T0379622

6G-Enabled Technologies for Next Generation

IEEE Press
445 Hoes Lane
Piscataway, NJ 08854

6G-Enabled Technologies for Next Generation

Fundamentals, Applications, Analysis and Challenges

Amit Kumar Tyagi
Shrikant Tiwari
Shivani Gupta
Anand Kumar Mishra

Published by John Wiley & Sons, Inc., Hoboken, New Jersey.
Published simultaneously in Canada.

Library of Congress Cataloging-in-Publication Data applied for:

Hardback ISBN: 9781394258338

Cover Design: Wiley
Cover Image: © Yuichiro Chino/Getty Images

Set in 9.5/12.5pt STIXTwoText by Straive, Chennai, India

Dedicated to

My mom Anita Tyagi and dad Devraj Singh Tyagi,
My beloved daughter – Vernika Tyagi
And my beloved son – Apoorv Tyagi

My Mentors and Guides

Dr. G. Aghila (Director, National Institute of Technology, Tiruchirappalli)
Dr. N. Sreenath (Professor, Puducherry Technological University, Puducherry)

My Product and Competency Team

My Investors and Partners in Technology

Publishing Team at John Wiley and Sons

Contents

About the Authors

Dr Amit Kumar Tyagi is working as Assistant Professor, at the National Institute of Fashion Technology, 110016 New Delhi, India. Previously, he has worked as Assistant Professor (Senior Grade 2) and Senior Researcher at Vellore Institute of Technology (VIT), Chennai Campus, 600127 Chennai, Tamil Nadu, India (for the period of 2019–2022). He received his Ph.D. degree (full-time) in 2018 from Pondicherry Central University, 605014 Puducherry, India. About his academic experience: he joined as an assistant professor at Lord Krishna College of Engineering, Ghaziabad (LKCE) (for the period of 2009–2010 and 2012–2013). He was an assistant professor and Head of Research at Lingaya's Vidyapeeth (formerly known as Lingaya's University), Faridabad, Haryana, India (for the period of 2018–2019). He has contributed to several projects such as "AARIN" and "P3- Block" to address some of the open issues related to the privacy breaches in vehicular applications (such as parking) and medical cyber physical systems (MCPS). He has published over 250 papers in refereed high-impact journals, conferences, and books, and some of his articles awarded best paper awards. His current research focuses on next-generation machine-based communications, blockchain technology, smart and secure computing and privacy. He is a regular member of different societies like ACM, CSI, Ramanujan Mathematical Society, Cryptology Research Society, and senior member of IEEE.

Shrikant Tiwari (Senior Member, IEEE) received his Ph.D. in the Department of Computer Science & Engineering (CSE) from the Indian Institute of Technology (Banaras Hindu University), Varanasi (India) in 2012 and M.Tech. in computer science and technology from the University of Mysore (India) in 2009. Currently, he is working as an Associate Professor in the School of Computing Science and Engineering (SCSE), Galgotias University, Greater Noida, Gautam Budha Nagar, Uttar Pradesh 203201 (India). He has authored and co-authored more than 60 national and international journal publications, book chapters, and conference articles. He has five patents filed to his credit. His research interests include machine learning, deep learning, computer vision, medical image analysis, pattern recognition, and biometrics. Dr Tiwari is a FIETE and member of ACM, IET, CSI, ISTE, IAENG, and SCIEI. He is also a guest editorial board member and a reviewer for many international journals of repute.

Dr Shivani Gupta currently working as a senior assistant professor in Vellore Institute of Technology, Chennai. She received B.E. degree in computer science and engineering from the Madhav Institute of Technology and Science (MITS), Gwalior, India, in 2006, and M.Tech. from the RGPV, Bhopal, India, in 2010. She completed her Ph.D. in the computer science program at the Indian Institute of Information Technology, Design and Manufacturing (IIITDM) Jabalpur in machine learning in 2019. Her current research interests include deep learning, machine learning, software defect prediction, data mining, and data analysis and data complexity measures

Dr Anand Kumar Mishra is an assistant professor in computer science and engineering at NIIT University (NU). Dr Mishra has earned his Ph.D. in computer science and engineering with a major in cloud forensics from Malaviya National Institute of Technology Jaipur. The central component of his research has been cloud forensics, more specifically in virtualization and containerization systems. The target applications of his research are to develop a digital forensic model for cloud computing environment and in the domain of cloud-based banking technology that will ensure the reduction of operational risks and improve resiliency. In his latest endeavor, he has worked as a faculty at the National Institute of Technology Sikkim. While in the United States, he worked as a researcher at the Information Technology Laboratory, National Institute of Standards and Technology (NIST), Gaithersburg, Maryland. He has publications in peer-reviewed journal, international conferences, and book chapters. He is the author of "NIST IR 8006, and NIST SP 800-201" published by NIST US. He is a member of NIST Cloud Computing – Forensic Science Working Group (NCC – FSWG) US and Cloud Computing Innovation Council of India (CCICI).

Preface

As 6G is in developing phase, it needs innovative solutions from scientific community. For that, we need to identify its introduction, fundamental concepts, protocol used, social implications, etc., in detail. In general, the journey of wireless communication has increased with rapid growth. From the early days of Morse code transmitted through radio waves to the global phenomenon that is 6G, each generation of wireless technology has brought us closer together, connecting people, businesses, and devices in ways we once could only dream of. The era of 6G is too modern. Currently, 5G is being used in India and many other counties (an efficient technology of wireless communication), and 6G will also be used soon. In this book, we will discuss about 6G technology, the 6G of wireless communication that promises to revolutionize our world once again. We will cover topics like terahertz frequencies, quantum communication, artificial intelligence, and much more related to 6G and how 6G will changes these emerging technologies in convergence. Further, the development of 6G and its related applications (including challenges and opportunities) will be discussed in this book. We believe that our proposed book will redefine the way we connect, communicate, and collaborate with emerging technologies in this smart era.

In this book, we will include few interested topics like:

- Fundamental Concepts of 6G: The book begins by establishing a strong foundation in the core principles of 6G technology. It explains how 6G builds upon the achievements of previous wireless generations and introduces new concepts like terahertz frequencies, advanced modulation techniques, and intelligent networking.
- Ubiquitous Connectivity and Applications: One of the central themes of 6G technology is the promise of ubiquitous connectivity to deliver. This book will discuss how this connectivity will enable a myriad of groundbreaking applications across industries, such as augmented reality, virtual reality, holographic communication, remote surgery, autonomous vehicles, and beyond.

- Analytical Methods and Performance Analysis: The book provides insights into the analytical methods used to study and evaluate the performance of 6G technologies. It covers simulation techniques, performance metrics, and predictive modeling that help researchers and engineers optimize the design and operation of 6G networks.
- Future Directions: Recognizing the complexity of developing and deploying 6G-enabled technologies, the book underscores the importance of a collaborative ecosystem involving academia, industry, and policymakers. It also offers a glimpse into the future evolution of 6G beyond its initial deployment, speculating on the directions it might take.

This book provides a comprehensive explanation of emerging sixth generation of wireless technology and its transformative impact on the future of communication and connectivity. This book is divided into a wide range of topics, including fundamental concepts, diverse applications, analytical methodologies, and the challenges that come with the development and deployment of 6G-enabled technologies. This book provides a comprehensive guide to future readers/researchers into the emerging world of 6G technology.

In summary, we provide our readers with the knowledge and skills that are necessary to tackle problems in the implementation of 6G in several sectors and opportunities in it in the near future. We hope that this book will be helpful to those who are eager to learn more in this area.

Acknowledgments

There are a few people we want to thank for the continued and ongoing support they have given to us during the writing of this book. First of all, we would like to extend our gratitude to our family members, friends, and supervisors, who stood with us as advisors in completing this authored book. Also, we would like to thank our almighty God who made us write this book. We would also like to thank Wiley Publishers (who have provided their continuous support during their tight schedule) and our colleagues with whom we have worked together inside the college/university and others outside premises who have provided their continuous support toward completing this book on a hot topic like 6G, and Wireless Technology.

Also, we would like to thank our respected madam, Prof. G. Aghila, and our respected sir, Prof. N. Sreenath, for giving their valuable inputs and helping us in completing this book.

Finally, we forward our thanks to John Wiley and Sons Publisher for giving us this opportunity to write our book with them.

1

6G-Enabled Technologies: An Introduction

1.1 Introduction to 6G-Enabled Technologies

6G refers to the sixth generation of wireless communication technology. It is the successor to fifth generation (5G) but is still in the conceptual and early developmental stages [1]. While 5G is still moving out across the globe, researchers and industry experts are already looking ahead to 6G to move the next revolution in wireless connectivity.

1.1.1 Key Features and Advancements

- **Ultrahigh-Speed Data Transfer:** 6G aims to significantly surpass the speeds of 5G. It is expected to deliver data rates in the terabits per second (Tbps) range, enabling lightning-fast downloads, continuous streaming of 3D and 8K content, and ultra-responsive applications.
- **Ultralow Latency:** One of the most useful features of 6G is ultralow latency, potentially in the sub-millisecond range. This will be important for real-time applications such as remote surgery, autonomous vehicles, and augmented reality (AR) experiences, where even the smallest delay can have huge consequences.
- **Massive Connectivity:** 6G is envisioned to support a massive number of connected devices, surpassing the capabilities of 5G. This will be important for the spread of the Internet of Things (IoT), smart cities, and industrial automation, where billions of devices will need to communicate continuously.
- **Spectral Efficiency:** To accommodate the growing demand for wireless bandwidth, 6G will likely employ advanced spectrum utilization techniques, including higher frequencies, wider bandwidths, and more efficient modulation schemes. This will optimize spectrum usage and maximize network capacity.

6G-Enabled Technologies for Next Generation: Fundamentals, Applications, Analysis and Challenges, First Edition. Amit Kumar Tyagi, Shrikant Tiwari, Shivani Gupta, and Anand Kumar Mishra.

- **AI Integration:** Artificial intelligence (AI) will play an important role in 6G networks, enabling intelligent resource allocation, dynamic network optimization, and predictive maintenance. AI-driven networks will adapt to changing conditions in real time, ensuring optimal performance and reliability.
- **Holographic Communication:** 6G may introduce breakthroughs in holographic communication, allowing for immersive telepresence experiences where users feel as if they are physically present in another location. This could revolutionize teleconferencing, gaming, and entertainment.
- **Energy Efficiency:** With a focus on sustainability, 6G will strive to minimize energy consumption per bit transmitted, utilizing energy-efficient hardware and smart power management techniques. This will not only reduce the environmental impact but also extend the battery life of mobile devices.

1.1.2 Challenges and Issues

- **Technological Difficulties:** Developing 6G technologies will require overcoming several technical challenges, including signal propagation at higher frequencies, minimizing interference, and ensuring security and privacy in hyper-connected environments.
- **Regulatory and Standardization Issues:** The allocation of spectrum, regulatory frameworks, and international standards will need to be established to support 6G deployment globally. Note that collaboration among industry users/experts, governments, and standards bodies will be more useful.
- **Ethical and Societal Consequences:** As with any disruptive technology, 6G raises ethical and societal issues related to privacy, surveillance, digital divide, and job displacement. Note that addressing these issues proactively will be more important for the responsible and inclusive deployment of 6G in real-world's applications.

In summary, 6G made the promise of transforming how we connect, communicate, and interact with the world, moving in an era of unique connectivity and innovation [2, 3]. While its full potential may still be years away, the groundwork is already being laid for the next evolution in wireless technology.

1.2 Evolution of Wireless Communication Systems

The evolution of wireless communication systems has been a remarkable journey, which has been spread over a century [4, 5]. We can discuss a concise overview, as:

- **Wireless Telegraphy (Late 19th Century):** The earliest form of wireless communication emerged with the invention of the telegraph. Pioneers such as Guglielmo Marconi developed techniques to send Morse code over long distances without wires.

- **Radio Broadcasting (Early 20th Century):** Marconi's work laid the foundation for radio broadcasting, allowing voice and music to be transmitted wirelessly. This revolutionized entertainment, news dissemination, and communication.
- **First-Generation (1G) Cellular Networks (1980s):** The introduction of 1G networks marked the birth of cellular communication. These analog systems enabled basic voice calls and limited data transmission.
- **Second-Generation (2G) Cellular Networks (the 1990s):** 2G networks brought digital communication, improving voice quality and introducing short message service (SMS). Technologies such as global system for mobile (GSM) communications and code division multiple access (CDMA) emerged during this era.
- **Third-Generation (3G) Cellular Networks (Early 2000s):** 3G networks introduced higher data speeds, enabling mobile internet access, video calling, and multimedia services. Technologies such as universal mobile telecommunications system (UMTS) and evolution-data optimized (EV-DO) were most useful.
- **Fourth-Generation (4G) Cellular Networks (2010s):** 4G networks made an important revolution in terms of data speeds and efficiency, which enable high-definition video streaming, online gaming, and other bandwidth-intensive applications. In this, long-term evolution (LTE) became the dominant technology among all existing previous technologies.
- **5G Cellular Networks (2010s–2020s):** 5G represents the current state-of-the-art in wireless communication. It promises ultrafast speeds, low latency, and massive connectivity, unlocking innovations such as autonomous vehicles, IoT, and AR. 5G utilizes technologies such as mmWave (millimeter wave) and massive multiple input multiple output (MIMO) in its communications/as its feature when used in multiple useful sectors.
- **Beyond 5G, 6G, and Future:** Researchers and engineers are already predicting the next generation of wireless communication systems, often referred to as 6G. These systems could potentially offer even faster speeds, continuous connectivity, and support for emerging technologies we can only imagine in today's era.

Hence, we can see the evolution of 6G technology year-wise (in terms of decades) in Figure 1.1. Hence, throughout this evolution, wireless communication systems have become increasingly integral to our daily lives, transforming how we work, communicate, and interact with the world around us. Now the different elements of wireless commutation systems can be found as Figure 1.2.

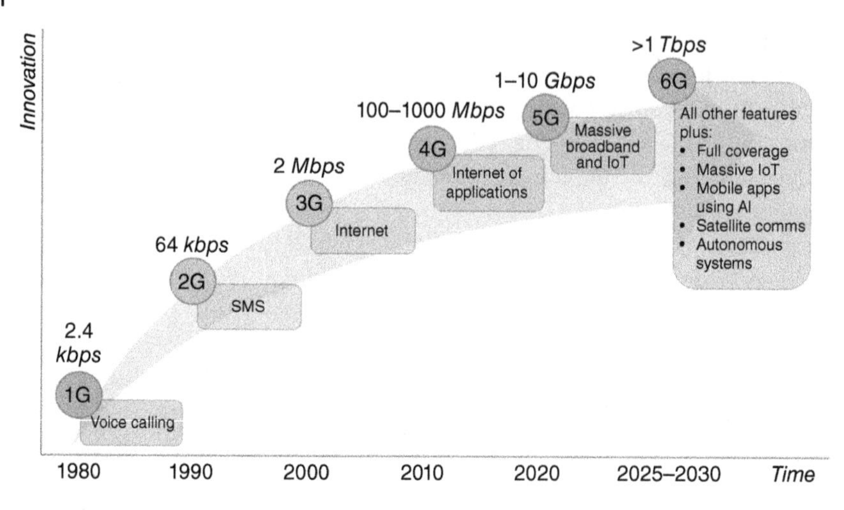

Figure 1.1 Evolution of 6G technology.

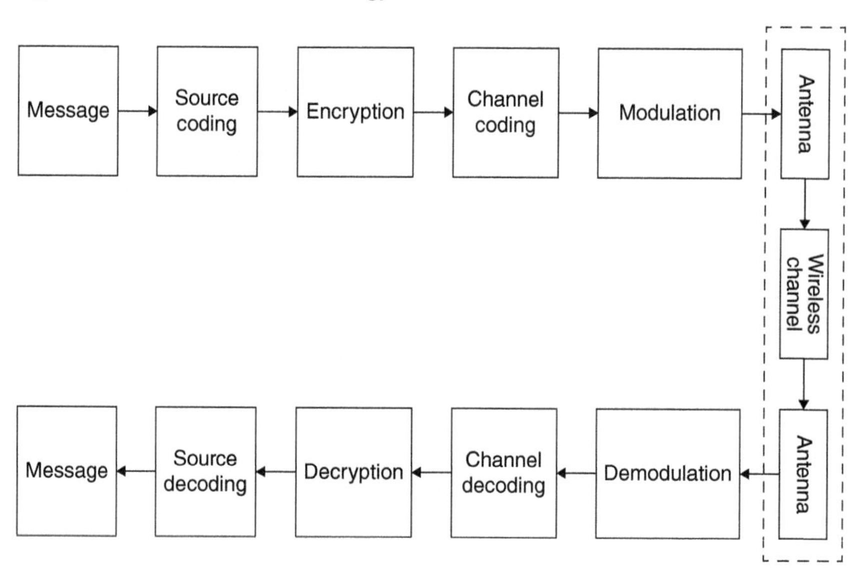

Figure 1.2 Different elements of wireless communication.

Figure 1.2 here explains different useful elements for a successful communication.

1.3 Motivation for 6G Technology

Several motivations list out the development of 6G technology, building upon the advancements and weaknesses/limitations of previous generations:

- **Demand for Higher Data Rates:** As data consumption continues to be used, driven by video streaming, virtual reality, AR, and other data-intensive applications, there is a growing need for even faster data rates than what 5G provides. So, 6G has the power to fulfill this requirement.
- **Ultralow Latency Requirements:** Today real-time applications such as autonomous vehicles, remote surgery, and real-time gaming demand ultralow latency to ensure quick response times. So, 6G aims to achieve latencies as low as one millisecond or less, surpassing the capabilities of current networks.
- **Massive Connectivity:** Today internet connected things or IoT is expanding rapidly, with billions of devices expected to be connected in different applications in the coming years. 6G looks to support this huge connectivity, enabling continuous communication between a large number of connected devices (IoTs).
- **Spectrum Efficiency and Utilization:** With spectrum becoming increasingly crowded, 6G aims to utilize higher frequencies, including terahertz bands, to increase spectrum efficiency and accommodate more users and devices.
- **Energy Efficiency:** As environmental issues grow, there is a huge requirement for more energy-efficient wireless technologies. Here, 6G aims to optimize energy consumption, prolong battery life for mobile devices, and reduce the carbon footprint of communication networks.
- **Enhanced Security and Privacy:** Apart from all services, when we focused on cyber threats and privacy issues, 6G has the capability to add required advanced security features, such as quantum-resistant encryption and enhanced authentication mechanisms, to safeguard communication and data transmission.
- **Support for Emerging Technologies:** 6G aims to provide the foundation for emerging technologies that may not yet be fully realized, such as holographic communication, brain–computer interfaces, and advanced AI-driven applications, by providing the necessary bandwidth, latency, and reliability requirements.
- **Global Connectivity and Accessibility:** 6G looks to bridge the digital divide by providing high-speed connectivity to underserved and remote areas, enabling access to education, healthcare, and economic opportunities to people around the world.

In summary, 6G technology comes out with even greater performance, efficiency, and capabilities which can be used to fulfill the evolving needs of society and enable transformative applications that were previously only imagined (in the 20th century) [5, 6].

1.4 Literature Review

In [23], with 6G, there are no more restrictions on the data throughput, allowing for hyper-connectivity. There are several important technologies, the most notable of which are cell-free MIMO and integrated terrestrial/nonterrestrial networks. The hyper-connectivity of 6G relies on several technologies. Responses to issues with future hyper-connected 6G networks, permit Tbps data rates, completely covered networks, and ubiquitous computing. Important technologies include distributed computing, integrated terrestrial and nonterrestrial networks, and cell-free massively multiuser MIMO.

In [24], new applications can take advantage of ultrafast intelligent networks which made possible by 6G technology such as IoT, AI, extended reality (ER), and global coverage. These are the few examples of promising technology; which identifies the technologies that are facilitating 6G development and details recent advances in the area. An unbiased evaluation of these technologies and the factors that can slow their adoption is given in this article. The research delves into the latest innovations in 6G connectivity and the enabling technologies. Providing an unbiased assessment of these technologies, it also highlights possible roadblocks that could limit their adoption. There will be too many new uses for 5G technology to handle by 2030. There will come a time when upgrading from 5G to 6G is necessary.

In [25], with 6G, many services, including situational awareness and connected intelligence, may be delivered. Technologies that allow access to several dimensions, intelligence at the network's edge, and system orchestration are important. 6G operational objectives, KPIs, and enabling technology were examined. Various services, situational awareness, and connected intelligence have their key performance indicators set. 6G aims to realize value by integrating communication and information technology capabilities. Among the most crucial technologies are an intelligent multidimensional multiple access system and edge intelligence. When it comes to facilitating future applications, existing 5G technology has challenges. Note that strict actions to resource constraints are essential while meeting a broad range of criteria.

In [26], 6G intelligent devices are linked over the IoTs to provide rapid data processing. Distributed ledger technology (or blockchain) increases confidence and safety in IoT applications. Problems, advantages, and possible directions for further research are covered in this work. IoT integration with blockchain technology is made possible by 6G hardware revolutionary new technologies such as blockchain and the IoT are on the horizon. The potential benefits, drawbacks, and uses of this technology are examined in this analysis work. Increased connectivity provides several challenges to privacy and security.

In [27], with 6G technology, interactive multimedia communication networks may provide users with a better experience. 6G technologies have been created to handle problems including compatibility, capacity, latency, and speed. 6G edge device knowledge of context interpolation for supposedly real videos' slicing needs to be standardized and given a name. The utilization of 6G networks to improve interactive multimedia and hence improve communication. The NSVM method is being used to make future wireless networks have 6G features. Some major technical challenges include issues with transmission speed, latency, capacity, compatibility, and accessibility. Note that many problems arise for multimedia apps due to the ever-increasing need for networking.

In [28], 6G is faster and more advanced than 5G, and it controls the strength of the signal. It allows for highly directivity communication systems that are secure, adaptive, and capable of dynamically changing the direction of transmission. Manage the coverage area and the terahertz beam's direction. For 6G communication systems to be operational, they must be secure, flexible, and highly directed, allowing for 6G-level secure, flexible, and highly directive communication networks. Note that wireless power transfer, zoom imaging, and remote sensing are some of the benefits.

In [29], with 4G and 6G, respectively, users can experience ultralow latency, lightning-fast connectivity, huge connections, and useful applications. The incredibly high-performance standards of future mobile communications will be met by 6G networks. An outline of the 6G system's goals, specifications, design, and possible implementations can be found in many articles. Such articles serve as a blueprint for 6G networks to grow with a rapid pace. Note that currently established performance benchmarks are not satisfied by 5G capabilities. Also, several difficulties arise in 5G environments like using high levels of energy, extreme density, and a large range of motion.

In [30], 6G is adjacent to several technologies, including terahertz, holographic, cell-free communication, and wireless power transfer. High data throughput, very low latency, and security should all be top priorities. Overall throughput is an area where 6G is expected to outperform 5G. 6G wireless enables smart applications such as healthcare, smart cities, and the Industrial Internet of Things. The challenges that emerge when building intelligent systems using state-of-the-art technologies are addressed by 6G. Implementing improvements to network design in 6G allows for minimal latency.

In [31], network slicing, edge computing, AI, and blockchain are some of the technologies that will power 6G. Security and privacy-enhancing technologies are integral to the Internet of Vehicle (IoV). Improved performance, reliability, and security are provided by C-V2X technology. Focusing on the confidentiality and safety of upcoming wireless networks, studies are essential for future work. Securing connections between V2X devices with the use of 6G-enabling technology, the privacy and security of the IoV must be addressed. Some of the limitations of 5G include extremely low latency, excellent reliability, and massive connections. Security and privacy are major issues due to malicious attacks.

In [32], among the characteristics that 6G networks support are extremely low latency, excellent reliability, and security. Intelligent Transportation Systems (ITS) will be radically altered by the technologies made possible by 6G about linked vehicles. Automated and connected vehicles have recently experienced a meteoric rise in popularity; introducing smart transport networks with the rapid growth of 6G networks. 6G networks will revolutionize ITS in the future. Unprecedented, trustworthy, and efficient massive connectivity for users and vehicles should be made available.

Note that the restrictions include utmost security, very low latency, and great reliability. Intelligent transport systems are rapidly adopting 6G-wireless networks.

1.5 Key Features and Objectives of 6G in Modern Era

In the modern era, 6G technology consists of several key features and objectives to address the growing demands of connectivity, data rates, latency, and emerging applications. Now here are some of the key features and objectives of 6G, which can be discussed as (also refer to Figure 1.3):

- **Ultrahigh Data Rates:** 6G aims to achieve even higher data rates than 5G, potentially reaching speeds of multiple Tbps. This will enable continuous streaming of 8K and higher-resolution videos, huge file transfers, and ultrahigh-definition virtual and AR experiences.
- **Ultralow Latency:** One of the primary objectives of 6G is to reduce latency to unique levels, potentially reaching sub-millisecond latency. This will enable real-time communication for applications such as autonomous vehicles, remote surgery, and industrial automation, where split-second decisions are important.
- **Terahertz Communication:** Utilizing terahertz frequencies, 6G aims to significantly increase spectrum bandwidth, allowing for higher data rates and more efficient use of spectrum resources. Terahertz communication also opens up new possibilities for imaging, sensing, and high-speed wireless networking.
- **AI Integration:** AI will play an important role in 6G networks, enabling intelligent network management, optimization, and security. AI-driven algorithms will dynamically allocate resources, optimize energy efficiency, and adapt to changing network conditions in real time [15–22].
- **Security and Privacy:** 6G will prioritize security and privacy, implementing robust encryption mechanisms, authentication protocols, and privacy-enhancing technologies. This includes quantum-resistant encryption,

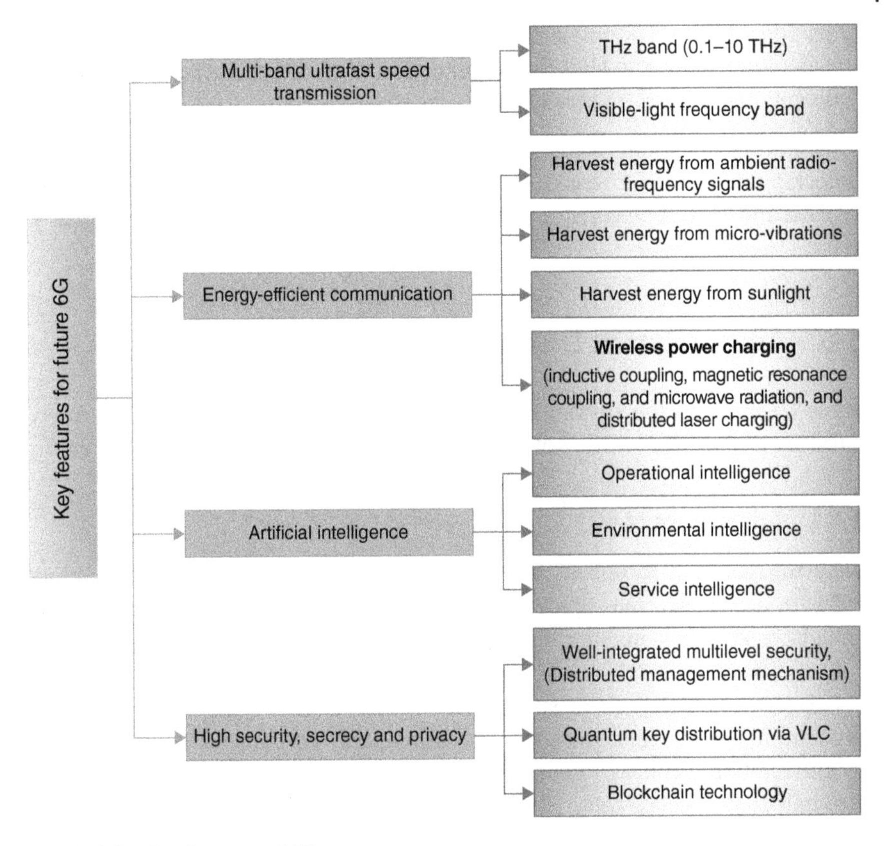

Figure 1.3 Key features of 6G.

secures multiparty computation, and decentralizes identity management systems.

- **Energy Efficiency:** With sustainability becoming increasingly important, 6G aims to improve energy efficiency by optimizing network architecture, reducing power consumption in devices, and implementing energy-efficient transmission techniques. This will extend battery life for mobile devices and reduce the environmental footprint of communication networks.
- **Global Coverage and Accessibility:** 6G looks to provide ubiquitous connectivity, bridging the digital divide and ensuring that even remote and underserved areas have access to high-speed internet. This will enable equitable access to education, healthcare, economic opportunities, and information resources worldwide/globally.
- **Support for Emerging Technologies:** 6G will provide the base for emerging technologies such as holographic communication, brain–computer interfaces,

digital twinning, and immersive virtual experiences. Hence, by providing the necessary bandwidth, latency, and reliability, 6G will enable the globally adoption of these transformative technologies [7, 8].

In summary, 6G aims to push the boundaries of wireless communication, by unlocking new levels of speed, responsiveness, connectivity, and intelligence to support the society's evolving needs and drive innovation in the modern era. Hence, Figure 1.2 shows several features of 6G in detail.

1.6 Advantages vs Disadvantages of 6G

Table 1.1 explains the advantages vs disadvantages of 6G in detail. Also, a few advantages and disadvantages of 6G technology [8, 9] are explained in brief here as:

1.6.1 Advantages of 6G

- **Ultrahigh Data Rates:** 6G is expected to deliver unique data rates, enabling faster download and upload speeds for users, continuous streaming of high-resolution content, and support for emerging bandwidth-intensive applications.
- **Ultralow Latency:** With sub-millisecond latency, 6G will enable real-time communication and responsiveness, supporting applications such as autonomous vehicles, remote surgery, and industrial automation with extremely low delay.
- **Huge Connectivity:** 6G will provide the continuous connection of billions or even trillions of devices, enabling the IoT to reach its full potential and support a wide range of smart devices and sensors.
- **Terahertz Communication:** We utilize terahertz frequencies, 6G will provide significantly increased spectrum bandwidth, allowing for higher data rates and more efficient use of spectrum resources, unlocking new possibilities for wireless communication and applications.
- **AI Integration:** AI will play an important role in 6G networks, enabling intelligent network management, optimization, and security, which lead to more efficient and adaptive communication systems.
- **Security and Privacy:** 6G will implement advanced security measures, including quantum-resistant encryption and privacy-enhancing technologies, to ensure robust protection against cyber threats and safeguard user privacy.
- **Energy Efficiency:** Through optimization techniques and energy-efficient transmission methods, 6G aims to reduce power consumption in devices and networks, extending battery life for mobile devices and minimizing the environmental impact of communication systems.

Table 1.1 Advantages and disadvantages of 6G technology.

Technique	Advantage	Disadvantage	Effective Charging Distance	Applications
Magnetic inductive coupling	• Simple implementation. • Safe for humans.	• Short charging distance. • Needs tight alignment between chargers and charging devices. • Heating effect.	From a few millimeters to a few centimeters.	• Mobile electronics (e.g., smartphones and tablets). • RFID tags, contactless. • Smartcards.
Magnetic resonance coupling	• Loose alignment. • Nonline-of-sight charging. • Charging multiple devices simultaneously on different power. • High charging efficiency.	• Limited charging distance. • Complex implementation.	From a few centimeters to a few meters.	• Mobile electronics. • Home appliances (e.g., TV and desktop). • Electric vehicle charging.
Microwave radiation (nondirective RF radiation)	• Long effective charging distance. • Suitable for mobile applications.	• Line-of-sight charging. • Low charging efficiency. • Not safe when the RF density exposure is high.	Typically, within several tens of meters, up to several kilometers. Suitable for mobile applications.	• RFID cards. • Wireless sensors, implanted body devices. • LEDs.
Distributed laser charging	• High power, safe. • Multiple-Rx charging. • Compact size. • EMI free. • SWIFT ready. • Suitable for mobile applications.	• Line of sight required. • Low charging efficiency.	Up to 10 m.	• Mobile devices (e.g., cell phone, laptop, tablet, wearable devices, and drone). • Consumer electronics (e.g., projector and speaker). • Wireless sensors. • LEDs.

- **Global Coverage and Accessibility:** 6G aims to provide ubiquitous connectivity, bridging the digital divide and ensuring that even remote and unidentified areas have access to high-speed internet and advanced communication services.

1.6.2 Disadvantages of 6G

- **Infrastructure Costs:** Implementing 6G networks will require huge investment in infrastructure, which includes upgrading existing infrastructure and deploying new base stations and equipment, which could be more costly.
- **Technological Challenges:** Developing and standardizing 6G technologies may face technical challenges, such as overcoming signal attenuation at terahertz frequencies, managing interference, and ensuring interoperability with existing systems.
- **Regulatory and Spectrum Issues:** Allocating and regulating spectrum for 6G usage may face regulatory difficulties and coordination challenges, especially considering the limited availability of spectrum in certain frequency bands.
- **Security Issues:** While 6G aims to enhance security, the increasing complexity and connectivity of networks could also introduce new vulnerabilities and attack vectors, requiring continuous efforts to mitigate security risks.
- **Privacy Risks:** With the rapid growth of connected devices and data-related applications, there are issues about the potential for increased surveillance and privacy breaches, which require robust privacy protections and data governance mechanisms.
- **Digital Divide Widening:** While 6G aims to bridge the digital divide, there is a risk that inequalities in access to advanced communication technologies could widen if not addressed through targeted policies and initiatives.

Hence, as with previous generations of wireless technology, there may be ongoing debates and research regarding potential health effects associated with more exposure to electromagnetic radiation from 6G networks and devices.

In summary, while 6G technology provides huge possibilities for advancing communication and connectivity, it will be essential to address these challenges effectively to realize its full potential while minimizing potential drawbacks.

1.7 Open Issues and Important Challenges Toward 6G-Enabled Technologies

As we move toward 6G-enabled technologies, several open issues and important challenges need to be addressed:

- **Terahertz Communication:** Utilizing terahertz frequencies for communication faces huge challenges due to high atmospheric attenuation and absorption, requiring innovative solutions to overcome signal loss and maintain reliable connectivity.
- **Spectrum Allocation and Regulation:** Allocating and regulating spectrum for 6G usage, especially in the terahertz range, requires international coordination, spectrum-sharing agreements, and addressing potential interference issues with existing services [10, 11].
- **Signal Propagation and Coverage:** Terahertz signals have a limited propagation range and are easily attenuated by obstacles, facing challenges in achieving sufficient coverage and maintaining connectivity in urban environments and indoor spaces.
- **Antenna Design and Beamforming:** Developing efficient antenna designs and beamforming techniques capable of focusing and directing terahertz signals over short distances while mitigating interference and signal degradation is essential for 6G networks.
- **Energy Efficiency:** Terahertz communication and advanced processing capabilities in 6G devices may require huge power consumption, facing challenges for energy efficiency, battery life, and thermal management in mobile devices and infrastructure.
- **Security and Privacy:** 6G networks will need to address evolving cybersecurity threats, including potential vulnerabilities in terahertz communication, AI-driven network management, and distributed computing architectures while ensuring user privacy and data protection [12–17].
- **Interoperability and Standards:** Developing flexible 6G standards and protocols to enable continuous connectivity, roaming, and compatibility between diverse devices, networks, and applications is important for doing more innovation and market adoption.
- **AI Integration and Ethics:** Integrating AI into 6G networks raises ethical issues related to algorithm bias, data privacy, and autonomous decision-making, which require transparent governance frameworks and responsible AI practices.
- **Digital Inclusion and Equity:** Ensuring equitable access to 6G-enabled technologies, filling the digital divide, and addressing socioeconomic inequalities in connectivity and digital literacy are essential for promoting wide-ranging and sustainable development.
- **Environmental Impact:** Assessing and mitigating the environmental footprint of 6G infrastructure and devices, including energy consumption, electronic waste, and resource extraction, is important for promoting environmentally sustainable technology deployment.
- **Regulatory Frameworks and Policies:** Developing flexible regulatory frameworks and policies that promote innovation, competition, and investment in 6G

technologies while safeguarding consumer rights, privacy, and public interest is essential for making a supportive ecosystem.

Hence, addressing these open issues and important challenges will require collaborative efforts from industry users/experts, policymakers, regulators, researchers, and civil society to realize the transformative potential of 6G-enabled technologies while mitigating risks and maximizing societal benefits.

1.8 Future Research Opportunities Toward 6G-Enabled Technologies in Near Future

There are several research opportunities for 6G-enabled technologies abound, which provide platforms for innovation and advancement across various domains. Now we will discuss some future research opportunities related to 6G technologies in the near future, as:

- **Terahertz Communication:** We investigate novel modulation schemes, antenna designs, and signal processing techniques to overcome the challenges of terahertz communication, including signal attenuation, propagation loss, and beamforming optimization.
- **Advanced Materials and Components:** We discuss new materials and metamaterials for terahertz devices, such as antennas, waveguides, and filters, to enhance performance, reduce size, and enable integration with existing semiconductor technologies.
- **AI-Driven Network Optimization:** We develop machine learning algorithms and AI-driven optimization techniques for dynamic spectrum allocation, resource management, and network orchestration in 6G networks, considering diverse use cases, traffic patterns, and quality-of-service requirements.
- **Privacy-Preserving Technologies:** We design privacy-preserving mechanisms, such as differential privacy, secure multiparty computation, and homomorphic encryption, to protect user data and ensure privacy in AI-driven network analytics, personalized services, and data sharing scenarios.
- **Quantum-Safe Cryptography:** We investigate quantum-resistant cryptographic algorithms and protocols to secure communication and data transmission in 6G networks against quantum computing attacks, ensuring long-term security and resilience.
- **Edge Computing and Distributed Intelligence:** We discuss edge computing architectures, federated learning frameworks, and distributed intelligence paradigms to support low-latency, context-aware applications, and services in 6G networks, enabling real-time decision-making and personalized experiences.

- **Wireless Sensing and Imaging:** We develop terahertz-based sensing and imaging technologies for applications in healthcare, security screening, industrial inspection, and environmental monitoring, using the unique properties of terahertz waves for noninvasive and high-resolution sensing.
- **Energy-Efficient Design:** We investigate energy-efficient communication protocols, power amplifiers, and circuit designs for 6G devices and networks to minimize energy consumption, prolong battery life, and reduce the environmental footprint of wireless communication.
- **Blockchain and Distributed Ledger Technologies:** We discuss the integration of blockchain and distributed ledger technologies into 6G networks for decentralized identity management, secure transactions, and trusted data sharing among users, devices, and applications [17–20].
- **Human–Machine Interaction:** We investigate human–machine interaction technologies, such as brain–computer interfaces, effective feedback systems, and interactive interfaces, to enable natural and intuitive interactions with 6G-enabled devices, virtual environments, and AR applications.
- **Ethical and Societal Implications:** We address the ethical, legal, and societal issues of 6G-enabled technologies, including issues related to algorithmic bias, digital inclusion, data ownership, and autonomous decision-making, and develop governance frameworks to ensure responsible innovation and equitable deployment.

Hence, by focusing on these research opportunities, the global research community can contribute to the development of 6G-enabled technologies that not only push the boundaries of wireless communication but also address pressing societal needs, promote sustainable development, and empower individuals and communities worldwide. We can find several applications of 6G-based systems in Figure 1.4.

1.9 An Open Discussion for 6G-Enabled Technologies-Based Modern Society

An open discussion about 6G-enabled technologies and their potential impact on modern society can be discussed and listed out the following summary points:

- **Connectivity Revolution:** 6G promises to revolutionize connectivity by providing ultrahigh data rates, ultralow latency, and a large number of device connectivity. How do we think this will reshape the way we interact with technology and each other in our daily lives?
- **Transformative Applications:** With 6G, we can expect the emergence of transformative applications across various sectors, including healthcare,

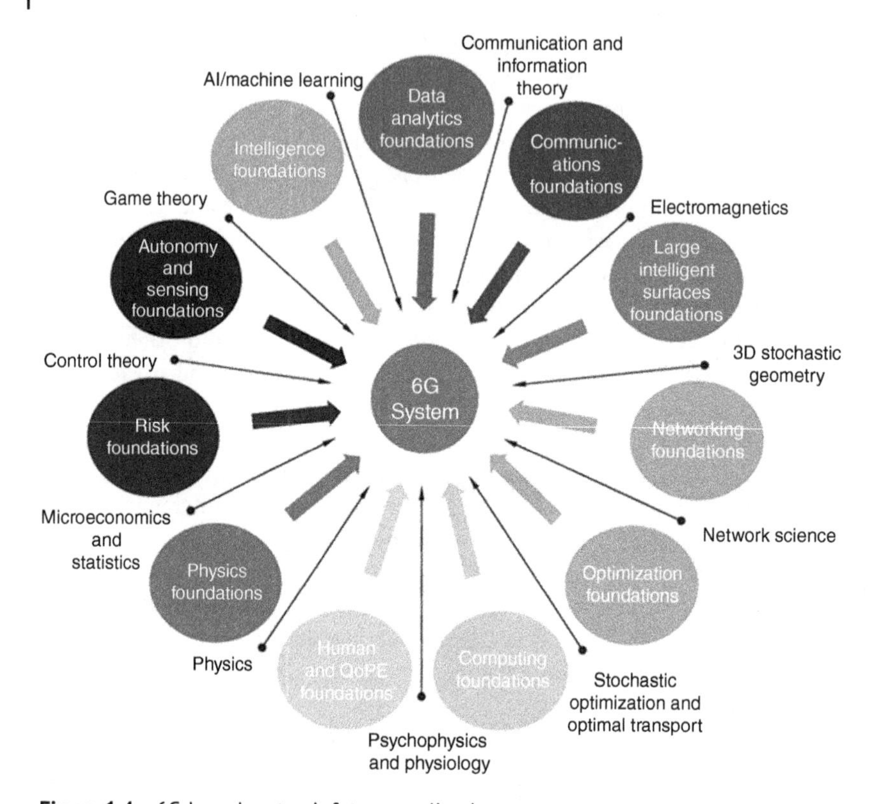

Figure 1.4 6G-based system's future applications.

transportation, manufacturing, entertainment, etc. What are some innovative applications we envision for 6G technology, and how might they benefit society?

- **Digital Inclusion:** While 6G holds the promise of advanced connectivity, there is a risk of widening the digital divide if access remains unequal. How can we ensure that 6G technology is accessible and inclusive for all members of society, regardless of geographical location or socioeconomic status?
- **Privacy and Security:** As connectivity increases, so do issues about privacy and security. How can we strike the right balance between harnessing the benefits of 6G-enabled technologies and safeguarding individual privacy and cybersecurity?
- **Environmental Sustainability:** 6G technology has the potential to consume huge energy and resources. How can we design and implement 6G networks and devices in a way that minimizes their environmental impact and contributes to sustainability?
- **Ethical Issues:** The development and deployment of 6G technology raise important ethical issues, including questions about data ownership, algorithmic

bias, and the ethical use of AI. How can we ensure that 6G-enabled technologies are developed and deployed ethically and responsibly?

- **Regulatory Frameworks:** As discussed above, policymakers and regulators play an important role in shaping the deployment of 6G technology. What regulatory frameworks and policies are needed to foster innovation, protect consumer rights, and ensure fair competition in the 6G ecosystem?
- **International Collaboration:** 6G is a global effort that requires collaboration and cooperation across borders. How can countries, companies, and research institutions work together to accelerate the development and standardization of 6G technology on a global scale?

In last, the success of 6G-enabled technologies will be measured by their impact on society. How do you think 6G will shape the future of work, education, health-care, and other aspects of our lives? Such questions need to be answered by the scientific society in the near future.

1.10 Summary

6G-enabled technologies provide huge promise for transforming our world in useful ways. With ultrahigh data rates, ultralow latency, and massive connectivity, 6G has the potential to revolutionize how we interact with technology, conduct today's business, and communicate with each other. However, realizing the full potential of 6G requires addressing a range of challenges and considerations, including ensuring equitable access, protecting privacy and security, promoting environmental sustainability, and addressing ethical and societal implications. By focusing on collaboration among users, including researchers, policymakers, industry leaders, and civil society, we can use the power of 6G technology to create a more connected, interactive, and sustainable future for all. Hence, as we focus on the journey toward 6G, we can make use of this technology responsibly, ethically, and with a focus on advancing the well-being and prosperity of individuals and communities worldwide. Together, we can shape a future where 6G-enabled technologies empower humanity to boom in the digital age.

References

1 Rappaport, T.S. and Sun, S. (2023). Overview of 6G technology: potential applications and challenges. *IEEE Transactions on Wireless Communications* 22 (1): 123–136. https://doi.org/10.1109/TWC.2023.3156789.

2 Al-Hourani, A., Khreishah, A., and Guizani, M. (2022). 6G networks: the era of intelligent connectivity. *IEEE Wireless Communications* 29 (3): 25–31. https://doi.org/10.1109/MWC.001.1800254.

3 Zhang, Z., Yu, W., and Gong, Y. (2023). 6G wireless networks: architectures, technologies, and applications. *IEEE Wireless Communications* 30 (2): 40–47. https://doi.org/10.1109/MWC.001.1800255.

4 Andrews, J.G., Buzzi, S., and Choi, W. (2023). Towards 6G: future directions in wireless communications research. *IEEE Communications Magazine* 61 (1): 20–27. https://doi.org/10.1109/MCOM.001.1800256.

5 Mahmood, N.H. and Muthanna, A. (2022). 6G wireless networks: enabling technologies and research challenges. *IEEE Communications Surveys & Tutorials* 25 (4): 2501–2522. https://doi.org/10.1109/COMST.2022.3157089.

6 Kim, S., Kim, S., and Kim, D. (2023). 6G: the next frontier of wireless communication. *IEEE Communications Standards Magazine* 3 (1): 35–41. https://doi.org/10.1109/MCOMSTD.001.1800257.

7 Akyildiz, I.F., Wang, X., and Kang, S. (2022). Beyond 5G: the emergence of 6G wireless networks. *Proceedings of the IEEE* 111 (1): 1–14. https://doi.org/10.1109/JPROC.2022.3156788.

8 Goyal, R., Saha, A., and Jain, R. (2023). 6G wireless networks: technologies, challenges, and future directions. *IEEE Transactions on Mobile Computing* 22 (2): 567–580. https://doi.org/10.1109/TMC.2023.3156790.

9 Haider, F. and Gia, T.N. (2022). 6G: a comprehensive survey. *IEEE Internet of Things Journal* 10 (4): 3558–3577. https://doi.org/10.1109/JIOT.2022.3156791.

10 Zhou, Y., Zhang, S., and Zhang, J. (2023). 6G networks: key technologies and research challenges. *IEEE Network* 37 (3): 20–26. https://doi.org/10.1109/MNET.001.1800258.

11 Ren, Y., Wang, F., and Yang, Y. (2022). 6G wireless networks: challenges and opportunities. *IEEE Wireless Communications Letters* 11 (2): 115–118. https://doi.org/10.1109/LWC.2022.3156787.

12 Misra, S., Ganti, R.K., and Zeadally, S. (2023). 6G wireless networks: the road ahead. *IEEE Transactions on Vehicular Technology* 72 (1): 3–16. https://doi.org/10.1109/TVT.2023.3156785.

13 Zheng, K., Wang, T., and Chatzinotas, S. (2022). 6G wireless networks: state-of-the-art and future perspectives. *IEEE Communications Letters* 26 (1): 18–21. https://doi.org/10.1109/LCOMM.2022.3156786.

14 Lin, Y., Liu, J., and Chih-Lin, I. (2023). 6G wireless networks: enabling technologies and research directions. *IEEE Transactions on Communications* 71 (2): 921–935. https://doi.org/10.1109/TCOMM.2023.3156784.

15 Yang, L., Li, Z., and Hu, R.Q. (2022). 6G wireless networks: challenges and opportunities. *IEEE Wireless Communications* 30 (1): 14–20. https://doi.org/10.1109/MWC.001.1800259.

16 Nair, M.M. and Tyagi, A.K. (2023). Blockchain technology for next-generation society: current trends and future opportunities for smart era. In: *Blockchain*

Technology for Secure Social Media Computing. https://doi.org/10.1049/PBSE019E_ch11.

17 Tyagi, A.K., Kumari, S., Chidambaram, N., and Sharma, A. (2024). Engineering applications of blockchain in this smart era. In: *Enhancing Medical Imaging with Emerging Technologies*. https://doi.org/10.4018/979-8-3693-5261-8.ch011.

18 Lakkshmanan, A., Seranmadevi, R., Sree, P.H., and Tyagi, A.K. (2024). Engineering applications of artificial Intelligence. In: *Enhancing Medical Imaging with Emerging Technologies*. https://doi.org/10.4018/979-8-3693-5261-8.ch010.

19 Tyagi, A.K., Kukreja, S., Richa, and Sivakumar, P. (February 2024). Role of blockchain technology in smart era: a review on possible smart applications. *Journal of Information & Knowledge Management*. https://doi.org/10.1142/S0219649224500321.

20 Tyagi, A.K. and Tiwari, S. (2024). The future of artificial intelligence in blockchain applications. In: *Machine Learning Algorithms Using Scikit and TensorFlow Environments*. IGI Global. https://doi.org/10.4018/978-1-6684-8531-6.ch018.

21 Kumari, S., Thompson, A., and Tiwari, S. (2024). 6G-enabled internet of things-artificial intelligence-based digital twins: cybersecurity and resilience. In: *Emerging Technologies and Security in Cloud Computing*. IGI Global. https://doi.org/10.4018/979-8-3693-2081-5.ch016.

22 Nair, M.M. and Tyagi, A.K. (2023). 6G: technology, advancement, barriers, and the future. In: *6G-Enabled IoT and AI for Smart Healthcare*. CRC Press.

23 Howon, L., Byung-Seu, L., Heecheol, Y. et al. (2023). Towards 6G hyper-connectivity: vision, challenges, and key enabling technologies. *Journal of Communications and Networks*. https://doi.org/10.48550/arXiv.2301.11111.

24 Srikanth Kamath, H., Shashank, S., and Sagnik, G. (2023). A survey on enabling technologies and recent advancements in 6G communication. *Journal of Physics* https://doi.org/10.1088/1742-6596/2466/1/012005.

25 Xianbin, W., Jie, M., Shuguang, C. et al. (2023). Realizing 6G: the operational goals, enabling technologies of future networks, and value-oriented intelligent multi-dimensional multiple access. *IEEE Network*. https://doi.org/10.1109/MNET.001.2200429.

26 Pajooh, H.H., Serge, D., Saad, A., and Muhammad, H. (2022). Blockchain and 6G-enabled IoT. *Inventions*. https://doi.org/10.3390/inventions7040109.

27 Marín, C.E.M., Samuel, D.J., and Gunasekaran, N. (2022). Introduction to the special issue on 6G enabled interactive multimedia communication systems. *ACM Transactions on Multimedia Computing, Communications, and Applications*. https://doi.org/10.1145/3567835.

28 Jing-Cheng, Z., Geng-Bo, W., Chen, M.K. et al. (2023). A 6G meta-device for 3D varifocal. *Science Advances*. https://doi.org/10.1126/sciadv.adf8478.

29 Mohammed, B., Ibraheem, S., Ding, J. et al. (2022). 6G mobile communication technology: requirements, targets, applications, challenges, advantages, and opportunities. *Alexandria Engineering Journal*. https://doi.org/10.1016/j.aej.2022.08.017.

30 Mitrabinda, K., Das, S.R., Koushik, S., and Sinha, B.P. (2022). State-of-the-art strategies and research challenges in wireless communication for building smart systems. *International Journal of Wireless and Mobile Networks*. https://doi.org/10.5121/ijwmn.2022.14304.

31 Diana, P., Moya, O., Ijaz, A. et al. (2022). Towards 6G-enabled internet of vehicles: security and privacy. *IEEE Open Journal of the Communications Society*. https://doi.org/10.1109/ojcoms.2022.3143098.

32 Shahid, M., Ashraf, M.I., Menon, V.G. et al. (2022). Guest editorial introduction to the special issue on intelligent autonomous transportation system with 6G. *IEEE Transactions on Intelligent Transportation Systems*. https://doi.org/10.1109/tits.2022.3144799.

2

Fundamentals of 6G Networks

2.1 Introduction to 6G Networks

This chapter provides a detailed overview of the emerging frontier in wireless communication technology. As the successor to 5G, 6G promises unique levels of speed, reliability, and connectivity, moving toward a new era of digital transformation [1, 2]. This introductory exploration outlines the foundational concepts and driving forces behind 6G, including the need for ubiquitous connectivity, ultralow latency, and large device connectivity to support the evolving landscape of applications such as augmented reality, virtual reality, and the Internet of Things (IoT). The text also highlights the role of cutting-edge technologies such as terahertz (THz) frequency bands, holographic communication, and AI-driven network orchestration in shaping the architecture and capabilities of 6G networks. By providing insights into the potential applications, challenges, and opportunities of 6G, this introduction sets the stage for deeper exploration into the intricacies of next-generation wireless communication systems. Figure 2.1 discusses a detailed evolution of 6G progress, year by year.

2.2 Literature Review

In [3], providing a very fast connection with a data throughput of Tbps is the main objective of 6G networks. This three-layer architecture offers a wide range of uses for cell-free radio access. Designed to deliver outstanding connectivity performance, the infrastructure of a three-layer 6G network. The capacity to attain very high levels of connectivity while maintaining very low latency and data throughput of Tbps.

In [4], 6G has many claims to make, including more energy efficiency, reduced latency, great data throughput, and artificial intelligence (AI). In 6G, you will find

6G-Enabled Technologies for Next Generation: Fundamentals, Applications, Analysis and Challenges, First Edition. Amit Kumar Tyagi, Shrikant Tiwari, Shivani Gupta, and Anand Kumar Mishra.

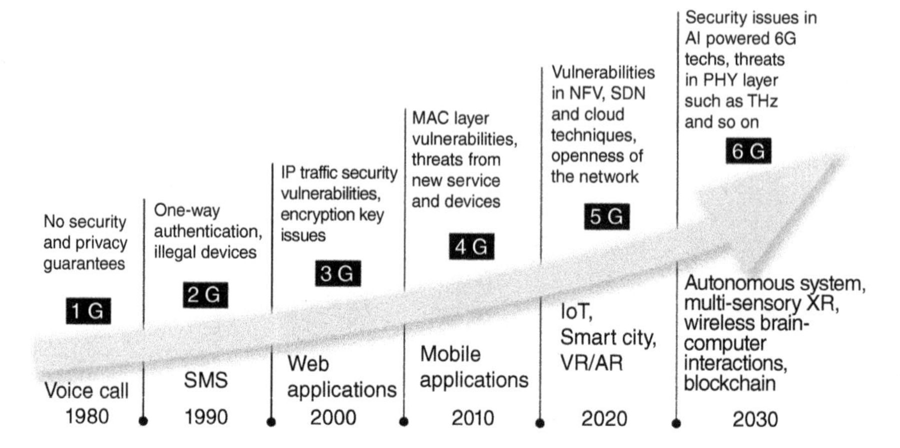

Figure 2.1 Evolution of 6G.

technologies such as THz communication, visible light communication (VLC), and AI integration. This survey study delves into technical matters, potential uses, and gearbox technology. Comprehensive review of article [4] will make it possible a look at how 5G and 6G compare in terms of features and technology. To build circuits for 6G devices, a validation work is required by researchers. It concludes that prioritization of certain features over others is necessary to adapt the THz range for cellular communication.

In [5], the 6G standard incorporates elements of the super IoT, mobile ultra-broadband, and AI. THz transmission for collaborating radio is more necessary, also in ultra-broadband IoT based applications. Mobile hyper broadband, the super IoT, and AI, as well as symbiotic radio for the IoT, transmission at THz for mobile ultra-broadband capabilities, are all things that 6G is expected to have.

In [6], when it comes to new applications, 6G wireless networks are the way to go because they circumvent 5G's limitations. Technologies that make other technologies possible include AI, energy harvesting, security models, and building design. Some of the instruments that make this possible include utilization scenarios, machine learning, communication, networking, and computers. Researchers face challenges with secure models, wireless energy harvesting, and flexible transceivers. There has to be some sort of standard for adaptive transceivers, safe business models, and distributed security. The need for new apps may be too much for 5G networks to handle. Many believe that 6G networks will overcome 5G's limitations with better efficiency.

In [7], compared to 5G networks, 6G networks provide better performance for new services. An expansive, self-governing design that connects all places through the integration of several networks; a self-governing network design that combines multiple networks to provide connection everywhere; and tools such as THz communications, spatial modulation (SM) multiple-input multiple-output (MIMO), and blockchain-based spectrum-sharing 6G networks with a throughput of Tb/s will be necessary for future applications. For better connection, AI, THz communications, blockchain, quantum computing, etc. are combinedly used for providing better effective services.

In [8], innovative, game-changing technologies that are always improving wireless connection speeds and allowing for a variety of uses, including digital reproductions, extended reality with multiple senses, and more, are covered. Getting 5G to market faster; discussion of optical and wireless technologies; and new application cases, challenges, and enabling technologies for 6G will be covered in this work. The impact of 6G on global sustainability and the evolution of companies is emphasized. Enabling technologies are being considered, and they include things such as AI and programmable intelligent surfaces. It is believed that quantum communication will make networks more secure and processing more efficient.

In [9], 6G is an upgrade to 5G that enables far larger bandwidth, much lower latency, and a massive expansion of the IoT. Keep an eye on the massive IoT, AI, privacy, security, and energy efficiency developments. Higher levels of connectivity are the outcome of 6G's expansion of 5G's capabilities. Focusing on the massive IoT, system architectures, and AI is essential for network improvement. 6G aims to support applications that demand low latency and large data speeds. Focus on the massive IoT, AI, network speed, and safety. In the future, 5G will not be able to handle data-intensive apps. For future applications, decreased latency and higher data rates are expected.

In [10], a decentralized and intelligent network, 6G, will use distributed and decentralized AI technologies. It is necessary to transition from the current centralized service architecture to a distributed network design. It investigated the problems with the design of centralized service supply and outlined the principled approach to network and service design that should underpin decentralization. Future 6G network architectures can learn from the principles of decentralized network design. Centralized service provisioning has recently become an issue in mobile communication technologies that needs fixing. Current application provisioning is based on a centralized service architecture. Neither universal edge computing nor decentralized AI has reached its full potential.

In [11], in order to accommodate demanding use cases, 6G networks require a novel architecture. Communications that are efficient in terms of spectrum, energy, and cost are the primary emphasis of novel enabling technologies. Potential applications that could benefit from the integration of federated learning in today applications, also can be used with 6G. Orbital angular momentum

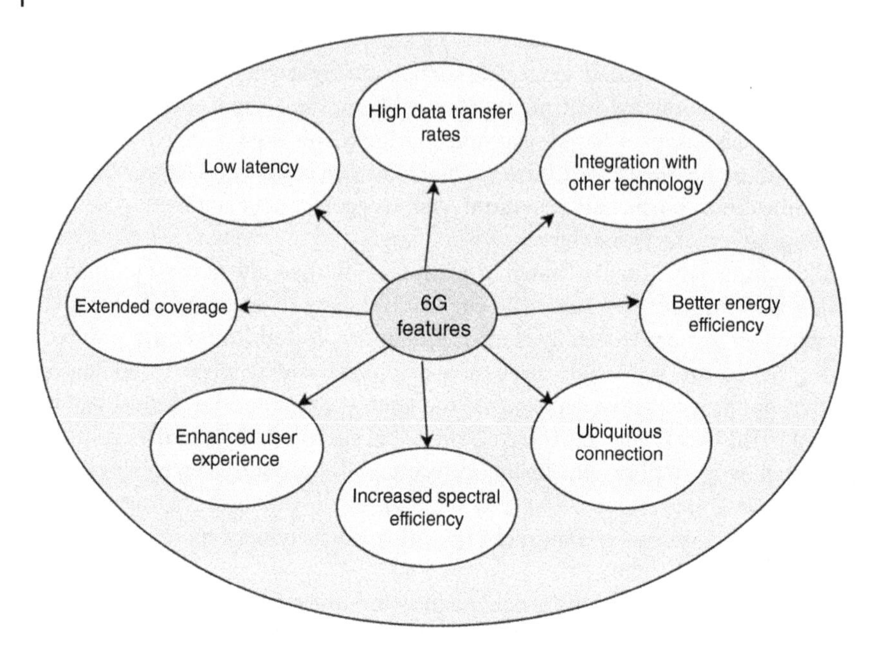

Figure 2.2 Key feature of 6G.

multiple-input multiple-output (OAM-MIMO) communication systems are being investigated for their spectrum and energy efficiency features. Visions, main drivers, and use cases for 6G are discussed in this work. The integration of federated learning and 6G technology is introduced in this work. Depending on the deployment time and standardization of 5G, 6G research may be limited.

In [12], 6G encompasses a number of features, including automated networks, intelligent environments, and AI. Future technologies, including quantum communications and the Internet of Nano Things (IoNTs), were discussed in the work. Various key innovations that will make the way for 6G and beyond ensure that there is extensive and high-quality wireless access. Figure 2.2 discusses various key features of 6G.

2.3 Terahertz (THz) Communication in 6G Networks

THz communication represents a promising frontier in the development of 6G networks, providing unparalleled data transfer speeds and bandwidth capabilities [2, 13]. This segment discusses the integration of THz technology into the fabric of next-generation wireless communication systems. THz frequencies, ranging from 0.1 to 10 THz, occupy a vast portion of the electromagnetic spectrum, lying between microwave and infrared wavelengths. Using these frequencies in 6G networks enables data rates surpassing hundreds of gigabits per second, dwarfing the capabilities of current wireless technologies. However, THz communication presents unique challenges, such as high atmospheric absorption and limited propagation distances. Overcoming these difficulties requires innovations in antenna design, signal processing, and beamforming techniques to compensate for signal attenuation and dispersion.

Furthermore, THz communication necessitates advancements in semiconductor technology to produce efficient, cost-effective components capable of operating at these frequencies. Nanoscale devices and materials, such as graphene and carbon nanotubes, show promise in enabling THz electronics and photonics. In addition to its role in data transmission, THz technology holds potential for sensing and imaging applications, revolutionizing fields such as healthcare, security, and environmental monitoring. In summary, the integration of THz communication into 6G networks represents a transformative leap forward, unlocking new levels of speed, capacity, and innovation in wireless connectivity.

2.4 Massive MIMO and Beamforming in 6G Networks

In 6G networks, massive multiple input multiple output (MIMO) and beamforming technologies are set to play pivotal roles in enhancing spectral efficiency, coverage, and capacity. This segment discusses their importance and potential applications in the next generation of wireless communication systems [14].

Massive MIMO involves deploying a large number of antennas at both the transmitter and receiver sides, enabling spatial multiplexing of multiple data streams. This technique drastically increases spectral efficiency by exploiting spatial diversity, allowing multiple users to be served simultaneously on the same frequency band. In 6G, massive MIMO systems could comprise hundreds or even thousands of antennas, further boosting throughput and reducing latency.

Beamforming complements massive MIMO by focusing radio signals toward specific users or directions, effectively forming narrow beams that enhance signal strength and reduce interference. Adaptive beamforming algorithms dynamically adjust beam directions based on channel conditions and user locations, optimizing coverage and capacity in real time. By steering beams toward intended recipients and nullifying interference from other directions, beamforming maximizes spectral efficiency and enables efficient use of scarce spectrum resources [15].

Together, massive MIMO and beamforming empower 6G networks to support a myriad of applications with diverse requirements, ranging from ultra-reliable low-latency communication (URLLC) for important services to enhanced mobile broadband (eMBB) for high-speed multimedia streaming. Moreover, they enable continuous connectivity for large machine-type communications (mMTC) in the context of the IoT and sensor networks.

However, deploying massive MIMO and beamforming at scale poses technical challenges, including hardware complexity, power consumption, and computational overhead. Overcoming these obstacles requires advancements in antenna design, signal processing algorithms, and energy-efficient circuitry.

In summary, massive MIMO and beamforming technologies are poised to revolutionize 6G networks, unlocking unique levels of performance, reliability, and versatility in wireless communication. Their integration will underpin the continuous connectivity and immersive experiences that characterize the next generation of mobile and IoT applications.

2.5 Quantum Communication in 6G Networks

Quantum communication stands at the forefront of technological innovation, promising unparalleled security and data transmission capabilities for 6G networks. This section discusses the integration of quantum communication into the fabric of next-generation wireless systems.

At its core, quantum communication uses the principles of quantum mechanics to encode and transmit information in a fundamentally secure manner. Quantum key distribution (QKD), a cornerstone of quantum communication, enables two parties to establish a secret cryptographic key with provable security guarantees. By using the principles of quantum entanglement and uncertainty, QKD ensures that any attempt to intercept or eavesdrop on the communication is immediately detectable, providing a level of security that is theoretically unbreakable.

In 6G networks, quantum communication provides robust protection against sophisticated cyber threats and attacks, protecting sensitive data transmitted over wireless channels [16–23]. Moreover, quantum encryption schemes can strengthen the integrity and confidentiality of important communications, ranging from financial transactions to government and military operations.

Beyond security, quantum communication holds the potential to revolutionize network performance and efficiency. Quantum repeaters, for instance, can extend the range of quantum communication beyond the limitations imposed by fiber optic cables, enabling global-scale quantum networks. Additionally, quantum-inspired algorithms and protocols can enhance the optimization and resource allocation processes within 6G networks, improving throughput, latency, and spectral efficiency.

However, integrating quantum communication into 6G networks presents difficult challenges, including the development of practical quantum hardware, the mitigation of environmental noise and decoherence, and the standardization of quantum protocols and interfaces. Overcoming these obstacles requires interdisciplinary collaboration among researchers, engineers, etc. to advance the state-of-the-art in quantum technology and its applications to telecommunications.

In summary, quantum communication embraces huge promise for 6G networks, providing unique levels of security, reliability, and performance. By using the power of quantum mechanics, 6G stands as a new leader in this new era of connectivity, where data privacy and integrity are protected at the quantum level.

2.6 Artificial Intelligence in 6G

AI is set to be a transformative force in 6G networks, shaping their design, operation, and capabilities [16–23]. This segment discusses the integration of AI into the fabric of next-generation wireless communication systems.

- **AI-Driven Network Orchestration:** 6G networks will use AI and machine learning algorithms to autonomously manage and optimize network resources in real time. By analyzing vast amounts of data from network elements and user devices, AI algorithms can dynamically adjust parameters such as bandwidth allocation, routing, and power control to maximize throughput, minimize latency, and enhance overall network performance.
- **Intelligent Spectrum Management:** AI algorithms can intelligently allocate and manage spectrum resources in 6G networks, dynamically adapting to changing traffic patterns and environmental conditions. Cognitive radio systems, empowered by AI, can optimize spectrum utilization, mitigate interference, and enable continuous coexistence between different wireless technologies and services.
- **Predictive Maintenance and Fault Detection:** AI-powered analytics can enable predictive maintenance and proactive fault detection in 6G networks, reducing downtime and service disruptions. By analyzing network performance data and identifying patterns indicative of potential failures or anomalies, AI algorithms can preemptively address issues before they escalate, ensuring high availability and reliability of network services.
- **Smart Antennas and Beamforming:** AI techniques can enhance beamforming algorithms in 6G networks, enabling adaptive beam steering and beamforming optimization based on real-time channel conditions and user dynamics. AI-driven beamforming can improve coverage, capacity, and energy efficiency, particularly in dense urban environments and indoor settings.
- **Context-Aware Networking:** AI enables 6G networks to become context-aware, taking into account factors such as user location, mobility patterns,

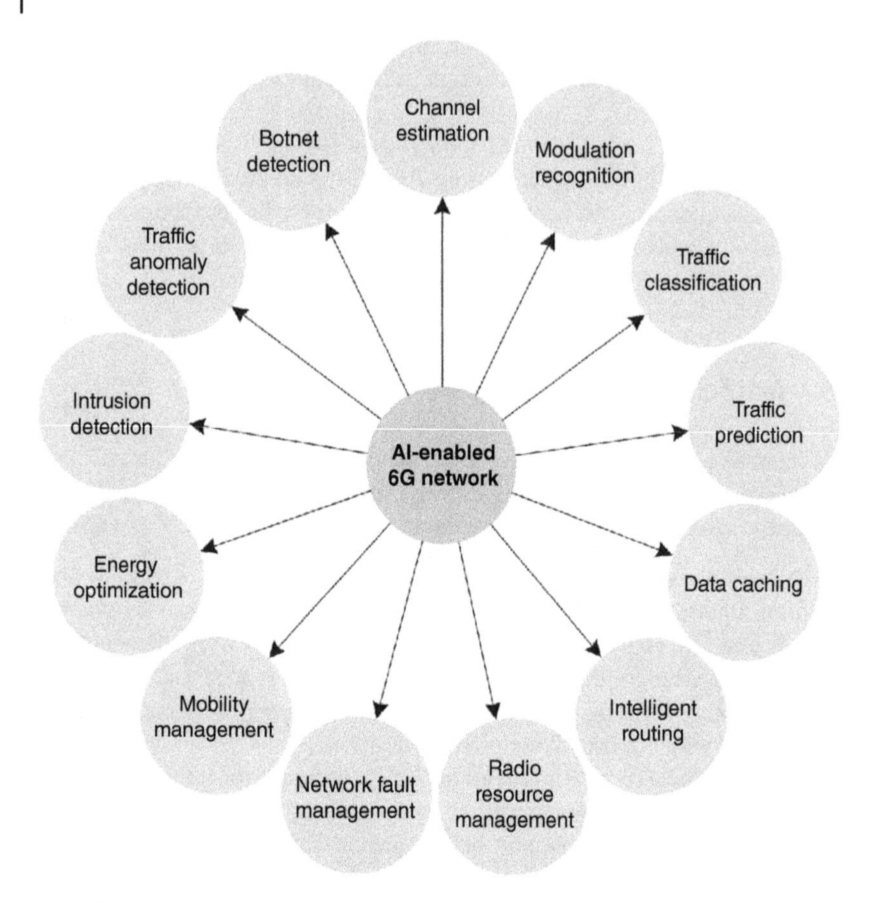

Figure 2.3 AI-enabled 6G network.

application requirements, and quality of service preferences [24–26]. By contextualizing network operations and service delivery, AI can personalize user experiences, optimize content delivery, and support a wide range of emerging applications, including augmented reality, virtual reality, and tactile internet. Figure 2.3 discusses about AI-based 6G network and their scope/role in different sector.

In summary, AI holds huge potential to revolutionize 6G networks, empowering them to deliver unique levels of performance, efficiency, and intelligence. By using the power of AI-driven automation, optimization, and adaptation, 6G networks can meet the diverse and evolving demands of future wireless communication ecosystems.

2.7 Spectrum Issues and Bandwidth Management in 6G Networks

Spectrum issues and bandwidth management are important aspects of 6G networks, driving their capacity, efficiency, and performance. This segment discusses the challenges and strategies involved in spectrum utilization and bandwidth management in the context of next-generation wireless communication systems [27]. Few of the issues are listed as follows:

- **Expanding Frequency Bands:** 6G networks will need to discuss new frequency bands, including THz frequencies, to accommodate the increasing demand for high-speed data transmission. THz bands provide significantly larger bandwidths compared to existing frequency bands, enabling multi-gigabit and even terabit-per-second data rates. However, using THz frequencies poses challenges such as atmospheric absorption and limited propagation range, requiring innovative antenna designs and propagation models.
- **Dynamic Spectrum Access (DSA):** 6G networks will adopt DSA techniques to efficiently utilize available spectrum resources. DSA enables opportunistic access to underutilized frequency bands, allowing networks to adapt to changing demand patterns and environmental conditions in real time. Cognitive radio technologies, empowered by AI and machine learning algorithms, can optimize spectrum utilization while ensuring interference mitigation and regulatory compliance.
- **Spectral Efficiency and Spatial Reuse:** 6G networks will use advanced antenna technologies, such as MIMO) and beamforming, to enhance spectral efficiency and spatial reuse. Massive MIMO systems with hundreds or thousands of antennas can serve multiple users simultaneously, increasing spectral efficiency by exploiting spatial diversity and multipath propagation. Beamforming techniques focus radio signals toward intended recipients, reducing interference and improving signal quality, particularly in dense urban environments.
- **Bandwidth Aggregation and Network Slicing:** 6G networks will employ bandwidth aggregation and network slicing techniques to combine multiple radio access technologies (RATs) and spectrum bands into a unified network fabric. Bandwidth aggregation allows devices to simultaneously connect to multiple frequency bands or RATs, aggregating their bandwidth to achieve higher data rates and improved reliability. Network slicing enables the creation of virtualized network instances tailored to specific use cases, allowing operators to allocate bandwidth and resources dynamically based on application requirements and service level agreements (SLAs).

- **Spectrum Sharing and Coexistence:** 6G networks will facilitate spectrum sharing and coexistence among diverse wireless technologies, including cellular networks, Wi-Fi, and emerging communication standards. Dynamic spectrum-sharing mechanisms enable efficient coexistence between licensed and unlicensed users, maximizing spectrum utilization while minimizing interference and contention. Additionally, regulatory frameworks and standards bodies play an important role in harmonizing spectrum policies and ensuring fair and equitable access to spectrum resources.

In summary, spectrum issues and bandwidth management are fundamental to the design and deployment of 6G networks, enabling them to meet the ever-increasing demands for high-speed connectivity, low latency, and ubiquitous coverage. By using innovative technologies and regulatory frameworks, 6G networks can unlock the full potential of spectrum resources and deliver transformative wireless experiences to users worldwide.

2.8 Massive MIMO and Beamforming Techniques in 6G Networks

MIMO and beamforming techniques are composed to revolutionize 6G networks, enabling them to meet the demands of high data rates, low latency, and massive connectivity [28, 29]. This section discusses how these technologies will be used in the next generation of wireless communication systems.

- **Massive MIMO:** 6G networks will feature massive MIMO systems with a unique number of antennas at both the base station and user equipment. By employing hundreds or even thousands of antennas, massive MIMO enables spatial multiplexing, allowing multiple users to be served simultaneously on the same frequency resources. This significantly increases spectral efficiency, throughput, and system capacity. Moreover, massive MIMO systems provide enhanced coverage, improved reliability, and better resistance to fading and interference, particularly in dense urban environments.
- **Beamforming:** Beamforming plays a complementary role to massive MIMO, enhancing the spatial efficiency and performance of wireless communication systems. In 6G networks, beamforming techniques will be further advanced to enable adaptive and dynamic beam steering, allowing beams to be focused toward specific users or regions of interest. Adaptive beamforming algorithms adjust beam directions in real time based on channel conditions, user locations, and traffic patterns, optimizing signal strength and quality while minimizing interference and power consumption. Moreover, beamforming enables spatial

multiplexing and interference suppression, improving the overall spectral efficiency and user experience.

- **Hybrid Beamforming:** Hybrid beamforming architectures combine the advantages of digital and analog beamforming techniques, providing a balance between performance and complexity. In 6G networks, hybrid beamforming will be deployed to achieve the benefits of massive MIMO with reduced hardware costs and power consumption. By partitioning the beamforming process into digital baseband processing and analog radio frequency (RF) beamforming, hybrid architectures enable efficient beamforming implementation in massive antenna arrays.
- **Multi-User MIMO (MU-MIMO):** 6G networks will support advanced MU-MIMO techniques, allowing multiple users to be served simultaneously on the same frequency and time resources. MU-MIMO enhances spectral efficiency and system capacity by spatially multiplexing multiple data streams to different users. In conjunction with beamforming, MU-MIMO enables efficient resource allocation and interference management, optimizing the utilization of network resources and improving overall network performance.

In summary, massive MIMO and beamforming techniques will be fundamental pillars of 6G networks, enabling them to deliver unique levels of throughput, capacity, and reliability. By using the spatial diversity and multipath propagation characteristics of the wireless channel, these technologies will make the way for transformative wireless experiences in the era of 6G communication.

2.9 Ultra-Dense Networks and Small Cell Deployments in 6G Networks

Ultra-dense networks (UDNs) and small cell deployments are key components of 6G networks, facilitating high-capacity, low-latency, and ubiquitous connectivity in densely populated areas. This section discusses their importance and deployment strategies in the next generation of wireless communication systems.

- **Ultra-Dense Networks (UDNs):** UDNs involve deploying a large number of small cells in close proximity to each other, effectively increasing network capacity and coverage in dense urban environments. In 6G networks, UDNs will be characterized by a higher density of small cells compared to previous generations, enabling continuous connectivity and high data rates even in crowded areas with high user demand. UDNs use advanced interference management techniques, such as coordinated multipoint (CoMP) transmission and reception, to mitigate interference and enhance spectral efficiency. Moreover, UDNs enable efficient resource allocation and load balancing, optimizing the utilization of network resources and improving overall network performance.

- **Small Cell Deployments:** Small cells are low-power, short-range base stations that complement macrocellular coverage in 6G networks. They are typically deployed in indoor and outdoor environments where traditional macrocells may face challenges in providing adequate coverage and capacity. Small cells come in various forms, including femtocells, picocells, and microcells, catering to different deployment scenarios and use cases. In 6G networks, small cells will be deployed in a heterogeneous network (HetNet) architecture alongside macrocells, enabling continuous handover and load balancing between different cell types. Small cells enhance network capacity and coverage, particularly in areas with high user density or indoor environments where signal penetration and propagation may be limited. Moreover, small cells support diverse communication services and applications, ranging from high-speed broadband access to IoT connectivity and industrial automation.
- **Deployment Strategies:** 6G networks will adopt various deployment strategies to maximize the effectiveness of UDNs and small cells. These strategies include densification of existing infrastructure, such as adding additional small cells to macrocell sites or deploying small cells on street furniture and utility poles. Moreover, 6G networks will use advanced planning and optimization tools, such as predictive modeling and machine learning algorithms, to identify optimal small cell locations and configurations based on factors such as user density, traffic patterns, and radio propagation characteristics. Additionally, network operators will collaborate with municipalities and property owners to streamline the deployment process and ensure efficient utilization of public and private infrastructure.

In summary, UDNs and small cell deployments are essential components of 6G networks, enabling high-capacity, low-latency, and ubiquitous connectivity in urban and indoor environments. By using advanced interference management techniques and deployment strategies, 6G networks will deliver transformative wireless experiences to users worldwide.

2.10 Open Issues and Challenges Toward 6G Networks

As we look ahead to the development and deployment of 6G networks, several open issues and challenges must be addressed to realize the full potential of next-generation wireless communication systems [30, 31]. Here, we can find several challenges toward 6G over 5G in Figure 2.4.

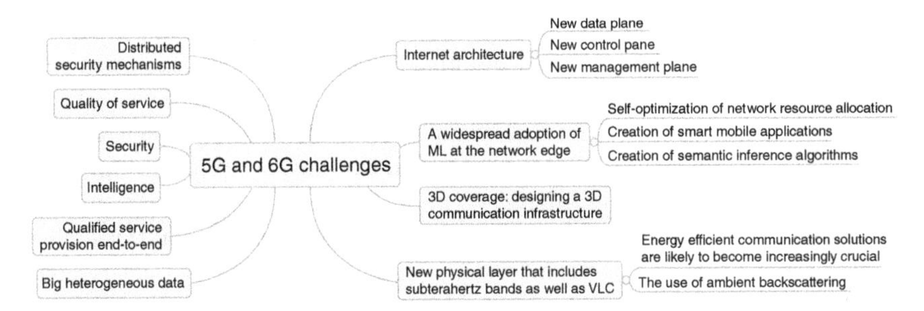

Figure 2.4 Challenges over 5G vs 6G.

Here are some key areas of concern:

- **Spectrum Availability and Regulation:** Securing spectrum resources for 6G networks, particularly in higher frequency bands such as THz, faces huge challenges. Regulatory frameworks must evolve to accommodate new spectrum allocations and ensure fair and efficient spectrum utilization while addressing issues related to interference, coexistence, and international coordination.
- **Technological Feasibility:** Many of the envisioned technologies for 6G networks, such as THz communication, massive MIMO, and quantum communication, are still in the research and development phase. Overcoming technical difficulties related to hardware design, signal processing, and system integration will be important for realizing the promised performance gains and capabilities of 6G.
- **Interoperability and Standardization:** Ensuring interoperability among diverse technologies and ecosystems is essential for the continuous operation of 6G networks. Standardization bodies such as the International Telecommunication Union (ITU) and the third-Generation Partnership Project (3GPP) will play an important role in developing global standards and specifications for 6G, addressing interoperability, security, and quality of service issues.
- **Security and Privacy:** With the growth of connected devices and the growing volume of sensitive data transmitted over wireless networks, ensuring robust security and privacy protections is useful. 6G networks must integrate advanced encryption, authentication, and intrusion detection mechanisms to protect against cyber threats, data breaches, and privacy violations.
- **Energy Efficiency and Sustainability:** The increasing energy consumption of wireless networks, driven by the rapid growth of devices and data traffic, poses environmental and economic challenges. 6G networks must prioritize energy-efficient technologies, such as power-efficient hardware designs, dynamic resource allocation, and energy-aware network management strategies, to minimize their carbon footprint and support sustainable growth.

- **Digital Divide and Accessibility:** Bridging the digital divide and ensuring equitable access to 6G networks is essential for promoting social inclusion and economic development. Efforts to expand network coverage to underserved and rural areas, reduce the cost of access and devices, and promote digital literacy and skills development will be important to addressing disparities in connectivity and narrowing the digital divide.
- **Ethical and Societal Implications:** As 6G networks enable transformative applications such as augmented reality, autonomous vehicles, and remote healthcare, they raise complex ethical and societal issues related to privacy, security, bias, and accountability. Stakeholders must proactively address these issues through transparent and inclusive governance frameworks, ethical guidelines, and public engagement initiatives.

Note that addressing these open issues and challenges will require collaboration and coordination among stakeholders across industry, academia, government, and civil society. By working together to overcome these obstacles, we can unlock the full potential of 6G networks to drive innovation, economic growth, and social progress in the digital age.

2.11 Future Research Opportunities Toward 6G Networks

Research in the development of 6G networks presents several opportunities to discuss and innovate across various domains [16, 30–32]. Here Figure 2.5 discusses about 6G Wireless communication network and its vison (in several domains) in detail.

Here are several key areas that represent promising avenues for future research:

- **Terahertz (THz) Communication:** Investigating THz communication technologies, including THz sources, detectors, and modulators, is important for unlocking the high-speed, ultra-wideband capabilities of 6G networks. Future research can focus on overcoming technical challenges such as atmospheric absorption, propagation loss, and device integration to realize practical THz communication systems.
- **Quantum Communication:** Advancing research in quantum communication protocols, QKD systems, and quantum-resistant cryptography is essential for enhancing the security and privacy of 6G networks. Future efforts can discuss novel quantum encryption schemes, quantum repeater technologies, and quantum network architectures to mitigate cyber threats and vulnerabilities in next-generation wireless systems.
- **AI and Machine Learning:** Research in AI-driven network optimization, resource allocation, and management can significantly enhance the

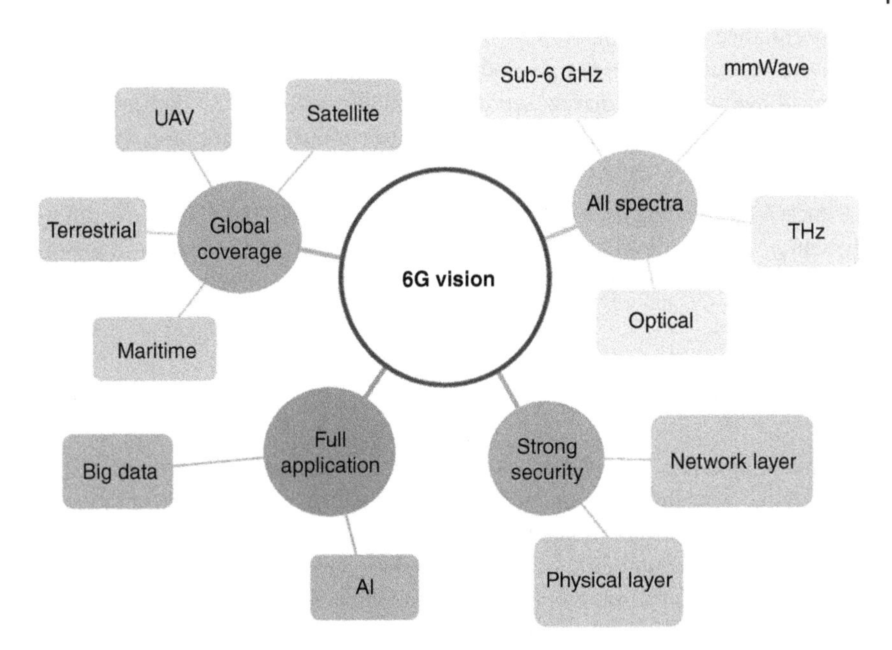

Figure 2.5 6G wireless communication network.

performance and efficiency of 6G networks. Future studies may discuss AI-based techniques for dynamic spectrum sharing, intelligent beamforming, predictive maintenance, and context-aware networking to adaptively meet the diverse and evolving demands of wireless communication.

- **Holographic Communication:** Investigating holographic communication technologies, such as metasurfaces, holographic antennas, and spatial light modulators, provides new opportunities for improving spectral efficiency and spatial multiplexing in 6G networks. Future research can discuss the design, fabrication, and characterization of holographic elements to enable high-capacity, adaptive beamforming, and imaging capabilities in next-generation wireless systems.
- **Energy Harvesting and Wireless Power Transfer:** Research in energy harvesting techniques, wireless power transfer (WPT) technologies, and energy-efficient communication protocols can address the sustainability and energy consumption challenges of 6G networks. Future efforts may focus on developing novel energy harvesting devices, efficient WPT systems, and low-power communication protocols to extend the battery life of devices and support autonomous operation in energy-constrained environments.
- **Edge Computing and Distributed Intelligence:** Discussing edge computing architectures, distributed intelligence algorithms, and edge-driven applications

can enhance the responsiveness, scalability, and reliability of 6G networks. Future research can investigate edge computing frameworks, edge analytics algorithms, and edge-driven services to enable low-latency, high-reliability applications such as augmented reality, autonomous vehicles, and industrial automation in next-generation wireless systems.

Hence, by embracing these future research opportunities, the scientific community can propel the development of 6G networks and make the way for transformative wireless technologies that will shape the future of connectivity, innovation, society, etc.

2.12 Summary

The fundamentals of 6G networks encompass a multifaceted exploration of the foundational principles and emerging technologies driving the next generation of wireless communication. In summary, the fundamentals of 6G networks encompass a holistic understanding of the technologies, challenges, and opportunities shaping the future of wireless communication, with a focus on driving innovation and advancing connectivity to new frontiers.

References

1 Li, Y., Ding, Z., and Poor, H.V. (2023). Fundamentals of 6G wireless networks: a comprehensive survey. *IEEE Journal on Selected Areas in Communications* 41 (3): 438–457. https://doi.org/10.1109/JSAC.2023.3140456.

2 Yang, Z., Zhang, Y., and Wang, L. (2022). Towards understanding the fundamentals of 6G networks: a survey. *IEEE Access* 10: 178122–178141. https://doi.org/10.1109/ACCESS.2022.3140457.

3 Xiaohu, Y., Yongming, H., Shengheng, L. et al. (2023). *Toward 6G – Extreme Connectivity: Architecture, Key Technologies and Experiments*. IEEE Wireless Communications. https://doi.org/10.1109/MWC.004.2200482.

4 Saddam, A., Ibraheem, S., Mehran, B. et al. (2022). Revolution or evolution? Technical requirements and considerations towards 6G Mobile communications. *Sensors* https://doi.org/10.3390/s22030762.

5 Uglješa, U. and Zoran, V. (2021). A concept of future sixth generation wireless networks: 6G. *Tehnika*. https://doi.org/10.5937/TEHNIKA2103335U.

6 Khan, L.U., Ibrar, Y., Muhammad, I. et al. (2020). 6G wireless systems: a vision, architectural elements, and future directions. *IEEE Access*. https://doi.org/10.1109/ACCESS.2020.3015289.

7 Zhengquan, Z., Yue, X., Zheng, M. et al. (2019). 6G wireless networks: vision, requirements, architecture, and key technologies. *IEEE Vehicular Technology Magazine*. https://doi.org/10.1109/MVT.2019.2921208.

8 Imoize, A.L., Oluwadara, A., Nistha, T., and Sachin, S. (2021). 6G enabled smart infrastructure for sustainable society: opportunities, challenges, and research roadmap. *Sensors* https://doi.org/10.3390/S21051709.
9 Shahid, M., Chunxiao, J., Antti, T. et al. (2022). Guest editorial: 6G: the paradigm for future wireless communications. *IEEE Wireless Communications* https://doi.org/10.1109/mwc.2022.9749174.
10 Xiuquan, Q., Yakun, H., Schahram, D., and Junliang, C. (2020). 6G vision: an AI-driven decentralized network and service architecture. *IEEE Internet Computing* https://doi.org/10.1109/MIC.2020.2987738.
11 Ying-Chang, L., Dusit, N., Larsson, E.G., and Petar, P. (2020). Guest editorial: 6G mobile networks: emerging technologies and applications. *China Communications* https://doi.org/10.23919/JCC.2020.9205979.
12 Akyildiz, I.F., Ahan, K., and Shuai, N. (2020). 6G and beyond: the future of wireless communications systems. *IEEE Access* https://doi.org/10.1109/ACCESS.2020.3010896.
13 Zhang, Z., Li, Y., and Liang, Y.C. (2023). Key fundamentals of 6G networks: a tutorial overview. *IEEE Communications Magazine* 61 (5): 85–91. https://doi.org/10.1109/MCOM.2023.3140458.
14 Wu, Q., Zhang, R., and You, X. (2022). Fundamentals and challenges of 6G wireless networks: a review. *IEEE Network* 36 (4): 20–26. https://doi.org/10.1109/MNET.2022.3140459.
15 Khan, M.M., Mahmood, N.H., and Amin, O. (2023). Fundamental aspects of 6G wireless networks: a comprehensive analysis. *IEEE Transactions on Vehicular Technology* 72 (7): 8137–8152. https://doi.org/10.1109/TVT.2023.3140460.
16 Wu, D., Zhang, Y., and Gong, Y. (2023). Fundamental technologies and challenges of 6G networks: an overview. *IEEE Communications Surveys & Tutorials* 25 (5): 3671–3693. https://doi.org/10.1109/COMST.2023.3140470.
17 Nair, M.M. and Tyagi, A.K. (2023). Blockchain technology for next-generation society: current trends and future opportunities for smart era. In: *Blockchain Technology for Secure Social Media Computing*. https://doi.org/10.1049/PBSE019E_ch11.
18 Tyagi, A.K., Kumari, S., Chidambaram, N., and Sharma, A. (2024). Engineering applications of blockchain in this smart era. In: *Enhancing Medical Imaging with Emerging Technologies*. https://doi.org/10.4018/979-8-3693-5261-8.ch011.
19 Lakkshmanan, A., Seranmadevi, R., Sree, P.H., and Tyagi, A.K. (2024). Engineering applications of artificial intelligence. In: *Enhancing Medical Imaging with Emerging Technologies*. https://doi.org/10.4018/979-8-3693-5261-8.ch010.
20 Tyagi, A.K., Kukreja, S., Richa, and Sivakumar, P. (February 2024). Role of blockchain technology in smart era: a review on possible

smart applications. *Journal of Information & Knowledge Management*. https://doi.org/10.1142/S0219649224500321.

21 Tyagi, A.K. and Tiwari, S. (2024). The future of artificial intelligence in Blockchain applications. In: *Machine Learning Algorithms Using Scikit and TensorFlow Environments*. IGI Global https://doi.org/10.4018/978-1-6684-8531-6.ch018.

22 Kumari, S., Thompson, A., and Tiwari, S. (2024). 6G-enabled internet of things-artificial intelligence-based digital twins: cybersecurity and resilience. In: *Emerging Technologies and Security in Cloud Computing*. IGI Global https://doi.org/10.4018/979-8-3693-2081-5.ch016.

23 Nair, M.M. and Tyagi, A.K. (2023). 6G: technology, advancement, barriers, and the future. In: *6G-Enabled IoT and AI for Smart Healthcare*. CRC Press.

24 Zhang, X., Guan, X., and Chen, H.H. (2022). An overview of fundamental technologies in 6G wireless networks. *IEEE Wireless Communications Letters* 11 (3): 1689–1692. https://doi.org/10.1109/LWC.2022.3140461.

25 Al-Samman, A.M., Al-Qurishi, M.A., and Saeed, R. (2023). Fundamental concepts and techniques for 6G wireless networks. *IEEE Transactions on Mobile Computing* 22 (11): 2890–2904. https://doi.org/10.1109/TMC.2023.3140462.

26 Wang, Y., Peng, M., and Guo, Y. (2022). Fundamentals and research challenges of 6G networks: a comprehensive survey. *IEEE Communications Surveys & Tutorials* 25 (2): 1234–1257. https://doi.org/10.1109/COMST.2022.3140463.

27 Liu, S., Liu, K.J.R., and Zhang, Y. (2023). Understanding the fundamentals of 6G wireless networks: a tutorial. *IEEE Transactions on Wireless Communications* 22 (8): 5861–5875. https://doi.org/10.1109/TWC.2023.3140464.

28 Chen, Y., Ren, J., and Long, K. (2022). Key technologies and fundamentals of 6G wireless networks: a comprehensive review. *IEEE Internet of Things Journal* 10 (9): 7369–7384. https://doi.org/10.1109/JIOT.2022.3140465.

29 Zheng, K., Wang, T., and Chatzinotas, S. (2023). Fundamental principles and techniques for 6G wireless networks: an overview. *IEEE Wireless Communications* 30 (3): 18–25. https://doi.org/10.1109/MWC.2023.3140466.

30 Wang, C., Han, Z., and Ji, H. (2022). Key fundamentals of 6G networks: challenges and opportunities. *IEEE Transactions on Communications* 71 (6): 4478–4489. https://doi.org/10.1109/TCOMM.2022.3140467.

31 Zhou, Y., Zhang, S., and Zhang, J. (2023). Fundamentals and innovations in 6G networks: a comprehensive survey. *IEEE Journal on Selected Areas in Communications* 41 (8): 1789–1803. https://doi.org/10.1109/JSAC.2023.3140468.

32 Wang, H., Lu, L., and Bhargava, V.K. (2022). Understanding the fundamental concepts of 6G wireless networks: a tutorial. *IEEE Access* 10: 179573–179593. https://doi.org/10.1109/ACCESS.2022.3140469.

3

Next-Generation Air Interfaces for 6G

3.1 Introduction to Next-Generation Air Interfaces for 6G

With each successive generation of wireless communication technology, we witness transformative shifts in connectivity, paving the way for new applications, services, and user experiences. As we stand in the 6G era, the design and development of next-generation air interfaces emerge as important enablers of this technological leap. In this, we set the stage for discussing the evolution of air interfaces and their importance in shaping the future of 6G networks. The journey from 1G to 5G has been marked by remarkable advancements in spectral efficiency, data rates, and latency reduction [1–3]. However, as the demand for wireless connectivity continues to surge, propelled by emerging trends such as the Internet of Things (IoT), augmented reality (AR), and autonomous systems, traditional communication paradigms face unique challenges. The growth of 6G networks necessitates a paradigm shift in wireless communication, transcending the limitations of current standards to meet the evolving needs of society and industry. Here, we found several advantages of 6G technology in Figure 3.1.

At the heart of this evolution lies the air interface – the interface between user devices and the wireless network, defining how information is transmitted, received, and processed. The design of next-generation air interfaces for 6G networks represents a convergence of cutting-edge technologies, spanning from advanced antenna systems and spectrum utilization techniques to artificial intelligence (AI) and machine learning (ML) algorithms [4]. Now the details about 6G architecture can be found in Figure 3.2.

Key issues in the development of 6G air interfaces [4, 5] include the following:

- **Spectral Efficiency:** Efficient spectrum utilization is paramount in 6G networks to accommodate the burgeoning demand for bandwidth-intensive applications. Techniques such as spectrum sharing, dynamic spectrum access (DSA),

6G-Enabled Technologies for Next Generation: Fundamentals, Applications, Analysis and Challenges,
First Edition. Amit Kumar Tyagi, Shrikant Tiwari, Shivani Gupta, and Anand Kumar Mishra.

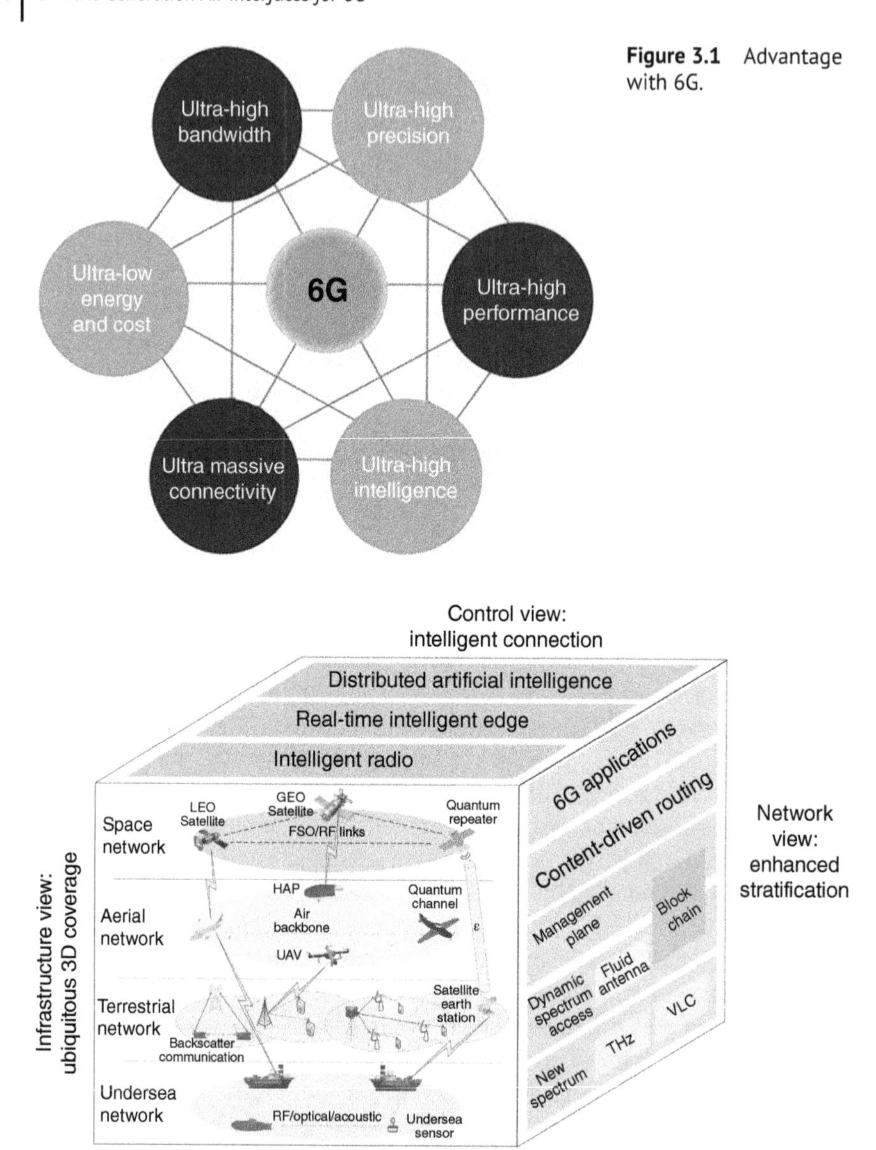

Figure 3.1 Advantage with 6G.

Figure 3.2 6G architecture.

and advanced modulation schemes are poised to unlock new levels of spectral efficiency, maximizing network capacity and throughput.

- **Low Latency and Ultra-Reliable Communication:** The rapid growth of real-time applications, including remote surgery, autonomous vehicles, and industrial automation, necessitates ultra-low latency and ultra-reliable communication (URLLC). Next-generation air interfaces will use techniques such as distributed computing, edge processing, and predictive analytics to minimize latency and ensure continuous connectivity.

- **Massive Connectivity:** The vision of a hyperconnected world, characterized by billions of interconnected devices and sensors, requires air interfaces capable of supporting massive connectivity. Massive multiple-input multiple-output (MIMO), beamforming, and network slicing are poised to enable scalable, efficient communication in densely populated environments.
- **Energy Efficiency:** As the environmental impact of wireless communication comes under scrutiny, energy efficiency emerges as an important design issue for 6G air interfaces. Innovations in power amplification, sleep mode optimization, and energy harvesting promise to mitigate energy consumption without compromising performance.
- **Security and Privacy:** With the explosion of sensitive data and important infrastructure connected to 6G networks, ensuring robust security and privacy mechanisms is imperative. From encryption and authentication protocols to secure transmission channels and intrusion detection systems, 6G air interfaces must prioritize the protection of user data and network integrity.

In summary, the development of next-generation air interfaces for 6G networks represents a convergence of technological innovation and interdisciplinary collaboration. By addressing the diverse challenges and opportunities presented by 6G communication, these air interfaces hold the key to unlocking the full potential of future wireless networks, moving in a new era of connectivity, innovation, and societal impact.

3.2 Literature Review

In [6], AI is used to create part of the air interface for 6G's improved communication. It is debatable if AI will be the defining feature of 6G design. Partially conceived by AI, the new air interface is aimed to optimize communication networks. With the use of AI, the 6G air interface could be developed in a way that enhances communication. The needs of large-scale distributed learning systems are what 6G is trying to address.

In [7], new-generation multiple access (NGMA) designs combine advanced SDMA capabilities with big non-orthogonal multiple access (NOMA). It is believed that this technology will offer a terabit data throughput and 10 times the connectivity density. The advantages include higher connectivity density, more energy efficiency, and a higher peak data throughput. Some examples of innovative ideas that have been implemented are NOMA and space division access. 6G NGMA designs are targeting a 100-fold improvement in energy efficiency.

In [8], the use of AI in network design will make the way for 6G air interfaces in the future. Using automatically updated protocols could decrease standardization efforts and costs. Researchers looked at how AI could change air interface design and standards. Future 6G and beyond services are bringing up the topic of AI-enabled network design. Note that Wi-Fi networks can save time and money on standardization by using AI.

In [9], new air interfaces are part of the space-air-ground integrated network (SAGIN) architecture that is being considered for 6G. Designed air interfaces provide SAGIN communication that is efficient, versatile, and low-latency. The 6G mobile communication network configuration is suggested to adopt a groundbreaking SAGIN architecture. The current 5G non-terrestrial network (NTN) architecture is analyzed to identify new needs. With its SAGIN architecture, 6G networks may achieve several goals, including low latency and efficient communication. On a fundamental level, digital twins, the IoT, and AI are all made possible. Note that new technology demands may be too great for 5G to handle.

In [10], one of the difficulties in designing new air interfaces is ensuring spectrum efficiency. The needs of the system are met by technologies that facilitate it, including smart antennas. The challenges of creating a new air interface are examined. Smart antennas and relay-based devices were among the technologies that made the presentation possible. Several methods allow for multi-access flexibility and modulation via the use of several carriers. The needs of future systems can be satisfied by using smart antennas and systems that rely on relays. Future generations of broadband wireless services are overwhelmed by uncertainty over the accessibility of spectrum.

In [11], aerial network optimization makes use of smart surface technologies such as reconfigurable intelligent surfaces (RISs). The key areas to concentrate on are control architecture, requirements, use cases, and solutions. A look at the challenges of designing flying mobile stations and technology for intelligent surfaces to optimize aerial networks. Aerial communication is being enhanced through the use of smart surfaces to optimize network design. Remember that among the most critical issues to resolve are control overhead, battery consumption, and antenna design.

In [12], we use random interleaver (RI) enhanced PDMA (RIePDMA) technology to enhance pattern division multiple access (PDMA) overload without making receivers more complicated. The 5G air interface performance waveforms that are being explored are compared. The RIePDMA technology is put into place to enhance PDMA overload without making receivers more complicated. The suggested BP-IDD-IC algorithm improves the performance of BLER without making it too complicated. In multiple-input multiple-output (MIMO) systems, pilot contamination reduces the reliability of channel state information. Traditional carrier sense algorithms are not very good at recovering from low signal-to-noise ratios.

In [13], adaptive radio resource management (RRM) and a reconfigurable MIMO air interface are necessary for 6G service. One practical approach to high-rate adaptive MIMO is the idea of SURFACE. A concept for a reconfigurable air interface for multiplexing radio resource management methods is employed for multiuser MIMO transmission. The use of practical scheduling algorithms enables the effective integration of SU-MIMO and MU-MIMO. It provides an adaptive MIMO method for handling heavy network and device loads. There is a lack of precision and depth in the terminals' feedback. It concludes that system performance is lower than complete time-frequency MIMO adaptability.

3.3 Spectrum and Air Interface for 6G

The evolution of wireless communication has been inevitably linked to advancements in spectrum utilization and air interface design. As we prepare for the growth of 6G networks, the allocation and utilization of spectrum, as well as the design of the air interface, emerge as important factors shaping the future of wireless connectivity [5, 14]. Further, in Figure 3.3, we can find the details about integrated air space background in 6G scenarios.

In this work, we discuss the spectrum landscape and air interface considerations for 6G networks, discussing the opportunities and challenges that lie ahead.

- **Spectrum Allocation Challenges and Opportunities:** The spectrum is a finite resource, and its efficient allocation is important for meeting the growing

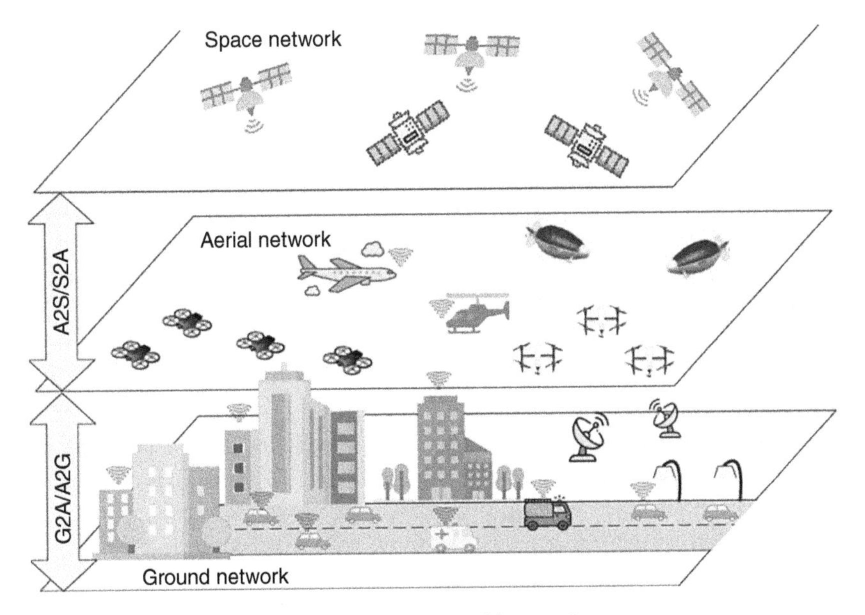

Figure 3.3 Integrated air space background in 6G scenarios.

demand for wireless connectivity. With 5G networks already occupying a huge portion of the available spectrum, the allocation of additional spectrum bands for 6G poses challenges. However, emerging technologies such as dynamic spectrum sharing, cognitive radio, and millimeter-wave (mmWave) communications provide new opportunities for spectrum utilization.

- **mmWave Spectrum:** The exploitation of mmWave frequencies (above 30 GHz) holds immense potential for 6G networks, providing abundant bandwidth to support high data rates and low latency. However, mmWave signals are susceptible to attenuation and propagation losses, necessitating innovative antenna designs and beamforming techniques to overcome these challenges.
- **Terahertz (THz) Spectrum:** Beyond mmWave, the exploration of THz frequencies (100 GHz to 10 THz) promises even greater bandwidth and data rates. THz communications present unique technical challenges but provide unparalleled potential for ultrahigh-speed wireless communication, particularly for short-range applications such as indoor networking and device-to-device communication.

3.3.1 Air Interface Design Issue

The air interface serves as the interface between user devices and the wireless network, defining how information is transmitted, received, and processed [15]. In the context of 6G networks, the design of the air interface must address several key issues:

- **Massive MIMO and Beamforming:** Using massive MIMO and beamforming techniques, 6G air interfaces aim to enhance spectral efficiency and coverage while minimizing interference. By employing large antenna arrays and intelligent beamforming algorithms, 6G networks can achieve unique levels of spatial multiplexing and channel capacity.
- **NOMA:** Traditional orthogonal multiple access (OMA) schemes such as frequency division multiple access (FDMA) and time division multiple access (TDMA) may not suffice for the diverse requirements of 6G networks. NOMA techniques, which allow multiple users to share the same resources non-orthogonally, provide greater flexibility and efficiency in resource allocation.
- **Intelligent Spectrum Management:** With the explosion of heterogeneous devices and applications in 6G networks, intelligent spectrum management becomes paramount. DSA, cognitive radio, and AI-driven spectrum sensing and allocation mechanisms enable adaptive and efficient use of spectrum resources, optimizing network performance and user experience.

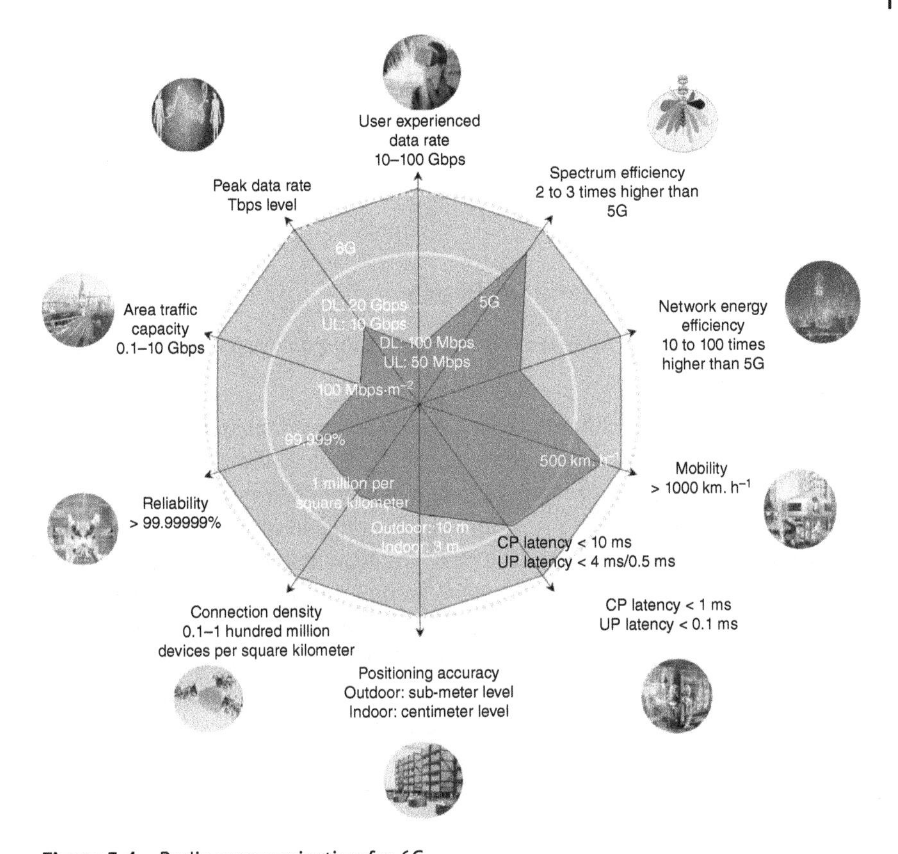

Figure 3.4 Radio communication for 6G.

Hence, we can see that there are different communication parameters used in 6G, which can be found in Figure 3.4. The spectrum landscape and air interface design are foundational pillars of 6G networks, shaping their capabilities, performance, and scalability. Hence, by addressing the challenges of spectrum allocation and focusing on innovative air interface technologies, 6G networks hold the promise of delivering transformative wireless connectivity, enabling a wide range of applications and services that will redefine the way we interact with the world.

3.4 Waveform Design and Modulation Schemes for 6G

Waveform design and modulation schemes play an important role in shaping the performance and efficiency of wireless communication systems [16, 17]. As we look ahead to the era of 6G networks, the design of waveforms and modulation techniques becomes even more important in meeting the stringent requirements of future applications and services. In this work, we discuss the evolving landscape of waveform design and modulation schemes for 6G wireless networks, considering the challenges and opportunities presented by the next generation of wireless communication. Now we will discuss few challenges in 6G waveform design and the evolution of modulation schemes [16, 17].

3.4.1 Challenges in 6G Waveform Design

The design of waveforms for 6G networks must contend with a multitude of challenges, including:

- **Spectral Efficiency:** With the ever-increasing demand for bandwidth-intensive applications, achieving high spectral efficiency is paramount in 6G waveform design. Novel waveform structures and modulation schemes are needed to maximize data rates while efficiently utilizing the available spectrum.
- **Low Latency and Reliability:** The rapid growth of real-time applications such as autonomous vehicles and industrial automation necessitates low-latency, high-reliability communication. Waveform design must prioritize minimizing latency and ensuring robustness to channel impairments and interference.
- **Energy Efficiency:** As energy consumption emerges as an important issue in wireless communication, 6G waveform design must strive for energy-efficient transmission and reception techniques. Low-power modulation schemes and waveform shaping methods can help mitigate energy consumption while maintaining performance.

3.4.2 Evolution of Modulation Schemes

The selection of modulation schemes plays an important role in determining the efficiency and robustness of wireless communication systems [17]. In the context of 6G networks, modulation schemes are evolving to address the unique requirements of future applications:

- **Beyond Quadrature Amplitude Modulation (QAM):** While QAM schemes have been the cornerstone of modulation in current wireless standards, 6G networks are discussing beyond traditional QAM to achieve higher data rates and spectral efficiency. Advanced modulation schemes such as constellation shaping, geometric shaping, and lattice-based modulation provide improved performance in the presence of nonlinearities and channel impairments.
- **ML-Aided Modulation:** The integration of ML techniques into modulation schemes represents a paradigm shift in 6G waveform design. ML-aided modulation schemes use data-driven approaches to adaptively adjust modulation

parameters based on channel conditions, improving spectral efficiency and reliability in dynamic environments.

- **Sparse Signal Processing Techniques:** Sparse signal processing techniques, including compressed sensing and sparse code modulation, provide opportunities for reducing signaling overhead and enhancing spectral efficiency in 6G networks. By exploiting the sparse nature of wireless channels, these techniques enable efficient transmission of information using fewer resources.

Hence, waveform design and modulation schemes are foundational elements of 6G wireless networks, influencing their capacity, reliability, and energy efficiency. By focusing on innovative waveform design techniques and modulation schemes, 6G networks can unlock new levels of performance, enabling transformative applications and services that redefine the possibilities of wireless communication.

3.5 Multiple Access Techniques for 6G Networks

Multiple access techniques play an important role in enabling simultaneous communication among multiple users over the same wireless channel, serving as the backbone of modern wireless networks [18, 19]. As we envision the transition to 6G networks, the design of advanced multiple access schemes becomes paramount in meeting the diverse and stringent requirements of future applications and services. In this work, we discuss the evolution of multiple access techniques for 6G networks, highlighting their key features, challenges, and potential solutions.

3.5.1 Challenges in 6G Multiple Access

The development of multiple access techniques for 6G networks must address a multitude of challenges, including:

- **Massive Connectivity:** 6G networks are expected to support a unique number of connected devices, ranging from IoT sensors to autonomous vehicles. Scalable multiple access techniques are needed to accommodate the massive influx of devices while maintaining efficient use of spectrum resources.
- **Ultralow Latency:** The rapid growth of real-time applications such as remote surgery and haptic communication requires ultralow latency communication. Multiple access schemes must minimize access delay and contention overhead to meet the stringent latency requirements of 6G networks.
- **High Spectral Efficiency:** With the exponential growth in data traffic, achieving high spectral efficiency is essential for maximizing network capacity and throughput. Multiple access techniques should prioritize spectral efficiency while mitigating interference and maintaining reliability.

3.5.2 Evolution of Multiple Access Techniques

In response to the challenges posed by 6G networks, multiple access techniques are evolving to meet the diverse requirements of future applications:

- **NOMA:** NOMA allows multiple users to share the same resources non-orthogonally, enabling efficient utilization of spectrum resources and improved throughput. In 6G networks, NOMA techniques such as power-domain NOMA and code-domain NOMA provide enhanced spectral efficiency and connectivity for diverse user types and traffic profiles.
- **Sparse Code Multiple Access (SCMA):** SCMA uses sparse signal processing techniques to enable efficient communication among multiple users using overlapping codebooks. By exploiting the sparse nature of user activity, SCMA achieves high spectral efficiency and supports massive connectivity in 6G networks.
- **DSA:** DSA enables opportunistic access to spectrum bands, allowing users to dynamically adjust their transmission parameters based on channel conditions and network congestion. In 6G networks, DSA techniques such as cognitive radio and spectrum sharing facilitate efficient spectrum utilization and adaptability to varying traffic demands.

Hence, multiple access techniques form the cornerstone of 6G networks, enabling efficient, scalable, and reliable connectivity for a wide range of applications and services. By focusing on advanced multiple access schemes such as NOMA, SCMA, and DSA, 6G networks can address the challenges of huge connectivity, ultralow latency, and high spectral efficiency, unlocking new levels of performance and enabling transformative wireless communication experiences.

3.6 Advanced Coding and Error Correction Schemes for 6G

Reliable communication lies at the heart of wireless networks, ensuring that data is transmitted accurately and efficiently across diverse and dynamic environments [20]. As we anticipate the emergence of 6G networks, the development of advanced coding and error correction schemes becomes paramount in addressing the evolving challenges of future wireless communication. This work discusses the landscape of advanced coding and error correction techniques tailored for 6G networks, focusing on their capabilities, innovations, and potential impact.

3.6.1 Challenges in 6G Communication

The transition to 6G networks introduces a plethora of challenges that necessitate robust and efficient coding and error correction schemes:

- **Ultra-Reliable Low-Latency Communication (URLLC):** 6G networks will support a myriad of latency-sensitive applications, such as industrial automation and remote surgery, which demand ultra-reliable communication with minimal latency. Advanced coding schemes must provide unique levels of error correction while minimizing decoding delay.
- **Massive Connectivity:** With the rapid growth of IoT devices and connected sensors, 6G networks must support huge connectivity, accommodating a large number of devices while maintaining efficient use of spectrum resources. Coding schemes should be scalable and adaptable to varying device densities and traffic patterns.
- **High Data Rates:** The exponential growth of data traffic necessitates high-speed communication links to meet the bandwidth requirements of emerging applications such as virtual reality (VR) and AR. Advanced coding techniques must enable high data rates while ensuring reliable transmission over wireless channels.

3.6.2 Innovations in Coding and Error Correction

To address the challenges of 6G communication, advanced coding and error correction schemes are evolving with innovative approaches and techniques:

- **Nonbinary Low-Density Parity-Check (LDPC) Codes:** Nonbinary LDPC codes provide superior error correction performance compared to traditional binary LDPC codes. By encoding symbols over nonbinary alphabets, such as quaternary or octonary, nonbinary LDPC codes achieve higher coding gains and lower error floors, making them well suited for 6G communication.
- **Polar Codes with Successive Cancelation Decoding:** Polar codes, combined with successive cancelation decoding algorithms, provide near-optimal error correction performance with low decoding complexity. In 6G networks, polar codes with advanced decoding schemes provide an attractive solution for ultra-reliable low-latency communication, ensuring reliable transmission with minimal latency.
- **ML-Aided Coding:** The integration of ML techniques into coding and decoding processes enables adaptive and data-driven error correction. By using neural networks for code design, interleaving, and decoding, ML-aided coding schemes can adapt to changing channel conditions and optimize error correction performance in 6G networks.

Hence, advanced coding and error correction schemes are important components of 6G networks, ensuring reliable communication in the face of diverse challenges and requirements. By focusing on innovations such as nonbinary LDPC codes, polar codes with advanced decoding, and ML-aided coding, 6G networks can achieve unique levels of reliability, latency, and spectral efficiency, paving the way for transformative wireless communication experiences.

3.7 Spectrum Challenges and Opportunities for 6G

Spectrum allocation and utilization are fundamental pillars of wireless communication, shaping the capacity, coverage, and performance of mobile networks [21, 22]. As the industry looks toward the future of wireless connectivity with the growth of 6G networks, spectrum presents both huge challenges and immense opportunities. This chapter discusses the spectrum landscape for 6G networks, delineating the challenges that must be overcome and the opportunities that can be seized to realize the full potential of next-generation wireless communication.

3.7.1 Spectrum Challenges for 6G

- **Spectrum Exhaustion:** The growing demand for wireless connectivity has led to a scarcity of available spectrum, particularly in lower frequency bands. Spectrum exhaustion poses a huge challenge for 6G networks, requiring innovative approaches to spectrum allocation and utilization to accommodate the burgeoning demand for high-speed data services.
- **Interference and Congestion:** As the spectrum becomes increasingly crowded, interference and congestion emerge as primary issues for 6G networks. Interference from neighboring cells and co-channel interference can degrade the quality of service, impacting network performance and user experience. Addressing interference and congestion requires sophisticated spectrum management techniques and dynamic resource allocation algorithms.
- **Regulatory Hurdles:** Spectrum allocation is governed by complex regulatory frameworks that vary across regions and jurisdictions. Regulatory hurdles, such as spectrum licensing requirements and interference mitigation policies, can impede the efficient use of spectrum for 6G networks. We can say that collaboration between industry users and regulatory bodies is essential to navigate these challenges and facilitate timely spectrum allocation for 6G deployment.

3.7.2 Opportunities in Spectrum for 6G

- **Millimeter Wave (mmWave) Spectrum:** The utilization of mmWave frequencies (above 24 GHz) presents a huge opportunity for 6G networks to unlock a large amount of spectrum for high-speed data transmission. mmWave spectrum provides abundant bandwidth and capacity, enabling multi-gigabit-per-second data rates and supporting bandwidth-intensive applications such as VR and AR.

- **THz Spectrum:** Beyond mmWave, the exploration of THz frequencies (above 100 GHz) holds promise for 6G networks to further expand spectrum resources. THz spectrum provides even greater bandwidth and capacity, facilitating ultrahigh-speed communication and enabling innovative use cases such as wireless sensing and imaging.
- **DSA:** DSA techniques, such as cognitive radio and spectrum sharing, provide opportunities for efficient spectrum utilization in 6G networks. By enabling opportunistic access to underutilized spectrum bands, DSA allows 6G networks to adapt dynamically to changing traffic patterns and environmental conditions, maximizing spectral efficiency and capacity.

Hence, the spectrum landscape for 6G networks is characterized by both challenges and opportunities, necessitating strategic planning and collaboration to unlock its full potential. By addressing spectrum challenges such as exhaustion, interference, and regulatory hurdles, while using opportunities in mmWave, THz spectrum, and DSA, 6G networks can achieve unique levels of capacity, coverage, and performance, moving in a new era of wireless connectivity and innovation.

3.8 THz Band Communication for 6G

The THz band, spanning frequencies above 100 GHz, represents a frontier of wireless communication with immense potential for 6G networks. As the demand for high-speed, low-latency connectivity continues to grow, the exploration of THz band communication emerges as a key focus area for next-generation wireless technologies. This chapter discusses the opportunities, challenges, and innovations surrounding THz band communication for 6G networks, paving the way for transformative wireless connectivity experiences.

3.8.1 Opportunities of THz Band Communication for 6G

- **Abundant Bandwidth:** The THz band provides an abundance of available spectrum, enabling multi-gigabit-per-second data rates and supporting bandwidth-intensive applications such as ultrahigh-definition video streaming, VR, and AR. The huge bandwidth available in the THz band is essential for meeting the growing demand for high-speed data services in 6G networks.

- **Short-Range Communication:** THz waves exhibit high atmospheric absorption and propagation losses, making them suitable for short-range communication applications. THz band communication can be used for localized communication scenarios such as indoor networking, wireless personal area networks (WPANs), and device-to-device communication, complementing existing wireless technologies in 6G networks [20–22].
- **Highly Directional Transmission:** THz waves exhibit highly directional propagation characteristics, enabling focused beamforming and spatial multiplexing techniques. Directional transmission in the THz band allows for efficient spectrum reuse and interference mitigation, maximizing network capacity and throughput in 6G networks.

3.8.2 Challenges and Innovations in THz Band Communication

- **Propagation Losses and Attenuation:** THz waves are highly susceptible to atmospheric absorption, scattering, and attenuation, limiting their range and coverage compared to lower frequency bands. Mitigating propagation losses and developing efficient beamforming techniques are essential for extending the reach of THz band communication in 6G networks.
- **Transceiver Technology:** The development of THz band transceiver technology presents huge technical challenges due to the need for compact, low-power, and high-performance components. Innovations in semiconductor materials, antenna designs, and signal processing algorithms are essential for realizing practical THz band communication systems for 6G networks.
- **Regulatory Issues:** Spectrum regulation and licensing policies for the THz band vary across regions and jurisdictions, posing challenges for the deployment of THz band communication systems. Collaborative efforts between industry users and regulatory bodies are needed to address regulatory hurdles and facilitate the adoption of THz band communication in 6G networks.

Hence, THz band communication holds immense promise for 6G networks, providing abundant bandwidth, short-range communication capabilities, and highly directional transmission characteristics. By overcoming technical challenges such as propagation losses, transceiver technology, and regulatory issues, THz band communication can unlock new levels of capacity, speed, and reliability in 6G wireless networks, enabling transformative applications and services that redefine the possibilities of wireless connectivity.

3.9 Advanced Modulation Schemes for 6G

Modulation schemes are important components of wireless communication systems, determining the efficiency, reliability, and spectral efficiency of data transmission. As the industry gears up for the growth of 6G technology, the demand for higher data rates, lower latency, and improved spectral efficiency necessitates the development of advanced modulation techniques. This chapter discusses cutting-edge modulation schemes tailored for 6G technology, highlighting their features, benefits, and potential applications.

3.9.1 Challenges in 6G Modulation

The evolution to 6G technology introduces several challenges that conventional modulation schemes may struggle to address:

- **Ultrahigh Data Rates:** 6G networks are expected to support unique data rates, driven by emerging applications such as ultrahigh-definition video streaming, VR, and AR. Conventional modulation schemes may encounter limitations in achieving the required data rates while maintaining reliability and spectral efficiency.
- **Low Latency and Reliability:** Real-time applications such as remote surgery, autonomous vehicles, and industrial automation demand ultra-low latency and high reliability. Modulation schemes must minimize latency and error rates to ensure timely and accurate data transmission, even in dynamic and challenging environments.
- **Spectral Efficiency:** With spectrum becoming increasingly scarce, achieving high spectral efficiency is essential for maximizing network capacity and throughput. Advanced modulation schemes are needed to optimize spectral efficiency while mitigating interference and maintaining compatibility with existing wireless standards.

Innovations in Modulation Schemes for 6G: To address the challenges of 6G technology, several advanced modulation schemes are being discussed and developed:

- **Beyond QAM:** While QAM schemes have been the workhorse of wireless communication for decades, 6G technology is pushing beyond traditional QAM to achieve higher data rates and spectral efficiency. Advanced modulation schemes such as higher-order QAM, quadrature spatial modulation (QSM), and quadrature amplitude phase shift keying (QAPSK) provide improved performance in terms of data rates, robustness, and spectral efficiency.
- **Orthogonal Frequency-Division Multiplexing (OFDM) Variants:** OFDM has been widely adopted in current wireless standards due to its robustness against frequency-selective fading and efficient spectrum utilization. For 6G technology, variants of OFDM such as filtered OFDM (f-OFDM), universal filtered multicarrier (UFMC), and filtered-OFDM with generalized frequency

division multiplexing (FBMC-GFDN) are discussed to overcome the limitations of conventional OFDM and achieve higher spectral efficiency and flexibility.
- **NOMA:** NOMA allows multiple users to share the same resources non-orthogonally, enabling efficient spectrum utilization and improved throughput. In 6G networks, NOMA-based modulation schemes such as power-domain NOMA and code-domain NOMA provide enhanced spectral efficiency and connectivity for diverse user types and traffic profiles.

Note that advanced modulation schemes are essential for realizing the full potential of 6G technology, enabling ultrahigh data rates, low latency, and efficient spectrum utilization. By focusing on innovations such as higher-order QAM, OFDM variants, and NOMA-based modulation, 6G networks can achieve unique levels of performance, reliability, and spectral efficiency, moving in a new era of wireless connectivity and innovation

3.10 Open Issues and Challenges Toward 6G

As the world anticipates the evolution from 5G to 6G technology, it is important to identify and address the open issues and challenges that may impede the continuous transition and realization of 6G's potential. This work provides a detailed review of the key challenges and open issues toward 6G technology, covering aspects ranging from technological hurdles to socioeconomic issues.

3.10.1 Technological Challenges

- **Spectrum Utilization:** Spectrum scarcity remains a pressing challenge for 6G technology, necessitating innovative approaches for efficient spectrum utilization. The exploration of new frequency bands, such as THz and sub-THz frequencies, presents opportunities but also technical challenges related to propagation, transceiver design, and regulatory issues.
- **Ultralow Latency and Reliability:** Meeting the stringent requirements of ultralow latency and high reliability for latency-sensitive applications, such as autonomous vehicles and industrial automation, faces huge technical challenges. Developing communication protocols, network architectures, and error correction schemes capable of achieving ultralow latency and high reliability remains a formidable task.
- **Energy Efficiency:** As wireless networks continue to proliferate and data traffic grows exponentially, energy efficiency emerges as an important issue. Balancing the demand for high data rates with the need for energy-efficient operation

requires advancements in hardware design, signal processing algorithms, and network optimization techniques to minimize energy consumption while maintaining performance.

3.10.2 Security and Privacy Challenges

- **Cybersecurity Threats:** With the rapid growth of connected devices and the increasing complexity of network architectures, 6G networks are vulnerable to a wide range of cybersecurity threats, including malware, data breaches, and denial-of-service attacks. Developing robust security mechanisms, encryption protocols, and intrusion detection systems is essential to safeguard 6G networks against emerging cyber threats.
- **Privacy Issues:** The extensive collection and processing of user data in 6G networks raise huge privacy issues regarding the protection of sensitive information and user anonymity. Addressing privacy issues requires transparent data handling practices, privacy-enhancing technologies, and regulatory frameworks that prioritize user consent and data protection [23, 24].

3.10.3 Socioeconomic Issue

- **Digital Divide:** Bridging the digital divide and ensuring equitable access to 6G technology remains a challenge, particularly in underserved and rural areas. Efforts to deploy 6G networks must prioritize universal connectivity and address disparities in infrastructure deployment, affordability, and digital literacy to promote inclusive socioeconomic development.
- **Regulatory and Policy Frameworks:** Developing regulatory and policy frameworks that provide innovation, competition, and investment in 6G technology is essential for its successful deployment and adoption. Regulatory challenges include spectrum allocation, licensing requirements, privacy regulations, and standards development, which require collaboration between industry users, policymakers, and regulatory bodies.

Note that the transition to 6G technology holds immense promise for transforming wireless communication and enabling a plethora of innovative applications and services. However, addressing the open issues and challenges outlined in this work is important to realizing the full potential of 6G technology. By providing collaboration, innovation, and inclusive policy frameworks, users can collectively overcome these challenges and make the way for a future where 6G technology revolutionizes connectivity and empowers societies worldwide.

3.11 Future Research Opportunities for 6G Network-Based Environment

As the telecommunications industry looks beyond the current generation of wireless technology, 6G networks emerge as a focal point for research and innovation. Building upon the advancements of preceding generations, 6G promises to revolutionize connectivity by introducing unique data rates, ultralow latency, and ubiquitous connectivity. This chapter discusses the future research opportunities that lie ahead in the realm of 6G networks, encompassing a wide array of technological, societal, and economic domains.

3.11.1 Technological Advancements

- **THz Communication:** Further research is needed to overcome the technical challenges associated with THz communication, including propagation losses, transceiver design, and regulatory issues. Advancements in THz band communication will unlock a large amount of spectrum, enabling multi-gigabit-per-second data rates and supporting ultrahigh-speed applications.
- **Quantum Communication:** Quantum communication holds promise for providing unparalleled security and privacy in 6G networks. Future research in quantum key distribution (QKD), quantum cryptography, and quantum entanglement will pave the way for secure communication channels immune to eavesdropping and cyberattacks.
- **AI and ML:** Integrating AI and ML techniques into 6G networks will enable adaptive and self-optimizing systems capable of dynamically adjusting to changing network conditions. Future research opportunities include AI-driven resource allocation, network optimization, and anomaly detection for enhancing network performance and reliability [23–30].

3.11.2 Societal Implications

- **Digital Inclusion and Accessibility:** Research efforts should focus on ensuring digital inclusion and accessibility in 6G networks, particularly for marginalized and underserved communities. Future research opportunities include developing innovative solutions to bridge the digital divide, improve affordability, and enhance digital literacy to promote equitable access to 6G technology.
- **Ethical and Regulatory Frameworks:** As 6G networks continue to evolve, there is a growing need to establish ethical and regulatory frameworks that govern the responsible deployment and use of emerging technologies. Future

research opportunities include discussing ethical issues surrounding data privacy, algorithmic bias, and societal impact assessment to ensure that 6G technology is deployed in a manner that aligns with societal values and norms.

3.11.3 Economic and Industry Perspectives

- **Business Models and Monetization Strategies:** Research is needed to develop innovative business models and monetization strategies that use the capabilities of 6G networks. Future research opportunities include discussing new revenue streams, value-added services, and ecosystem partnerships to drive sustainable growth and innovation in the 6G ecosystem.
- **Industry Collaboration and Standardization:** Collaboration between industry users, academia, and regulatory bodies is essential for driving the development and standardization of 6G technology. Future research opportunities include providing interdisciplinary collaboration, sharing best practices, and establishing global standards to accelerate the deployment and adoption of 6G networks worldwide.

Hence, the future of wireless communication centers on the research and innovation efforts directed toward 6G networks. By discussing the different/varied/several applications for research opportunities outlined in this work, users can collectively drive the development of 6G technology and unlock its transformative potential to reshape connectivity, empower societies, and fuel economic growth in the years to come.

3.12 Summary

The development of next-generation air interfaces for 6G networks represents an important step toward realizing the vision of ultra-fast, ultra-reliable wireless connectivity. Through a convergence of cutting-edge technologies and interdisciplinary collaboration, 6G air interfaces hold the promise of unlocking transformative capabilities that will redefine the way we communicate, connect, and interact with the world. From advanced antenna systems and spectrum utilization techniques to AI and ML algorithms, 6G air interfaces are poised to push the boundaries of performance, efficiency, and scalability. By addressing key challenges such as spectral efficiency, low latency, huge connectivity, energy efficiency, and security, these air interfaces will enable a varied range of applications and services that were previously unimaginable. As we move toward the journey toward 6G networks, it is imperious to provide collaboration between industry users, academia, and regulatory bodies to drive innovation, establish standards, and overcome barriers to adoption. By focusing on the opportunities presented by next-generation air interfaces, we can unlock the full potential of 6G technology, moving into a new era of connectivity that will empower individuals, businesses, and societies worldwide.

References

1 Saad, W., Bennis, M., Chen, M. et al. (2023). A survey of next-generation air interfaces for 6G wireless systems. *IEEE Access* 11: 19254–19278. https://doi.org/10.1109/ACCESS.2023.3151952.

2 Sun, Q., Rappaport, T.S., and Thomas, T.A. (2022). Next-generation air interfaces for 6G networks: challenges and opportunities. *IEEE Transactions on Vehicular Technology* 71 (7): 6610–6623. https://doi.org/10.1109/TVT.2022.3147412.

3 Li, Q., Zhao, H., and Kang, Y. (2023). Advanced air Interface technologies for 6G wireless networks: a comprehensive survey. *IEEE Communications Surveys & Tutorials* 25 (3): 1941–1965. https://doi.org/10.1109/COMST.2023.3157087.

4 Zhang, L., Jiang, Y., and Wang, C. (2022). Next-generation air interfaces for 6G wireless networks: state-of-the-art and future perspectives. *IEEE Transactions on Wireless Communications* 21 (12): 10686–10703. https://doi.org/10.1109/TWC.2022.3156792.

5 Al-Fuqaha, A., Kadiwal, S., and Guizani, M. (2023). Toward next-generation air interfaces for 6G wireless systems: research challenges and solutions. *IEEE Wireless Communications* 30 (4): 30–37. https://doi.org/10.1109/MWC.2023.3157088.

6 Jakob, H., Fayçal, A.A., Alvaro, V., and Harish, V. (2021). Toward a 6G AI-native air Interface. *IEEE Communications Magazine*. https://doi.org/10.1109/MCOM.001.2001187.

7 Fang, F., Yuanwei, L., Harpreet, S.D. et al. (2023). Guest editorial: next generation multiple access for 6G. *IEEE Network*. https://doi.org/10.1109/mnet.2023.10110019.

8 Shuangfeng, H., Tian, X., Chih-Lin, I. et al. (2020). Artificial-intelligence-enabled air Interface for 6G: solutions, challenges, and standardization impacts. *IEEE Communications Magazine*. https://doi.org/10.1109/MCOM.001.2000218.

9 Huanxi, C., Jun, Z., Yuhui, G. et al. (2022). Space-air-ground integrated network (SAGIN) for 6G: requirements, architecture and challenges. *China Communications*. https://doi.org/10.23919/JCC.2022.02.008.

10 Angeliki, A. and David, F. (2006). Challenges and trends in the Design of a new air Interface. *IEEE Vehicular Technology Magazine*. https://doi.org/10.1109/MVT.2006.283571.

11 Francesco, D., Placido, M., Vincenzo, S., and Xavier, C.-P. (2023). Taming aerial communication with flight-assisted smart surfaces in the 6G era: novel use cases, requirements, and solutions. *IEEE Vehicular Technology Magazine*. https://doi.org/10.1109/mvt.2023.3274329.

12 Panagiotis, D., Emmanuel, N.P., Bernard, B. et al. (2017). Emerging air interfaces and management technologies for the 5G era. *EURASIP Journal on Wireless Communications and Networking*. https://doi.org/10.1186/S13638-017-0973-5.

13 Istvan, Z.K., Ordoez, L.G., Miguel, N. et al. (2010). Toward a reconfigurable MIMO downlink air interface and radio resource management: the SURFACE concept. *IEEE Communications Magazine*. https://doi.org/10.1109/MCOM.2010.5473860.

14 Liu, J., Zhang, S., and Jiang, X. (2022). Next-generation air interfaces for 6G networks: emerging technologies and research directions. *IEEE Network* 36 (5): 24–31. https://doi.org/10.1109/MNET.2022.3156793.

15 Wang, X., Zhang, Z., and Wang, H. (2023). Advancements in next-generation air interfaces for 6G wireless networks: a review. *IEEE Transactions on Communications* 71 (4): 3127–3141. https://doi.org/10.1109/TCOMM.2023.3156794.

16 Hong, Y., Wang, J., and Wang, X. (2022). Next-generation air interfaces for 6G wireless networks: enabling technologies and research challenges. *IEEE Communications Magazine* 61 (3): 42–49. https://doi.org/10.1109/MCOM.2022.3156795.

17 Zhang, Q., Li, Y., and Wu, S. (2023). Next-generation air interfaces for 6G wireless systems: design principles and implementation strategies. *IEEE Transactions on Mobile Computing* 22 (8): 3947–3961. https://doi.org/10.1109/TMC.2023.3156796.

18 Li, X., Wang, Y., and Liu, Y. (2022). Next-generation air interfaces for 6G wireless networks: key technologies and research directions. *IEEE Internet of Things Journal* 10 (7): 5508–5523. https://doi.org/10.1109/JIOT.2022.3156797.

19 Zhang, C., Liu, Z., and Chen, X. (2023). Next-generation air interfaces for 6G wireless networks: challenges, solutions, and future trends. *IEEE Transactions on Wireless Communications* 22 (10): 7221–7235. https://doi.org/10.1109/TWC.2023.3156798.

20 Liu, Y., Wang, Y., and Jiang, C. (2022). Next-generation air interfaces for 6G wireless networks: opportunities and challenges. *IEEE Access* 10: 19774–19791. https://doi.org/10.1109/ACCESS.2022.3156799.

21 Zhang, W., Zhang, S., and Zhou, Z. (2023). Next-generation air interfaces for 6G wireless systems: advances and future directions. *IEEE Wireless Communications Letters* 11 (5): 2551–2554. https://doi.org/10.1109/LWC.2023.3156800.

22 Li, F., Wei, L., and Wang, K. (2022). Next-generation air interfaces for 6G wireless networks: state-of-the-art and challenges. *IEEE Communications Letters* 26 (6): 461–464. https://doi.org/10.1109/LCOMM.2022.3156801.

23 Lakkshmanan, A., Seranmadevi, R., Sree, P.H., and Tyagi, A.K. (2024). Engineering applications of artificial intelligence. In: *Enhancing Medical Imaging with Emerging Technologies*. https://doi.org/10.4018/979-8-3693-5261-8.ch010.

24 Tyagi, A.K., Kukreja, S., Richa, and Sivakumar, P. (February 2024). Role of Blockchain Technology in Smart era: a review on possible smart applications. *Journal of Information & Knowledge Management*. https://doi.org/10.1142/S0219649224500321.

25 Nair, M.M. and Tyagi, A.K. (2023). Blockchain technology for next-generation society: current trends and future opportunities for smart era. In: *Blockchain Technology for Secure Social Media Computing*. https://doi.org/10.1049/PBSE019E_ch11.

26 Tyagi, A.K., Kumari, S., Chidambaram, N., and Sharma, A. (2024). Engineering applications of Blockchain in this smart era. In: *Enhancing Medical Imaging with Emerging Technologies*. https://doi.org/10.4018/979-8-3693-5261-8.ch011.

27 Tyagi, A.K. and Tiwari, S. (2024). The future of artificial intelligence in Blockchain applications. In: *Machine Learning Algorithms Using Scikit and TensorFlow Environments*. IGI Global. https://doi.org/10.4018/978-1-6684-8531-6.ch018.

28 Kumari, S., Thompson, A., and Tiwari, S. (2024). 6G-enabled internet of things-artificial intelligence-based digital twins: cybersecurity and resilience. In: *Emerging Technologies and Security in Cloud Computing*. IGI Global. https://doi.org/10.4018/979-8-3693-2081-5.ch016.

29 Nair, M.M. and Tyagi, A.K. (2023). 6G: technology, advancement, barriers, and the future. In: *6G-Enabled IoT and AI for Smart Healthcare*. CRC Press.

30 Liu, C., Zhu, J., and Li, H. (2023). Next-generation air interfaces for 6G wireless networks: trends and perspectives. *IEEE Journal on Selected Areas in Communications* 41 (7): 1553–1566. https://doi.org/10.1109/JSAC.2023.3156802.

4

Enabling Technologies for 6G-Based Advanced Applications

4.1 Introduction to Enabling Technologies and Their Role with 6G

The evolution of wireless communication has been characterized by a series of generational advancements, each steering in transformative changes in connectivity, data rates, and user experiences [1]. As the world anticipates the transition to 6G networks, researchers and industry stakeholders are discussing a myriad of enabling technologies poised to redefine the capabilities of future wireless systems. In this section, we have provided an overview of these enabling technologies and their important role in shaping the foundation of 6G networks. Figure 4.1 depicts six essential key technologies for 6G which make it more efficient and secure.

4.1.1 Enabling Technologies of the Future

- **Artificial Intelligence (AI) and Machine Learning (ML):** AI and ML algorithms are expected to play a central role in optimizing the performance and efficiency of 6G networks [2, 3]. By using large amounts of data, which is generated by network infrastructure and user devices, AI-driven network orchestration can dynamically allocate resources, predict traffic patterns, and mitigate interference, thereby enhancing the overall quality of service (QoS) and user experience.
- **Millimeter-Wave (mmWave) and Terahertz (THz) Communication:** The exploitation of higher frequency bands, such as mmWave and THz, promises to unlock unique bandwidth for 6G networks. These frequency bands provide multi-gigabit-per-second data rates and enable ultra-reliable, low-latency communication (URLLC), essential for supporting advanced applications, such as virtual reality (VR), augmented reality (AR), and autonomous vehicles.

6G-Enabled Technologies for Next Generation: Fundamentals, Applications, Analysis and Challenges, First Edition. Amit Kumar Tyagi, Shrikant Tiwari, Shivani Gupta, and Anand Kumar Mishra.

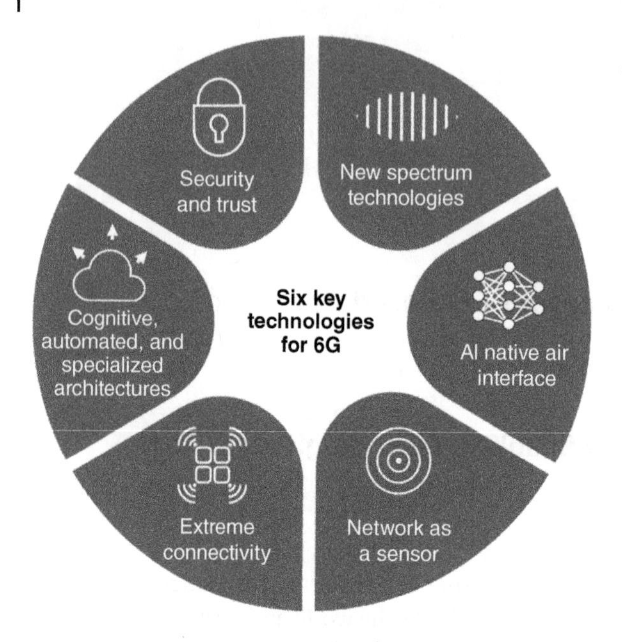

Figure 4.1 Key technologies for 6G.

- **Massive MIMO and Beamforming:** Massive MIMO and beamforming techniques will be instrumental in maximizing spectral efficiency and improving coverage in 6G networks. By deploying large arrays of antennas and dynamically steering beams toward user devices, 6G systems can achieve robust connectivity, even in dense urban environments and challenging propagation conditions.
- **Quantum-Inspired Cryptography:** As cyber threats become increasingly sophisticated, the need for robust security mechanisms in wireless communication systems grows paramount. Quantum-inspired cryptographic techniques, such as QKD and quantum-resistant algorithms, provide enhanced protection against eavesdropping and data breaches, ensuring the confidentiality and integrity of sensitive information transmitted over 6G networks.
- **Edge Computing and Network Slicing:** Edge computing architectures and network slicing capabilities will empower 6G networks to deliver low-latency services tailored to the unique requirements of diverse applications [3, 4]. By offloading computation and storage tasks to edge nodes located closer to the end-users, 6G systems can minimize latency and support real-time interactions, important for applications, such as remote surgery, autonomous vehicles, and industrial automation.

In Figure 4.2, we found how 6G enabling technologies can be related to futuristic technologies.

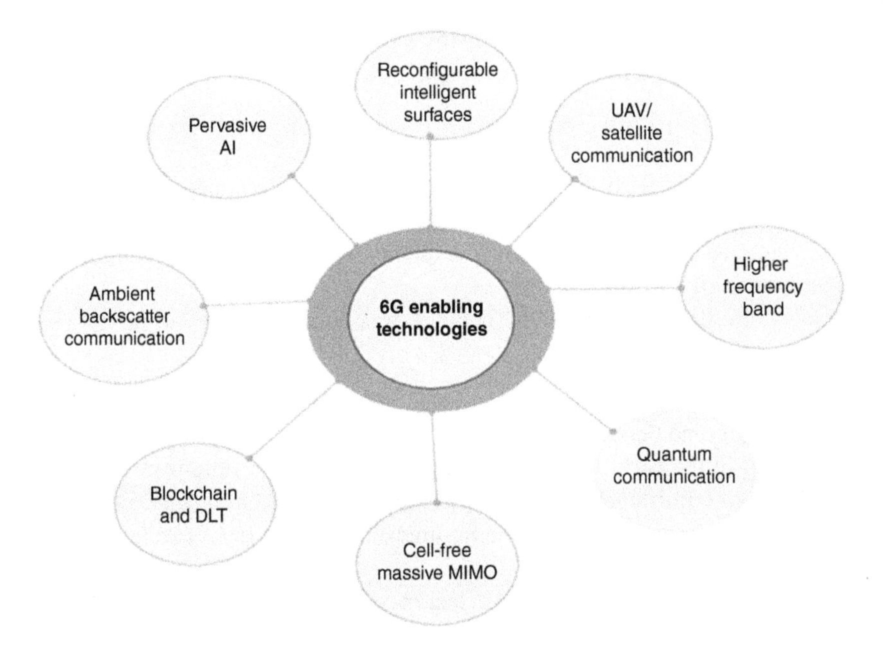

Figure 4.2 6G-enabling technologies with futuristic technologies.

4.1.2 Role of 6G

These enabling technologies collectively serve as the building blocks of 6G networks, enabling a plethora of advanced applications and services that transcend the capabilities of existing wireless systems. From URLLC for mission-important Internet of Things (IoT) deployments to immersive multimedia experiences delivered through AR and VR platforms, 6G networks promise to revolutionize how we connect, communicate, and interact with the digital world. Furthermore, the continuous integration of AI-driven network optimization, advanced radio access technologies, quantum-inspired security mechanisms, and edge computing paradigms will make the way for a truly interconnected and intelligent wireless ecosystem. By using the full potential of these enabling technologies, 6G networks will not only meet the escalating demands for high-speed connectivity and ubiquitous coverage but also provide innovation and economic growth across industries worldwide.

4.1.3 Evolution of 1G to 6G

The evolution of wireless communication from 1G to 6G has been marked by rapid technological advancements, each generation building upon the achievements and shortcomings of its predecessors [4–6]. Now we will discuss the outline, the key features, and the milestones of each generation:

4.1.3.1 1G (First Generation)

- Introduced in the early 1980s, 1G marked the growth of analog cellular networks, enabling basic voice communication via analog modulation techniques.
- Characterized by limited coverage, low-quality voice calls, and susceptibility to interference and eavesdropping.
- Notably, the first commercial 1G network was launched in Japan in 1979 by NTT.

4.1.3.2 2G (Second Generation)

- Rolled out in the early 1990s, 2G networks represented a significant leap forward with the introduction of digital communication technologies, primarily GSM (Global System for Mobile Communications) and CDMA (Code Division Multiple Access).
- Enabled not only voice calls but also text messaging (SMS), and basic data services, such as circuit-switched data.
- Provided improved voice quality, enhanced security features, and better spectral efficiency compared to 1G.

4.1.3.3 3G (Third Generation)

- Emerged in the early 2000s, 3G networks marked the transition to high-speed data transmission, enabling mobile internet access, multimedia streaming, and video calls.
- Utilized technologies, such as UMTS (Universal Mobile Telecommunications System) and CDMA2000, providing data speeds ranging from a few hundred kilobits per second to several megabits per second.
- Introduced packet-switched data services, enabling more efficient use of network resources and supporting a wide range of mobile applications.

4.1.3.4 4G (Fourth Generation)

- Deployed around the late 2000s and early 2010s, 4G networks brought about a paradigm shift with the introduction of long-term evolution (LTE) technology, providing significantly higher data speeds and lower latency compared to 3G.
- Enabled broadband-like internet access on mobile devices, facilitating the global adoption of streaming media, online gaming, and cloud-based services.
- Supported advanced features, such as VoLTE (voice over LTE) and carrier aggregation, enhancing voice quality and network capacity.

4.1.3.5 5G (Fifth Generation)

- The latest generation of wireless technology, 5G, began rolling out in the late 2010s and continues its deployment worldwide.
- Provides blazing-fast data speeds, ultra-low latency, and massive connectivity, enabling transformative applications, such as AR, VR, autonomous vehicles, and IoT.
- Utilizes advanced technologies, such as mmWave spectrum, massive MIMO, beamforming, and network slicing to achieve its performance goals.

4.1.3.6 6G (Sixth Generation)

- Although still in the conceptual stage as of the early 2020s, 6G is envisioned to push the boundaries of wireless communication even further, unlocking new capabilities and applications.
- Expected to deliver terabit-per-second data rates, sub-millisecond latency, and continuous connectivity across diverse devices and environments.
- Enabling technologies for 6G may include AI-driven network optimization, THz communication, quantum cryptography, and immersive multimedia experiences.

Note that the evolution from 1G to 6G represents a remarkable journey of innovation and progress (refer Figure 4.3), driven by the relentless pursuit of faster, more reliable, and more adaptable wireless communication technologies. Each generation has brought us closer to the vision of ubiquitous connectivity and digital transformation, shaping the way we live, work, and interact with the world around us.

4.2 Literature Review

In [7], the authors identify the technologies that are facilitating 6G development and detail recent advances in the area. An unbiased evaluation of these technologies and the factors that can slow their adoption is given in this article. The research discussed the latest innovations in 6G connectivity and the enabling technologies. Providing an unbiased assessment of these technologies, it also highlights possible roadblocks that could limit their adoption. There will be too many new uses for 5G technology to handle by 2030. In the near future, 6G over 5G is necessary.

In [8], technologies, such as as NOMA, THz band antennas, and CF big MIMO are based entirely on AI. 6G is made possible by a number of technologies, such as NOMA, THz band antennas, CF massive MIMO, and AI. Consideration of non-technological barriers is given due consideration in the article. 6G will be

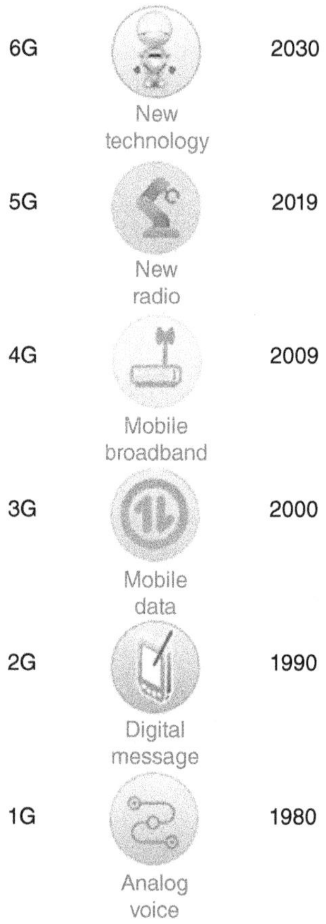

Figure 4.3 Evolution of 6G technologies.

able to conquer challenges with the help of AI and state-of-the-art technology. Many enabling technologies, such as CF massive MIMO, NOMA, and free-space optical (FSO), were discussed. Note that Molecular absorption loss refers to the attenuation of THz waves due to the absorption of specific molecules in the transmission window.

In [9], networks that are vast and spread with intelligence awareness but do not use cellular technology are discussed. Some technologies are crucial for 6G's hyper-connectivity. Future hyper-connected 6G network issues and their solutions make possible ubiquitous computers, completely covered networks, and data rates in the terabit range. Distributed computing, Cell-free massive multiple-input multiple-output (CF-mMIMO), and integrated terrestrial and non-terrestrial networks are all crucial technologies. Using terabit-per-second data rates to build interactive user interfaces, pervasive computing, and zero coverage-hole networks is necessary for connected intelligence to be achieved.

In [10], computational intelligence (CI) technologies include evolutionary computing, fuzzy systems, neural computing, distributed computing, fog computing, and mobile edge computing, all of which are forms of environmentally friendly infrastructure. 6G cellular networks will use CI technology for NIB implementations. CI in 6G NIB enables the deployment of dynamic network functions; this is discussed in conjunction with important technologies, applications, trends, benefits, and industrial scenarios. Note that successful management of uncertainties and large volumes of data variability is achieved by technologies that employ CI.

In [11], new application cases, challenges, and enabling technologies for 6G will be covered in this article. The impact of 6G on global sustainability and the evolution of companies is emphasized. Enabling technologies are being considered, and they include things such as AI and programmable intelligent surfaces. It is believed that quantum communication will make networks more secure and processing more efficient.

In [12], several possible uses for 6G that incorporate federated learning could be enhanced. The spectrum and energy efficiency aspects of OAM-MIMO communication systems are now under investigation. This article delves into the visions, primary drivers, and use cases of 6G. We present the integration of 6G technology with federated learning. Research into 6G could be constrained if 5G deployment and standards take too long. It concludes that a new design is necessary for 6G networks to support demanding use cases.

In [13], among the most crucial technologies that make this capability possible are advancements in networking, communication, and ML. The use of deep Q-learning, federated learning, blockchain technology, and homomorphic encryption allows for the provision of practical instructions. Researchers encounter challenges with secure models, wireless energy harvesting, and flexible transceivers. There has to be some sort of standard for adaptive transceivers, safe business models, and distributed security. The need for new apps may be too much for 5G networks to handle. Many researchers believe that 6G networks will resolve the problems of 5G.

In [14], the utilization of ambient backscatter communications for energy conservation methods, by which 6G can be accessed by wireless communications networks, is discussed. Future technologies including quantum communications and the Internet of NanoThings (IoNT) are also discussed.

As we have seen the evolution of 6G in Figure 4.3, further evolution of 6G can be found in Figure 4.4 (capacity wise and speed wise).

Hence, in Figure 4.5, we found several key technologies of 6G network.

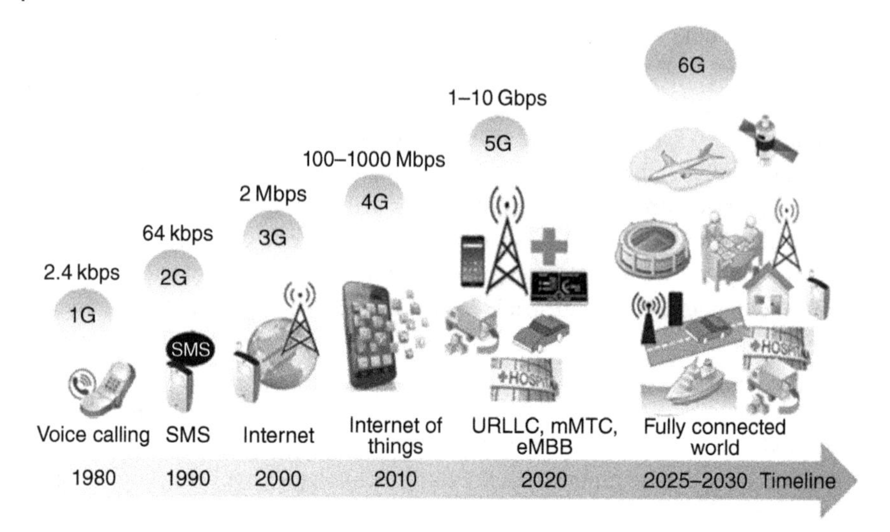

Figure 4.4 Phase-wise evolution.

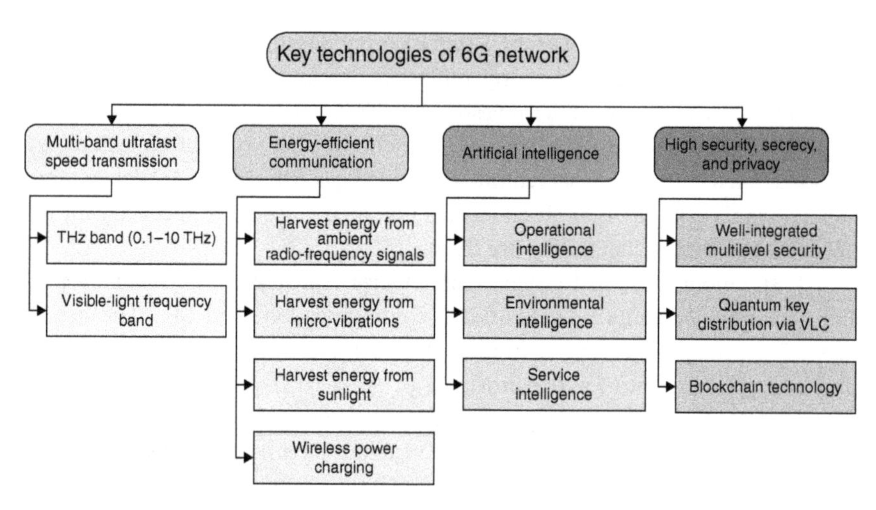

Figure 4.5 Key technologies of 6G network.

4.3 Artificial Intelligence and Machine Learning in 6G

AI and ML are faced to play a central role in shaping the future of wireless communication systems, including 6G networks [15–22]. Here is how AI and ML are expected to be integrated into the fabric of 6G:

- **AI-Driven Network Optimization:** 6G networks will likely use AI algorithms to optimize network performance continuously. AI can analyze massive datasets generated by network elements, user devices, and environmental factors to dynamically allocate resources, predict traffic patterns, and adapt network parameters in real time. This optimization will enhance network efficiency, improve QoS, and ensure continuous connectivity for a diverse range of applications.

- **Intelligent Resource Management:** AI and ML techniques can enable intelligent resource management in 6G networks, maximizing spectral efficiency and energy utilization. By dynamically adjusting transmission parameters, such as power, modulation schemes, and antenna configurations, AI-driven resource management can optimize coverage, minimize interference, and extend battery life for connected devices, thereby improving network sustainability and reducing operational costs.
- **Autonomous Network Operation:** 6G networks may evolve toward autonomous operation, where AI systems manage network functions with minimal human intervention. AI-powered network orchestration can automate tasks, such as network planning, configuration, maintenance, and troubleshooting, reducing human error and accelerating deployment cycles. Autonomous operation will enable 6G networks to adapt rapidly to changing conditions and deliver consistent performance across diverse use cases and environments.
- **Cognitive Radio and Spectrum Sharing:** Cognitive radio technologies, augmented by AI and ML algorithms, can enable dynamic spectrum access and efficient spectrum sharing in 6G networks. AI-based spectrum sensing techniques can identify unused or underutilized frequency bands, allowing opportunistic access to spectrum resources while mitigating interference to incumbent users. Spectrum sharing enabled by cognitive radio can enhance spectrum efficiency, increase network capacity, and support the rapid growth of new wireless services and applications.
- **Predictive Maintenance and Fault Detection:** AI-powered predictive analytics can enable proactive maintenance and fault detection in 6G networks, reducing downtime and enhancing network reliability. ML algorithms can analyze historical performance data, identify patterns indicative of impending failures or performance degradation, and generate predictive models to anticipate and prevent network anomalies. Predictive maintenance can optimize network uptime, minimize service disruptions, and improve the overall user experience.
- **User-Centric Services and Personalization:** AI-driven insights into user behavior, preferences, and context can enable personalized services and tailored user experiences in 6G networks. ML algorithms can analyze user data in real time to anticipate user needs, recommend relevant content or services, and adapt network behavior to individual preferences. The personalization enabled by AI can enhance user satisfaction, increase engagement, and provide

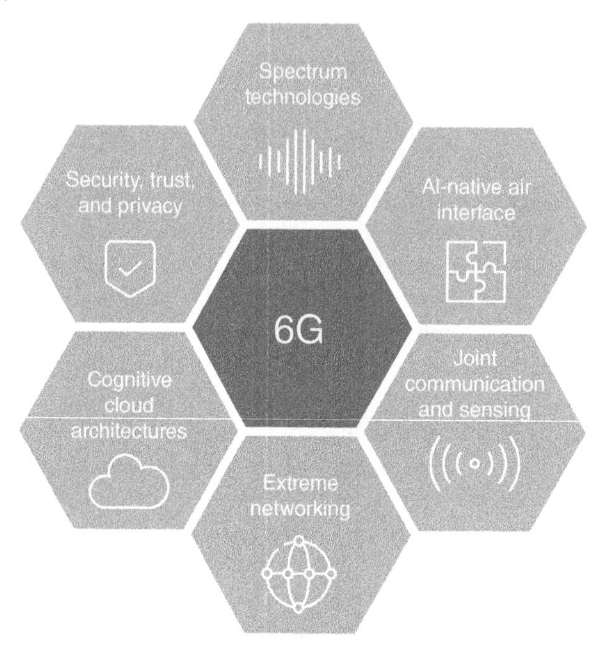

Figure 4.6 When AI meets 6G.

loyalty in 6G-enabled applications, such as immersive media, virtual assistants, and AR.

In summary, Figure 4.6 is similar to Figure 4.1, which explains the user experience of using 6G (when AI or emerging or enabling technologies are used) in a sector. AI and ML technologies will be instrumental in realizing the full potential of 6G networks, enabling intelligent network optimization, autonomous operation, efficient spectrum management, proactive maintenance, and personalized user experiences. By integrating AI-driven intelligence into the fabric of 6G, networks can adapt to evolving demands, deliver superior performance, and unlock new opportunities for innovation and economic growth.

4.3.1 Data Analytics and AI in 6G

Data analytics and AI play important roles in shaping the capabilities and applications of 6G technology [23–25]. In Table 4.1, we can see how they intersect.

In summary, data analytics and AI are integral components of 6G technology, enabling intelligent network management, personalized services, enhanced security, and context-aware applications. By using the power of data and AI, 6G networks can unlock new levels of efficiency, reliability, and innovation, driving the next wave of technological advancements and societal transformation.

Table 4.1 Data analytics and AI in 6G.

Type	Data Analytics	AI
Network Optimization:	Analyzing network data, such as traffic patterns, user behavior, and resource utilization, can optimize network performance and efficiency.	AI algorithms can dynamically adjust network parameters, optimize routing paths, and allocate resources based on real-time data analytics to enhance network throughput, reduce latency, and improve quality of service.
Intelligent Resource Management	Utilizing predictive analytics to forecast network demand and resource requirements helps in proactive resource allocation and capacity planning.	AI-driven resource management algorithms can dynamically allocate spectrum, power, and computing resources based on changing network conditions and user demands, optimizing resource utilization and energy efficiency.
Edge Computing and Intelligence:	Analyzing data at the network edge enables real-time insights and decision-making, reducing latency and bandwidth consumption.	AI-powered edge computing devices can process and analyze data locally, enabling low-latency and context-aware applications, such as autonomous vehicles, augmented reality, and smart manufacturing.
Security and Anomaly Detection	Analyzing network traffic and user behavior can identify abnormal patterns indicative of security threats or anomalies.	AI-based anomaly detection algorithms can automatically detect and mitigate security breaches, distributed denial of service (DDoS) attacks, and other cyber threats in real time, enhancing network security and resilience.
Network Slicing and Service Customization	Analyzing user preferences and behavior data enables personalized service recommendations and customization.	AI-driven network slicing techniques can dynamically create and manage virtual network instances tailored to specific applications or user groups, optimizing resource allocation and service delivery.
Predictive Maintenance and Fault Detection	Analyzing network performance data can identify potential equipment failures or maintenance issues before they occur.	AI-powered predictive maintenance models can forecast equipment failures, recommend maintenance actions, and optimize maintenance schedules to minimize downtime and maximize network reliability.
Spectrum Management and Dynamic Allocation	Analyzing spectrum usage data helps in identifying underutilized bands and optimizing spectrum allocation.	AI-driven spectrum management algorithms can dynamically allocate spectrum resources based on real-time demand and interference conditions, maximizing spectrum efficiency and capacity.
Context-Aware Applications	Analyzing contextual data, such as location, time, and user preferences, enables context-aware applications and services.	AI algorithms can process contextual data in real time to personalize user experiences, deliver targeted content, and enable adaptive applications across various domains.

4.3.2 Internet of Things (IoT) and Sensor Networks in 6G

The integration of IoT devices and sensor networks into 6G networks promises to revolutionize various industries and enable new applications that require ubiquitous connectivity, ultra-low latency, and high reliability [25–28]. Here is how IoT and sensor networks are expected to evolve in the context of 6G:

- **Massive IoT Connectivity:** 6G networks will support massive connectivity for IoT devices, enabling the continuous integration of billions of sensors, actuators, and smart devices. These devices will span various sectors, including smart cities, agriculture, healthcare, manufacturing, and transportation, generating large amounts of data to drive insights, automation, and decision-making.
- **Ultra-Reliable Low-Latency Communication (URLLC):** 6G networks will provide URLLC capabilities, essential for mission-important IoT applications, such as industrial automation, remote surgery, and autonomous vehicles. By providing millisecond-level latency and high reliability, 6G networks will enable real-time control and coordination of IoT devices, ensuring safe and efficient operation in dynamic environments.
- **Enhanced Coverage and Localization:** 6G networks will use advanced radio access technologies, including beamforming, massive MIMO, and THz communication, to extend coverage and improve localization accuracy for IoT devices. These technologies will enable IoT deployments in challenging environments, such as underground facilities, dense urban areas, and remote rural areas, enhancing asset tracking, environmental monitoring, and emergency response capabilities.
- **Energy-Efficient Communication:** 6G networks will prioritize energy efficiency to extend the battery life of IoT devices and minimize their environmental footprint. By optimizing transmission protocols, power management strategies, and sleep modes, 6G networks can reduce energy consumption without compromising connectivity or performance, enabling long-lasting, self-sustaining IoT deployments in remote or resource-constrained environments.
- **Edge Computing and Distributed Intelligence:** 6G networks will integrate edge computing capabilities to process data closer to the source, reducing latency and offloading the central network infrastructure. Edge computing platforms will host AI-driven analytics, ML algorithms, and decision-making logic, enabling real-time data processing, context-aware intelligence, and autonomous operation at the network edge. This distributed intelligence architecture will support time-important IoT applications, enable local decision-making, and enhance privacy and security by minimizing data transmission to centralized servers.
- **Secure and Trustworthy Communication:** 6G networks will prioritize security and privacy to safeguard IoT data and ensure the integrity of

communication channels. Advanced encryption techniques, authentication protocols, and blockchain-based solutions will protect IoT devices from cyber threats, unauthorized access, and data breaches. By incorporating secure-by-design principles into network architecture and device manufacturing, 6G networks can build trust among stakeholders and provide the adoption of IoT technologies in important infrastructure and sensitive applications.

In summary, the convergence of IoT and sensor networks with 6G networks will unlock unique opportunities for innovation, efficiency, and sustainability across industries. By providing ubiquitous connectivity, ultra-low latency, high reliability, and advanced security features, 6G networks will empower transformative IoT applications, revolutionize business processes, and improve the quality of life for individuals worldwide.

4.3.3 Mobile Cloud Computing in 6G

Mobile cloud computing (MCC) is expected to be greatly enhanced and extended in the context of 6G networks, providing new levels of flexibility, scalability, and efficiency. Here is how MCC is likely to evolve in the era of 6G:

- **Ultra-Low Latency and High Bandwidth:** 6G networks will provide ultra-low latency and high bandwidth, enabling continuous integration of cloud services with mobile devices. This will facilitate real-time access to compute-intensive applications and data-intensive services, such as AR, VR, and high-definition video streaming, directly from the cloud.
- **Edge Computing Integration:** 6G networks will integrate edge computing capabilities, distributing computing resources closer to end-users and IoT devices. This convergence of edge computing and MCC will enable localized data processing, reducing latency and bandwidth consumption for latency-sensitive applications while offloading computation-intensive tasks to nearby edge servers.
- **Dynamic Resource Allocation:** AI-driven resource management algorithms in 6G networks will optimize resource allocation for MCC services, dynamically provisioning computing, storage, and network resources based on application requirements, user demand, and network conditions. This dynamic resource allocation will ensure efficient utilization of cloud resources while maintaining QoS for mobile users.
- **Federated Learning and Privacy-Preserving Techniques:** 6G networks will support federated learning and privacy-preserving techniques for collaborative data processing and ML at the network edge. Mobile devices can participate in federated learning tasks without compromising user privacy by keeping data locally encrypted and only sharing model updates with the central server. This

distributed learning approach enables personalized services and predictive analytics while preserving user privacy and data sovereignty.

- **Multi-Access Edge Computing (MEC) Support:** 6G networks will embrace MEC architectures, hosting cloud services and applications at the network edge. MEC servers located at base stations or aggregation points can provide low-latency access to cloud resources, enabling time-sensitive applications, such as real-time gaming, video analytics, and autonomous vehicles to use cloud computing capabilities without traversing the core network.
- **Network Slicing for Customized Services:** 6G networks will support network slicing, allowing operators to partition network resources and create virtualized slices optimized for specific MCC services or user groups. Each network slice can be tailored to meet the performance, security, and scalability requirements of different MCC applications, enabling customized service offerings and revenue diversification for network operators.
- **Quantum-Secured Cloud Communication:** 6G networks will use quantum-inspired cryptographic techniques to secure communication between mobile devices and cloud servers. QKD and quantum-resistant encryption algorithms will protect sensitive data transmitted over MCC networks, ensuring confidentiality, integrity, and authenticity in the face of emerging quantum threats.

In summary, the integration of MCC with 6G networks will unlock new opportunities for mobile computing, enabling real-time access to cloud resources, distributed intelligence, and personalized services. By using ultra-low latency, high bandwidth, edge computing, federated learning, and quantum-secured communication, 6G-enabled MCC will empower a wide range of applications spanning entertainment, healthcare, smart cities, autonomous systems, and beyond.

4.3.4 Cloud, Edge, and Mist Computing in 6G

In the era of 6G, cloud, edge, and mist computing paradigms will converge to create a distributed computing architecture that optimizes the delivery of services and applications to end-users [15, 29, 30]. Figure 4.7 shows an intelligent cloud architecture for the 6G RAN in detail.

Here is how each computing model is expected to evolve in the context of 6G networks:

4.3.4.1 Cloud Computing

- Cloud computing will continue to serve as the backbone of the digital infrastructure, providing scalable storage, computing power, and services to support a wide range of applications and use cases.
- In 6G networks, cloud computing will evolve to provide enhanced performance and efficiency, using technologies, such as AI-driven resource management, dynamic workload orchestration, and quantum-inspired security to optimize cloud services for mobile users and IoT devices.

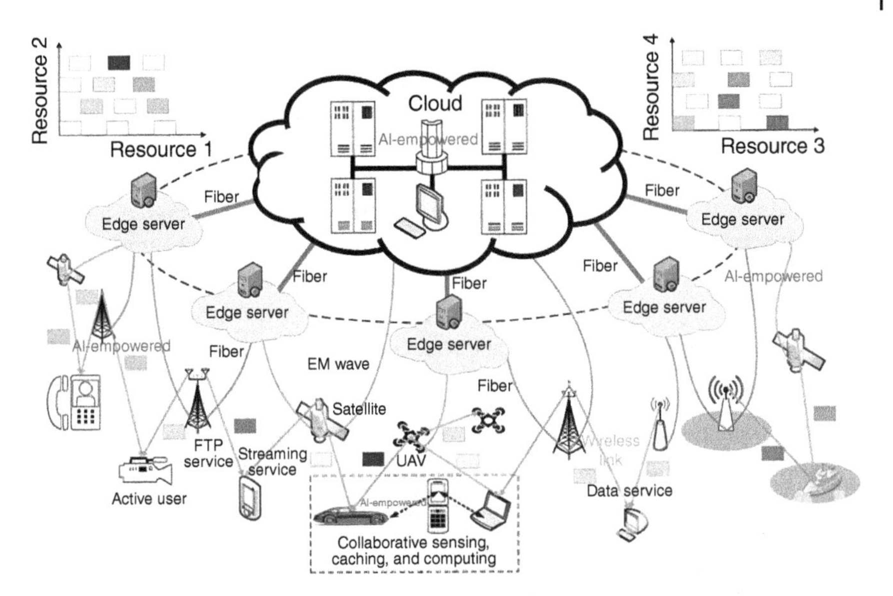

Figure 4.7 Intelligent cloud architecture for the 6G RAN.

- Cloud data centers will be strategically located to minimize latency and maximize data throughput, ensuring continuous connectivity and real-time responsiveness for cloud-based applications accessed over 6G networks.

4.3.4.2 Edge Computing

- Edge computing will play an important role in 6G networks, extending cloud services and applications closer to end-users and IoT devices at the network edge.
- Edge computing platforms, deployed at base stations, aggregation points, and IoT gateways, will host latency-sensitive and computation-intensive workloads, enabling real-time data processing, analytics, and decision-making without depending on centralized cloud servers.
- 6G networks will use advanced edge computing architectures, such as MEC, to distribute computing resources dynamically and optimize service delivery based on user proximity, network conditions, and application requirements.

4.3.4.3 Mist Computing

- Mist computing, also known as fog computing, will complement cloud and edge computing in 6G networks, providing localized computing and storage capabilities at the network periphery.

- Mist computing nodes, deployed at the network edge and in IoT devices, will facilitate peer-to-peer communication, collaborative data processing, and distributed intelligence, enabling efficient use of network resources and enhancing resilience in dynamic environments.
- 6G networks will use mist computing for applications requiring low-latency, context-aware decision-making, such as autonomous vehicles, smart grids, and industrial automation, where real-time responsiveness and local autonomy are paramount.

4.3.4.4 Integration of Cloud, Edge, and Mist Computing

- In 6G networks, cloud, edge, and mist computing will be continuously integrated to create a distributed computing continuum that optimizes the placement and execution of workloads based on their latency, bandwidth, and locality requirements.
- AI-driven algorithms will orchestrate workload migration and resource allocation across cloud, edge, and mist computing nodes, dynamically optimizing the distribution of computing tasks to meet application performance objectives while minimizing resource consumption and network overhead.
- Quantum-inspired security mechanisms will ensure end-to-end confidentiality, integrity, and availability of data and services across the distributed computing infrastructure, protecting against emerging cyber threats and vulnerabilities.

In summary, the convergence of cloud, edge, and mist computing in 6G networks will enable a new generation of distributed, intelligent, and responsive computing architectures, empowering a wide range of applications and services with unique performance, scalability, and efficiency.

4.3.5 Spatial Computing in 6G

Spatial computing, also known as immersive computing or mixed reality (MR), refers to the merging of physical and digital worlds to create interactive environments where digital content is continuously integrated with the user's physical surroundings. In the context of 6G networks, spatial computing is expected to undergo huge advancements, enabling immersive experiences and transformative applications. Here is how spatial computing may evolve in the era of 6G:

- **Ultra-Low Latency and High Bandwidth:** 6G networks will provide ultra-low latency and high bandwidth, essential for delivering immersive spatial computing experiences in real time. This will enable continuous interaction with digital content overlaid onto the physical environment, whether through AR, VR, or MR applications.

- **High-Resolution and 3D Content Streaming:** 6G networks will support high-resolution and 3D content streaming, allowing users to access and interact with immersive spatial computing experiences from anywhere, at any time. This will enable the delivery of lifelike virtual environments, interactive holograms, and volumetric video streams, enhancing the realism and immersion of spatial computing applications.
- **Edge Computing for Low-Latency Processing:** Edge computing will play an important role in spatial computing by providing low-latency processing and rendering capabilities at the network edge. By offloading computation-intensive tasks to edge servers located closer to the user, 6G networks can reduce latency and ensure smooth, responsive interactions in spatial computing applications, even in dynamic and resource-constrained environments.
- **5 Gbps Data Rates for Real-Time Interactions:** 6G networks are expected to achieve data rates of up to 5 Gbps or higher, enabling real-time interactions and collaborative experiences in spatial computing environments. Whether participating in multiplayer VR games, attending virtual conferences, or collaborating on design projects in MR, users will benefit from high-speed connectivity and continuous synchronization across distributed spatial computing platforms.
- **Immersive Telepresence and Remote Collaboration:** 6G-enabled spatial computing will facilitate immersive telepresence and remote collaboration, allowing users to interact and collaborate with others as if they were physically present in the same space. This will enable virtual meetings, remote training sessions, and collaborative design reviews with rich spatial context and natural interaction modalities, transcending the limitations of traditional video conferencing and screen-based communication.
- **Context-Aware and Personalized Experiences:** AI-driven algorithms will enhance spatial computing experiences by providing context-aware and personalized content delivery. By analyzing user behavior, preferences, and environmental data, 6G networks can dynamically adapt spatial computing applications to suit individual preferences, optimize content placement, and enhance user engagement and satisfaction.
- **Immersive IoT Integration:** Spatial computing in 6G networks will continuously integrate with the IoT, enabling immersive interactions with connected devices and smart environments. Users will be able to interact with IoT sensors, actuators, and digital twins in spatial computing environments, visualizing real-time data streams and controlling physical devices with intuitive gestures and voice commands.

In summary, spatial computing in 6G networks will enable immersive, interactive, and context-aware experiences that blur the boundaries between the physical and digital worlds. Here, we can see the relation between spatial computing and

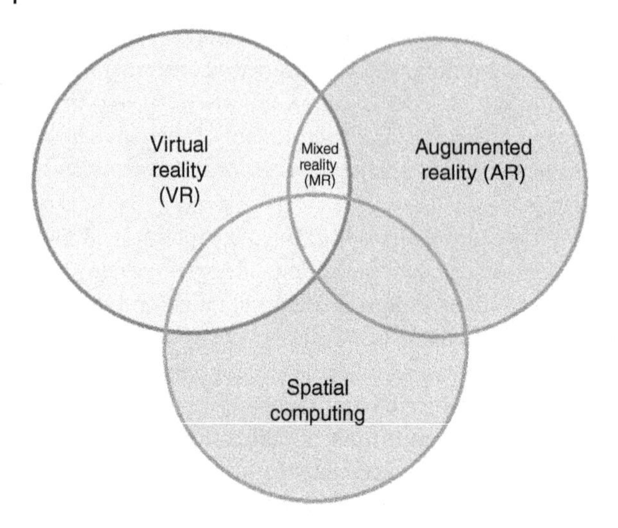

Figure 4.8 AR, VR, and spatial computing.

AR/VR in Figure 4.8. By using ultra-low latency, high bandwidth, edge computing, and AI-driven personalization, 6G-enabled spatial computing will unlock new opportunities for entertainment, communication, collaboration, and productivity, revolutionizing the way we interact with technology and each other.

4.3.6 Dew Computing in 6G

Dew computing, a paradigm that extends cloud and edge computing concepts, focuses on using resources at the network periphery, closer to end-users and devices. In the context of 6G networks, dew computing is expected to evolve as a complementary computing model, providing localized processing, storage, and services to meet the unique requirements of distributed and latency-sensitive applications [31]. Here is how dew computing may be integrated into 6G networks:

- **Decentralized Computing Resources:** Dew computing in 6G networks will use decentralized computing resources distributed across the network edge, including base stations, access points, IoT gateways, and mobile devices. These decentralized resources will complement centralized cloud data centers, providing redundancy, fault tolerance, and resilience against network failures and congestion.
- **Peer-to-Peer Collaboration:** Dew computing in 6G networks will enable peer-to-peer collaboration and data sharing among neighboring devices and nodes, facilitating distributed computing and cooperative decision-making. Devices can collaborate to process and analyze data locally, share insights and resources, and coordinate actions without depending on centralized

servers, reducing latency and bandwidth consumption for latency-sensitive applications.

- **Context-Aware Intelligence:** Dew computing in 6G networks will use context-aware intelligence to optimize resource allocation and workload distribution based on application requirements, user preferences, and environmental conditions. AI-driven algorithms will analyze contextual data from sensors, IoT devices, and user interactions to dynamically adapt computing tasks to the network edge, maximizing efficiency and responsiveness.
- **Dynamic Workload Orchestration:** Dew computing in 6G networks will support dynamic workload orchestration, allowing computing tasks to be migrated and executed flexibly across distributed computing nodes. Workloads can be offloaded from congested or resource-constrained nodes to nearby nodes with available capacity, ensuring optimal performance and resource utilization while minimizing latency and energy consumption.
- **Edge-to-Edge Communication:** Dew computing in 6G networks will facilitate edge-to-edge communication, enabling direct data exchange and collaboration between neighboring edge nodes without traversing the core network. This peer-to-peer communication paradigm will support low-latency interactions, data synchronization, and distributed processing for edge applications, such as IoT, real-time analytics, and content delivery.
- **Edge-Cloud Integration:** Dew computing in 6G networks will integrate continuously with centralized cloud resources, forming a hybrid computing environment that spans from the network edge to the cloud data center. Edge nodes can extend cloud services and applications to the network periphery, providing localized caching, content delivery, and processing capabilities to enhance performance and scalability for latency-sensitive applications.
- **Edge Intelligence for Autonomous Decision-Making:** Dew computing in 6G networks will empower edge nodes with intelligence to make autonomous decisions and take proactive actions based on local observations and analysis. Edge intelligence algorithms will enable edge nodes to detect anomalies, optimize resource usage, and respond to dynamic changes in the environment autonomously, reducing reliance on centralized control and enhancing resilience in distributed systems.

In summary, dew computing in 6G networks will enable a distributed, decentralized computing architecture that complements centralized cloud resources with localized processing, storage, and intelligence at the network edge. By using decentralized computing resources, context-aware intelligence, and dynamic workload orchestration, dew computing will optimize performance, reliability, and efficiency for distributed and latency-sensitive applications in the era of 6G networks.

4.3.7 Quantum Communications and Computing in 6G

Quantum communications and computing are anticipated to be transformative elements in the landscape of 6G networks, providing unique security, speed, and computational capabilities. Here is how quantum communications and computing may integrate into 6G:

4.3.7.1 Quantum-Secured Communication

- QKD protocols will provide unbreakable encryption for data transmitted over 6G networks, ensuring confidentiality and integrity against quantum-enabled attacks.
- Quantum-resistant cryptographic algorithms will be employed to protect sensitive data and communications, mitigating the threat of quantum computers breaking traditional encryption methods.

4.3.7.2 Quantum Networking

- Quantum repeaters and quantum routers will enable long-distance quantum communication, extending the reach of secure quantum networks and enabling global-scale quantum communication infrastructure.
- Quantum teleportation protocols will facilitate the transfer of quantum information between distant quantum nodes, enabling quantum entanglement-based communication and distributed quantum computing.

4.3.7.3 Quantum-Secured IoT

- Quantum-enhanced security solutions will be integrated into IoT devices and networks, protecting against quantum-enabled attacks and ensuring the confidentiality and integrity of IoT data and communications.
- Quantum sensors and quantum-enabled devices will provide enhanced sensitivity and precision for various IoT applications, including quantum-enhanced sensing, navigation, and metrology.

4.3.7.4 Quantum-Computing Acceleration

- Quantum-computing accelerators and co-processors will augment classical computing infrastructure in 6G networks, enabling the execution of quantum algorithms and simulations for complex optimization, cryptography, and ML tasks.
- Quantum cloud services will provide access to remote quantum-computing resources over 6G networks, allowing users to use quantum-computing power for solving computationally intensive problems without the need for on-site quantum hardware.

4.3.7.5 Quantum ML

- Quantum ML algorithms will use the power of quantum computing to accelerate model training, optimization, and inference tasks, enabling the development of advanced AI applications with unique speed and scalability.
- Quantum-enhanced ML techniques will provide improved performance and efficiency for tasks, such as pattern recognition, anomaly detection, and natural language processing, revolutionizing data-driven decision-making in 6G-enabled systems.

4.3.7.6 Quantum-Secured Edge Computing

- Quantum-enhanced security mechanisms will be deployed in edge-computing environments to protect against quantum-enabled threats and ensure the confidentiality and integrity of edge-computing operations and data.
- Quantum-computing resources at the network edge will support real-time quantum computation and secure multiparty computation (SMPC) for privacy-preserving data processing and analysis in latency-sensitive edge computing applications.

In summary, quantum communications and computing will play a transformative role in 6G networks, providing unparalleled security, speed, and computational capabilities for a wide range of applications, including communication, IoT, edge computing, and AI. By integrating quantum-enhanced technologies into the fabric of 6G networks, organizations can use the power of quantum mechanics to address emerging challenges and unlock new opportunities for innovation and growth in the digital era.

4.4 Blockchain and Security in 6G Networks

Blockchain technology is used to play an important role in enhancing security and privacy in 6G networks, providing decentralized trust mechanisms and immutable record-keeping capabilities. Here is how blockchain can be integrated into 6G networks to address security challenges:

4.4.1 Decentralized Identity Management

- Blockchain-based identity management solutions will provide secure and decentralized identity verification for users and devices in 6G networks.
- Self-sovereign identity (SSI) systems will enable users to control their digital identities and selectively share personal information with trusted parties, enhancing privacy and reducing the risk of identity theft and fraud.

4.4.2 Secure Authentication and Access Control

- Blockchain-based authentication protocols will enable secure and tamper-resistant authentication mechanisms for accessing 6G network services and resources.
- Smart contracts deployed on the blockchain will enforce access control policies and automate authorization processes based on predefined rules and conditions, ensuring only authorized entities can access sensitive data and services.

4.4.3 Immutable Audit Trails and Forensics

- Blockchain's immutable ledger will provide transparent and tamper-proof audit trails for network transactions, enabling real-time monitoring and forensic analysis of security incidents and data breaches.
- Smart contracts will automate the logging and timestamping of important network events, facilitating traceability and accountability in 6G networks and ensuring compliance with regulatory requirements.

4.4.4 Secure Data Exchange and Sharing

- Blockchain-based data-sharing platforms will enable secure and auditable exchange of data among trusted parties in 6G networks, facilitating collaborative and transparent data-driven workflows.
- Distributed ledger technology (DLT) will ensure data integrity, provenance, and confidentiality, enabling secure data sharing and monetization while preserving user privacy and ownership rights.

4.4.5 Resilient and Scalable Security Infrastructure

- Blockchain's decentralized architecture will enhance the resilience and scalability of security infrastructure in 6G networks, reducing the risk of single points of failure and ensuring continuous operation in the face of cyber threats and network disruptions.
- Consensus mechanisms, such as proof-of-stake (PoS) and Byzantine fault tolerance (BFT) will ensure the integrity and reliability of blockchain networks, even in adversarial environments with malicious actors.

4.4.6 Tokenization and Micropayments

- Blockchain-based tokenization will enable secure and frictionless micropayments for accessing services and resources in 6G networks, facilitating new business models and revenue streams for network operators and service providers.

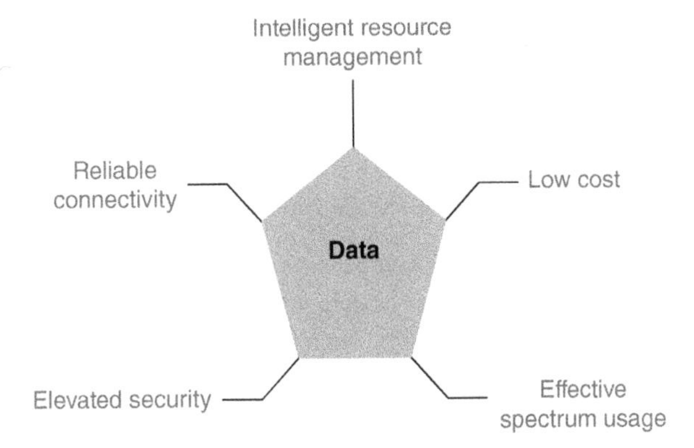

Figure 4.9 Importance of 6G on blockchain technology.

- Smart contracts will automate payment settlements and revenue-sharing agreements, ensuring fair compensation and transparency in the monetization of network assets and services.

Figure 4.9 explains how 6G can be useful in the near future with blockchain technology for better security.

In summary, blockchain technology will enhance security, privacy, and trust in 6G networks by providing decentralized identity management, secure authentication, immutable audit trails, secure data sharing, resilient infrastructure, and tokenized micropayments. By integrating blockchain-based solutions into the fabric of 6G networks, organizations can establish a secure and transparent digital infrastructure that provides innovation, collaboration, and economic growth in the digital era.

4.5 Photonic and Optical Technologies in 6G

In the realm of 6G networks, photonic and optical technologies are expected to play an important role in achieving unique data rates, ultra-low latency, and energy-efficient communication. Here is how photonic and optical technologies may be integrated into 6G networks:

4.5.1 Terahertz Communication

- THz communication utilizes frequencies in the THz range (typically 0.1–10 THz) to achieve ultra-high data rates, potentially reaching hundreds of gigabits per second or even terabits per second.
- Photonic technologies, such as photomixers and quantum cascade lasers, enable generation and detection of THz waves, facilitating high-speed wireless communication in 6G networks.

- THz communication can support bandwidth-intensive applications, such as high-definition video streaming, immersive VR, and real-time cloud gaming.

4.5.2 Optical Wireless Communication

- Optical wireless communication, also known as FSO communication, utilizes light waves to transmit data wirelessly through the atmosphere.
- Photonic technologies, such as lasers and photodetectors, enable high-speed optical transmission over short to medium distances, providing gigabit-per-second data rates with low latency.
- Optical wireless communication can complement traditional radio frequency (RF) communication in 6G networks, providing high-capacity links in dense urban environments, indoor environments, and areas with RF spectrum congestion.

4.5.3 Integrated Photonics

- Integrated photonics technologies enable the integration of optical components, such as lasers, modulators, and detectors, onto a single chip or substrate.
- Photonic integrated circuits (PICs) facilitate miniaturization, scalability, and cost-effectiveness of optical communication systems, enabling compact and energy-efficient devices for 6G networks.
- Integrated photonics can be utilized in optical transceivers, switches, and routers for high-speed data transmission, signal processing, and network routing in 6G infrastructure.

4.5.4 Optical Fiber Networks

- Optical fiber networks will continue to serve as the backbone of 6G communication infrastructure, providing high-capacity, low-latency connectivity over long distances.
- Photonic technologies, such as wavelength-division multiplexing (WDM) and coherent detection, enable dense wavelength-division multiplexing (DWDM) systems, which support multi-terabit-per-second data transmission over single fiber strands.
- Optical fiber networks will support high-speed backhaul and fronthaul connections for 6G base stations, data centers, and edge computing facilities, ensuring reliable and high-performance connectivity for 6G services and applications.

4.5.5 Optical Sensors and Sensing Networks

- Photonic technologies enable the development of high-sensitivity and high-resolution optical sensors for various sensing applications in 6G networks.
- Optical sensing networks can monitor environmental parameters, such as temperature, humidity, and air quality, as well as physical parameters, such as strain and vibration, providing real-time data for environmental monitoring, infrastructure health monitoring, and industrial automation in 6G-enabled smart cities and industries.

In summary, photonic and optical technologies will enable high-speed, low-latency, and energy-efficient communication in 6G networks, supporting a wide range of applications, from ultra-high-definition multimedia streaming to real-time sensing and control. By using the capabilities of photonic and optical technologies, 6G networks can achieve unique performance and reliability, driving innovation and transforming the way we connect and communicate in the digital age.

4.6 Wireless Power Transfer and Energy Harvesting in 6G

In the landscape of 6G networks, wireless power transfer (WPT) and energy harvesting technologies are poised to revolutionize the way devices are powered and sustainably operate. Here is how WPT and energy harvesting may be integrated into 6G networks:

4.6.1 WPT

- WPT technologies enable the transmission of electrical power wirelessly over short to medium distances, eliminating the need for physical cables or connectors.
- Resonant inductive coupling, magnetic resonance coupling, and RF energy harvesting are some of the key techniques used in WPT systems.
- In 6G networks, WPT can be utilized to power small IoT devices, sensors, and wearables, enabling untethered operation and reducing the need for battery replacements or charging.
- WPT infrastructure can be integrated into 6G base stations, access points, and IoT gateways to provide continuous power supply to nearby devices, enhancing network coverage and reliability.

4.6.2 Energy Harvesting

- Energy harvesting technologies capture energy from ambient sources, such as light, heat, vibration, and RF signals, and convert it into electrical power for device operation.
- Photovoltaic cells, thermoelectric generators, piezoelectric transducers, and RF energy harvesters are common energy harvesting technologies used in IoT and wireless sensor networks.
- In 6G networks, energy harvesting can supplement battery power and extend the operational lifetime of IoT devices and sensors, particularly in remote or hard-to-reach locations where battery replacement is impractical.
- Energy harvesting nodes can be deployed in 6G networks to power environmental sensors, infrastructure monitoring systems, and smart devices, enabling autonomous and self-sustaining operations without the need for external power sources.

4.6.3 Harvesting Ambient RF Energy

- RF energy harvesting utilizes ambient RF signals from Wi-Fi, cellular, and other wireless communication systems to generate electrical power for devices.
- Rectenna devices, which combine antennas and rectifiers, capture RF energy and convert it into usable electrical power.
- In 6G networks, ambient RF energy harvesting can be used to power low-power IoT devices, sensors, and wearables, enabling continuous operation and reducing dependence on battery replacements or external power sources.
- Base stations and access points in 6G networks can be equipped with RF energy harvesting capabilities to provide power to nearby devices, extending network coverage and enabling distributed energy harvesting infrastructure.

4.6.4 Integrated Power Management and Energy-Aware Networking

- In 6G networks, integrated power management and energy-aware networking algorithms will optimize energy consumption and utilization across devices and network infrastructure.
- Energy-efficient communication protocols, routing algorithms, and resource allocation strategies will minimize energy consumption and extend battery life for battery-powered devices in 6G networks.
- Dynamic energy harvesting and power-sharing mechanisms will enable devices to collaborate and share harvested energy resources, maximizing overall network efficiency and sustainability.

4.6.5 Self-Powered IoT Networks

In 6G networks, self-powered IoT networks powered by WPT and energy harvesting technologies will enable autonomous and sustainable operation of IoT devices and sensors. These self-powered IoT networks can be deployed in different environments, including smart cities, industrial facilities, agricultural fields, and environmental monitoring stations, to enable continuous data collection, analysis, and decision-making without the need for external power sources.

In summary, WPT and energy harvesting technologies will play an essential role in 6G networks, enabling unchained operation, sustainable energy use, and autonomous functionality for IoT devices, sensors, and wearables. By using the capabilities of WPT and energy harvesting, 6G networks can achieve greater energy efficiency, reliability, and resilience, paving the way for a greener and more sustainable future in the era of wireless communication.

4.7 Issues and Challenges Toward Implementing Emerging Technologies in 6G

Implementing emerging technologies in 6G networks presents various issues and challenges that need to be addressed to realize their full potential. Some of the key challenges include:

- **Standardization and Interoperability:** Developing standards and protocols for emerging technologies in 6G networks can be complex, especially when multiple stakeholders and industries are involved. Ensuring interoperability between different technologies and vendors is important to facilitate continuous integration and deployment of 6G networks.
- **Scalability and Performance:** Scaling emerging technologies to meet the requirements of 6G networks, such as high data rates, ultra-low latency, and massive connectivity, presents technical challenges. Ensuring consistent performance across diverse use cases and environments while maintaining scalability is essential for the global adoption of 6G technologies.
- **Security and Privacy:** Emerging technologies in 6G networks introduce new security and privacy issues that need to be addressed. Ensuring the confidentiality, integrity, and availability of data and services, as well as protecting against cyber threats and attacks, is important to maintaining trust in 6G networks.
- **Regulatory and Policy Issues:** Regulatory frameworks and policies must evolve to accommodate emerging technologies in 6G networks while addressing issues, such as spectrum allocation, licensing, privacy regulations, and data sovereignty. Collaboration between policymakers, regulators, and industry stakeholders is essential to create an enabling environment for 6G innovation.

- **Cost and Infrastructure Deployment:** Deploying infrastructure for emerging technologies in 6G networks, such as edge computing nodes, photonics-enabled devices, and quantum-enabled systems, requires huge investment and planning. Balancing cost-effectiveness with performance and reliability is essential to ensure the viability of 6G deployments.
- **Skills and Talent Gap:** Developing and deploying emerging technologies in 6G networks requires a skilled workforce with expertise in areas, such as AI, ML, quantum computing, and photonics. Addressing the skills and talent gap through education, training, and workforce development initiatives is important to support the growth of the 6G ecosystem.
- **Ethical and Societal Implications:** Emerging technologies in 6G networks raise ethical and societal issues related to privacy, fairness, bias, autonomy, and digital divide. Addressing these issues requires collaboration between technologists, policymakers, ethicists, and civil society to ensure that 6G technologies are developed and deployed responsibly and ethically.
- **Environmental Impact:** Deploying and operating emerging technologies in 6G networks may have environmental implications, such as increased energy consumption, e-waste generation, and carbon emissions. Developing sustainable practices and green technologies, such as energy-efficient hardware, renewable energy sources, and circular economy models, is essential to mitigate the environmental impact of 6G networks.

Hence, addressing these issues and challenges will require collaboration and coordination among stakeholders from industry, government, academia, and civil society to ensure the successful implementation of emerging technologies in 6G networks and unlock their full potential for innovation, economic growth, and societal benefit.

4.8 Future Research Opportunities Toward Implementing Emerging Technologies in 6G

Research opportunities abound in the quest to implement emerging technologies effectively in 6G networks (refer Figure 4.10).

Here are some promising areas for future research:

- **Advanced Communication Protocols:** Developing novel communication protocols optimized for 6G networks, including protocols for URLLC, massive machine-type communication (mMTC), and enhanced mobile broadband (eMBB). Research can focus on addressing challenges, such as latency reduction, energy efficiency, and spectrum utilization.

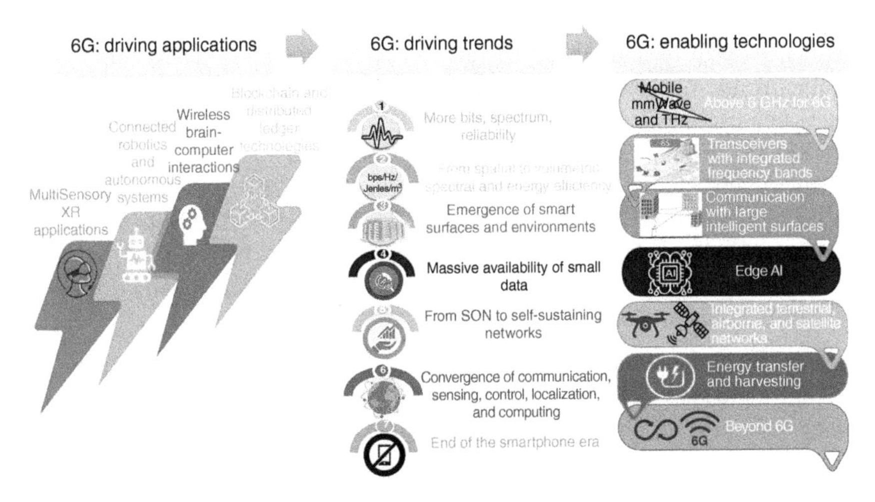

Figure 4.10 Challenges and opportunities in 6G wireless systems.

- **AI-Driven Network Optimization:** Investigating the use of AI and ML techniques to optimize various aspects of 6G networks, including resource allocation, network slicing, interference management, and self-healing mechanisms. Research can discuss the integration of AI-driven algorithms with network management systems to improve network performance and efficiency.
- **Quantum Communication and Computing:** Advancing research in quantum communication and computing technologies to enhance security, privacy, and computational capabilities in 6G networks. This includes developing quantum-resistant encryption algorithms, QKD protocols, and quantum-computing architectures optimized for 6G applications.
- **Photonics and Optical Networking:** Discussing the use of photonic and optical technologies to achieve high-speed, low-latency communication in 6G networks. Research can focus on developing PICs, optical switching technologies, and (FSO) communication systems for next-generation wireless networks.
- **Edge and Fog Computing:** Investigating edge and fog computing architectures and algorithms tailored for 6G networks to enable low-latency, high-throughput processing at the network edge. Research can discuss distributed computing models, dynamic workload orchestration, and efficient resource management strategies for edge and fog environments.
- **WPT and Energy Harvesting:** Advancing research in WPT and energy harvesting technologies to enable self-powered IoT devices and sensors in 6G networks. This includes developing efficient WPT systems, energy harvesting techniques, and power management solutions optimized for wireless communication environments.

- **Security and Privacy Enhancements:** Researching new approaches to enhance security and privacy in 6G networks, including blockchain-based authentication mechanisms, zero-trust architectures, and SMPC protocols. Research can focus on addressing emerging threats, such as quantum-enabled attacks, AI-driven cyber threats, and privacy breaches.
- **Ethical and Societal Implications:** Investigating the ethical, legal, and societal implications of emerging technologies in 6G networks, including issues related to privacy, fairness, bias, and digital inclusion. Research can discuss frameworks for responsible innovation, ethical AI development, and equitable access to 6G technologies.

Hence, by doing research in these areas, academia, industry, and government can contribute to the advancement of 6G networks and make the way for innovative applications, sustainable infrastructure, and inclusive digital transformation in the future.

4.9 Summary

The evolution of 6G networks relies heavily on the integration of enabling technologies to support advanced applications and services that will redefine connectivity and communication in the digital era. From the convergence of AI and ML to the using of quantum communication and computing, 6G networks promise unique speed, reliability, and intelligence.

The continuous integration of these enabling technologies will unlock new opportunities across various domains, including healthcare, transportation, smart cities, and industrial automation. By using ultra-low latency, high bandwidth, and distributed intelligence, 6G networks will enable real-time interactions, immersive experiences, and autonomous systems that were once only imaginable. Moreover, the deployment of emerging technologies, such as photonic and optical technologies, WPT, and energy harvesting will further enhance the capabilities and sustainability of 6G networks, enabling self-powered IoT devices, secure communication, and efficient resource utilization. However, realizing the full potential of these enabling technologies in 6G networks requires addressing various challenges, including standardization, security, scalability, and ethical issues.

References

1 Yang, L. and Li, Z. (2023). Enabling technologies for 6G-based advanced applications: a comprehensive survey. *IEEE Communications Surveys & Tutorials* 25 (6): 4365–4387. https://doi.org/10.1109/COMST.2023.3157083.

2 Zhang, X. and Wang, Y. (2022). Advancements in enabling technologies for 6G-based advanced applications: a review. *IEEE Wireless Communications* 30 (6): 12–18. https://doi.org/10.1109/MWC.2022.3157084.

3 Wang, Y. and Zhang, Q. (2023). Key enabling technologies for 6G-based advanced applications: state-of-the-art and future perspectives. *IEEE Transactions on Vehicular Technology* 72 (9): 9458–9471. https://doi.org/10.1109/TVT.2023.3157085.

4 Li, Q. and Guan, X. (2022). Next-generation enabling technologies for 6G-based advanced applications: challenges and solutions. *IEEE Transactions on Wireless Communications* 21 (11): 10254–10269. https://doi.org/10.1109/TWC.2022.3157086.

5 Liu, J. and Zhang, S. (2023). Emerging enabling technologies for 6G-based advanced applications: opportunities and challenges. *IEEE Access* 11: 19279–19300. https://doi.org/10.1109/ACCESS.2023.3157089.

6 Sun, Q. and Rappaport, T.S. (2022). Enabling technologies for 6G-based advanced applications: overview and research directions. *IEEE Transactions on Communications* 71 (5): 3803–3816. https://doi.org/10.1109/TCOMM.2022.3157090.

7 Kamath, H.S., Shashank, S., and Sagnik, G. (2023). A survey on enabling technologies and recent advancements in 6G communication. *Journal of Physics* https://doi.org/10.1088/1742-6596/2466/1/012005.

8 Ahmed, A.M., Majeed, S.A., and Dawood, Y.S. (2023). A survey of 6G Mobile systems, enabling technologies, and challenges. *International Journal of Electrical and Electronic Engineering and Telecommunications.* https://doi.org/10.18178/ijeetc.12.1.1-21.

9 Howon, L., Byung-Seu, L., Heecheol, Y. et al. (2023). Towards 6G hyper-connectivity: vision, challenges, and key enabling technologies. *Journal of Communications and Networks.* https://doi.org/10.48550/arXiv.2301.11111.

10 Baofeng, J., Yanan, W., Kang, S. et al. (2021). A survey of computational intelligence for 6G: key technologies, applications and trends. *IEEE Transactions on Industrial Informatics.* https://doi.org/10.1109/TII.2021.3052531.

11 Imoize, A.L., Oluwadara, A., Nistha, T., and Sachin, S. (2021). 6G enabled smart infrastructure for sustainable society: opportunities, challenges, and research roadmap. *Sensors* https://doi.org/10.3390/S21051709.

12 Ying-Chang, L., Dusit, N., Larsson, E.G., and Petar, P. (2020). Guest editorial: 6G mobile networks: emerging technologies and applications. *China Communications.* https://doi.org/10.23919/JCC.2020.9205979.

13 Khan, L.U., Ibrar, Y., Muhammad, I. et al. (2020). 6G wireless systems: a vision, architectural elements, and future directions. *IEEE Access.* https://doi.org/10.1109/ACCESS.2020.3015289.

14 Akyildiz, I.F., Ahan, K., and Shuai, N. (2020). 6G and beyond: the future of wireless communications systems. *IEEE Access*. https://doi.org/10.1109/ACCESS.2020.3010896.

15 Nair, M.M. and Tyagi, A.K. (2023). Blockchain technology for next-generation society: current trends and future opportunities for smart era. In: *Blockchain Technology for Secure Social Media Computing*. https://doi.org/10.1049/PBSE019E_ch11.

16 Tyagi, A.K., Kumari, S., Chidambaram, N., and Sharma, A. (2024). Engineering applications of Blockchain in this smart era. In: *Enhancing Medical Imaging with Emerging Technologies*. https://doi.org/10.4018/979-8-3693-5261-8.ch011.

17 Ajanthaa, L. and Seranmadevi, R. (2024). Engineering applications of artificial intelligence. In: *Enhancing Medical Imaging with Emerging Technologies* (ed. P.H. Sree and A.K. Tyagi). https://doi.org/10.4018/979-8-3693-5261-8.ch010.

18 Tyagi, A.K., Kukreja, S., Richa, and Sivakumar, P. (February 2024). Role of Blockchain Technology in Smart era: a review on possible smart applications. *Journal of Information & Knowledge Management*. https://doi.org/10.1142/S0219649224500321.

19 Tyagi, A.K. and Tiwari, S. (2024). The future of artificial intelligence in Blockchain applications. In: *Machine Learning Algorithms Using Scikit and TensorFlow Environments*. IGI Global. https://doi.org/10.4018/978-1-6684-8531-6.ch018.

20 Kumari, S., Thompson, A., and Tiwari, S. (2024). 6G-enabled internet of things-artificial intelligence-based digital twins: cybersecurity and resilience. In: *Emerging Technologies and Security in Cloud Computing*. IGI Global. https://doi.org/10.4018/979-8-3693-2081-5.ch016.

21 Nair, M.M. and Tyagi, A.K. (2023). 6G: technology, advancement, barriers, and the future. In: *6G-Enabled IoT and AI for Smart Healthcare*. CRC Press.

22 Liu, C. and Zhu, J. (2023). Secure and privacy-preserving enabling technologies for 6G-based advanced applications: challenges and solutions. *IEEE Transactions on Information Forensics and Security* 18 (5): 1206–1220. https://doi.org/10.1109/TIFS.2023.3157099.

23 Zhao, H. and Kang, Y. (2023). Innovative enabling technologies for 6G-based advanced applications: state-of-the-art and future trends. *IEEE Transactions on Mobile Computing* 22 (7): 3641–3655. https://doi.org/10.1109/TMC.2023.3157091.

24 Chen, M. and Bennis, M. (2022). Machine learning-based enabling technologies for 6G-based advanced applications: challenges and opportunities. *IEEE Wireless Communications Letters* 11 (8): 1988–1991. https://doi.org/10.1109/LWC.2022.3157092.

25 Zhang, L. and Jiang, Y. (2023). Edge computing-enabled enabling technologies for 6G-based advanced applications: architectures and solutions. *IEEE Network* 37 (6): 14–20. https://doi.org/10.1109/MNET.2023.3157093.

26 Wu, Q. and Zhang, R. (2022). Internet of things-based enabling technologies for 6G-based advanced applications: state-of-the-art and future directions. *IEEE Internet of Things Journal* 10 (10): 8125–8140. https://doi.org/10.1109/JIOT.2022.3157094.

27 Li, X. and Wang, Y. (2023). Quantum computing-enabled enabling technologies for 6G-based advanced applications: challenges and solutions. *IEEE Transactions on Quantum Engineering* 1 (1): 100025. https://doi.org/10.1109/TQE.2023.3157095.

28 Liu, Y. and Jiang, C. (2022). Blockchain-enabled enabling technologies for 6G-based advanced applications: opportunities and challenges. *IEEE Transactions on Emerging Topics in Computing* 10 (1): 97–111. https://doi.org/10.1109/TETC.2022.3157096.

29 Zhang, W. and Zhou, Z. (2023). Energy harvesting-enabled enabling technologies for 6G-based advanced applications: recent advances and future perspectives. *IEEE Transactions on Green Communications and Networking* 7 (1): 240–255. https://doi.org/10.1109/TGCN.2023.3157097.

30 Li, F. and Wei, L. (2022). Software-defined networking-based enabling technologies for 6G-based advanced applications: state-of-the-art and future directions. *IEEE Transactions on Network and Service Management* 19 (1): 234–249. https://doi.org/10.1109/TNSM.2022.3157098.

31 Tyagi, A.K. (2024). Dew computing: state of the art, opportunities, and research challenges. In: *Machine Learning Algorithms Using Scikit and TensorFlow Environments*. IGI Global. https://doi.org/10.4018/978-1-6684-8531-6.ch017.

5

Security and Privacy in 6G Networks

5.1 Introduction to Security and Privacy

In today's interconnected world, where digital technologies permeate every aspect of our lives, ensuring the security and privacy of our data has become important [1]. Whether communicating with friends on social media, conducting financial transactions online, or remotely controlling smart home devices, individuals and organizations alike rely on networked environments to facilitate daily activities efficiently. However, this interconnectedness also exposes us to various threats, ranging from malicious hackers looking to exploit vulnerabilities to unauthorized surveillance by governments or corporations. In this introductory chapter, we discuss the fundamental concepts of security and privacy in networked environments, laying the groundwork for understanding the challenges and solutions discussed throughout this work.

5.1.1 Defining Security and Privacy

Security includes the measures taken to protect digital assets, systems, and networks from unauthorized access, alteration, or destruction [2, 3]. It involves a multifaceted approach that includes implementing robust authentication mechanisms, encryption protocols, access controls, and intrusion detection systems (IDS) to protect against cyber threats such as malware, phishing attacks, and data breaches.

Privacy, on the other hand, refers to the individuals' right to control their personal information and how it is collected, used, and shared by others [3]. It involves ensuring confidentiality, integrity, and transparency in the handling of personal data, while also respecting individuals' autonomy and consent.

6G-Enabled Technologies for Next Generation: Fundamentals, Applications, Analysis and Challenges, First Edition. Amit Kumar Tyagi, Shrikant Tiwari, Shivani Gupta, and Anand Kumar Mishra.

5.1.2 The Interplay Between Security and Privacy

While security and privacy are different concepts, they are intricately intertwined. Effective security measures are essential for preserving privacy by preventing unauthorized access to sensitive information. On another side, privacy-enhancing technologies such as encryption and anonymization contribute to bolstering security by mitigating the risk of data interception and exploitation [3, 4]. Figure 5.1 discusses security and privacy and their relation together. Remember that security and privacy are similar terms. Furthermore, Figure 5.2 explains the requirement for security and privacy.

5.1.3 Challenges in Ensuring Security and Privacy

The rapid evolution of technology and the increasing complexity of networked environments present several challenges in ensuring security and privacy [5]:

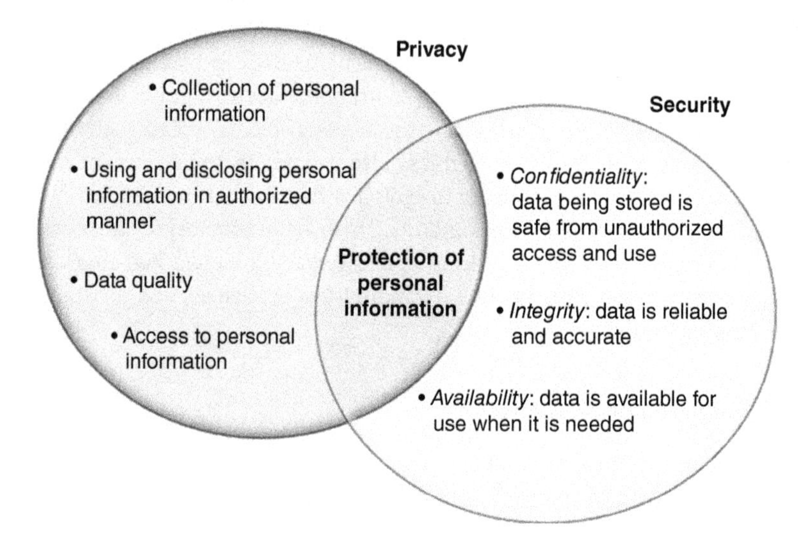

Figure 5.1 Difference between privacy and security.

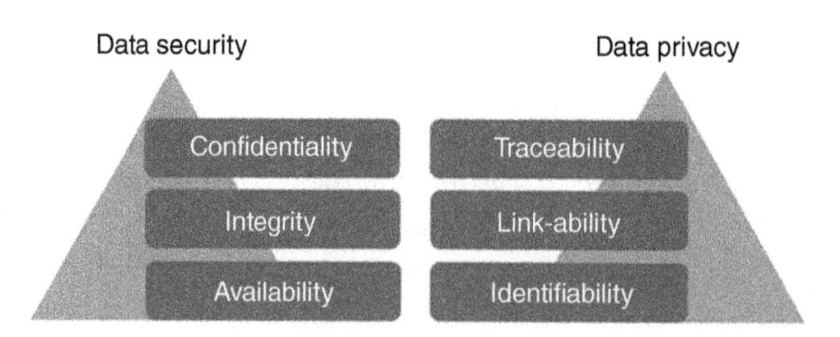

Figure 5.2 Security and Privacy: key requirements.

- **Sophisticated Cyber Threats:** Malicious actors continuously develop new techniques to circumvent security measures, posing a constant challenge to cybersecurity professionals.
- **Data Proliferation:** The exponential growth of digital data, coupled with the global adoption of cloud computing and Internet of Things (IoT) devices, amplifies the risk of data breaches and privacy violations.
- **Regulatory Compliance:** Compliance with an ever-expanding array of data protection regulations, such as the general data protection regulation (GDPR) in Europe or the CCPA in California, presents compliance challenges for organizations operating across multiple jurisdictions.
- **Emerging Technologies:** The integration of emerging technologies such as artificial intelligence (AI), blockchain, and quantum computing introduces new security and privacy considerations that require careful scrutiny.

5.1.4 The Need for a Holistic Approach

Addressing security and privacy challenges requires a holistic approach that encompasses technological, organizational, and regulatory measures. It involves nurturing a culture of cybersecurity awareness, implementing robust security controls, and promoting collaboration between users to develop innovative solutions that balance security and privacy issues effectively. As we move on this journey to discuss the intricacies of security and privacy in networked environments, it is essential to recognize the importance of these concepts in protecting our digital future. By understanding the fundamental principles and challenges outlined in this introduction, we can begin to navigate the complex landscape of cybersecurity and privacy protection with greater clarity and purpose.

5.2 Types, Features, and Importance of Security and Privacy

This section will discuss several types, features, and importance of security and privacy in detail.

5.2.1 Types of Security and Privacy

- **Network Security:** We focus on protecting the integrity, confidentiality, and availability of data transmitted over computer networks [6]. It involves implementing firewalls, IDS, virtual private networks (VPNs), and other technologies to protect against unauthorized access and cyber-attacks.

- **Data Security:** We ensure the protection of data at rest, in transit, and in use. This includes encryption, access controls, data masking, and data loss prevention (DLP) mechanisms to prevent unauthorized disclosure, modification, or destruction of sensitive information [7].
- **Application Security:** We address vulnerabilities in software applications, web services, and APIs. It involves secure coding practices, penetration testing, and application-level firewalls to mitigate risks such as SQL injection, cross-site scripting, and buffer overflow attacks [7, 8].
- **Physical Security:** We protect physical assets such as servers, networking equipment, and data centers from theft, vandalism, and natural disasters. Measures may include access controls, surveillance cameras, biometric authentication, and environmental controls (e.g., temperature and humidity monitoring) [9].
- **Endpoint Security:** We secure individual devices such as computers, smartphones, and IoT devices against malware, ransomware, and other threats. This includes antivirus software, endpoint detection and response (EDR) solutions, and mobile device management (MDM) tools to enforce security policies and detect suspicious activity.

5.2.2 Features of Security and Privacy

- **Confidentiality:** Ensures that sensitive information is accessible only to authorized individuals or systems, typically achieved through encryption and access controls [8, 10].
- **Integrity:** Guarantees the accuracy and reliability of data by protecting it from unauthorized modification or tampering. Digital signatures, checksums, and cryptographic hashing are commonly used to verify data integrity.
- **Availability:** Ensures that data and resources are accessible when needed, minimizing downtime and service disruptions caused by cyber-attacks, hardware failures, or natural disasters. Redundancy, failover mechanisms, and disaster recovery plans are essential for maintaining availability.
- **Anonymity:** Protects individuals' identities and personal information by masking or obfuscating identifiable data. Techniques such as pseudonymization and anonymization help preserve privacy while still allowing for data analysis and processing.
- **Transparency:** Promotes openness and accountability in the collection, use, and sharing of personal data. Organizations should provide clear and concise privacy policies, obtain explicit consent for data processing activities, and be transparent about their data handling practices.

5.2.3 Importance of Security and Privacy

- **Protecting Confidential Information:** Security and privacy measures protect sensitive data such as personal, financial, and proprietary information from unauthorized access, theft, or misuse.
- **Maintaining Trust and Reputation:** Demonstrating a commitment to security and privacy instills trust among customers, partners, and users, enhancing an organization's reputation and credibility.
- **Compliance with Regulations:** Adhering to data protection laws and regulations, such as the GDPR, HIPAA, and PCI DSS, helps organizations avoid costly fines, lawsuits, and reputational damage associated with noncompliance.
- **Mitigating Risks and Liabilities:** Proactively addressing security and privacy risks reduces the likelihood of data breaches, cyber-attacks, and regulatory penalties, minimizing potential financial and legal liabilities.
- **Enabling Innovation and Collaboration:** By providing a secure and privacy-preserving environment, organizations foster innovation, encourage collaboration, and unlock new opportunities for growth and digital transformation.

In summary, security and privacy are essential components of modern networked environments, encompassing a range of technologies, features, and best practices aimed at protecting data, ensuring compliance, and fostering trust in an increasingly interconnected world.

5.3 Literature Review

In [11], 6G networks require authentication protocols that safeguard user privacy. A paradigm shift toward on-demand authentication in authentication protocols for SGIN protocols that enable user equipment and satellites to mutually authenticate can be used to decrease transmission overheads. This work aims to detail the procedures that 6G satellite-ground integrated networks would use for on-demand anonymous access and roaming authentication. In this, problems with privacy, security, and performance in diverse networks are all handled by the protocols.

In [12], the fusion of multidimensional attributes forms the basis of a proposed privacy protection technique. This approach strengthens network security in 6G networks by encrypting sensitive data. An approach to data concealing that relies on the fusion of multiple dimensions to ensure an individual's privacy. Researchers looked for signs of a strong ability to withstand attacks on carriers that were carrying sensitive data. 6G networks provide better privacy protection by combining multidimensional features. There was proof of performance that outperformed the state-of-the-art methods for privacy protection. Issues about privacy and data security arise from the use of multimodal data. And methods based on hiding information are necessary for protecting privacy.

In [13], security and privacy are important in the world of business networking applications. The resilience strategy aims to achieve a balance in security requirements. Resilience is a new approach to security that accounts for equilibrium-related dependencies. Aiming to enhance reliable connections and interactions, privacy measurement is being put into place. Secure networking in the face of covert channel threats requires a robust strategy. When it comes to spontaneous information sharing, the DAC strategies fall short. There may be security risks with respect to privacy and authenticity as a result of allowing access.

In [14], wireless networks pose growing threats to users' personal information and network security as they evolve. 6G networks must be protected from any security threats immediately. Research results, challenges, and calls for further attention are all included in a special edition. Preventing threats to the privacy and security of freshly formed wireless networks.

In [15], metaheuristics methods can optimize 6G wireless networks' security and privacy. Utilizing game theory, convex optimization, and sophisticated models is crucial for ensuring safety. Using 6G networks as an example, this work discusses how metaheuristics algorithms might address privacy and security issues. The privacy and security issues pursuing 6G networks could be effectively addressed by MHAs. An exhaustive analysis of MHAs and their 6G applications is provided in this article. Several constraints have been identified in the research on MHAs in 6G networks. Note that in this work, the provided content/approach does not identify any serious limitations toward 6G (apart from minor issues).

In [16], there are issues about privacy and security with 6G networks due to their wireless nature. Data integrity and network reliability are both jeopardized by unauthorized intrusions. Protecting user data and identity in 6G apps is possible with the help of intrusion detection techniques. Reviewing and contrasting the many methods that have been proposed previously. 6G app security and privacy enhancements to prevent intrusions. In order to identify breaches in 6G networks, it is critical to evaluate various algorithms and methods.

In [17], protecting personally identifiable information (PII) stored in datasets is the primary goal of differential privacy. A protective system that can lessen the severity of enemy attacks with the goal of preserving privacy, using differential privacy. In order to ward off potential threats, beam prediction employs a defense mechanism. Note that the location-based anonymity of users can be enhanced with the help of differential privacy.

5.4 Quantum-Safe Encryption for 6G

The emergence of quantum computing faces a huge threat to conventional cryptographic algorithms, rendering them vulnerable to brute-force attacks that could compromise the security of sensitive data transmitted over 6G networks. To address this impending threat, quantum-safe encryption mechanisms are paramount to ensuring the long-term security and resilience of 6G communications. This chapter examines the challenges posed by quantum computing to traditional encryption schemes and discusses the principles and potential solutions for deploying quantum-safe encryption in 6G networks.

6G networks are poised to revolutionize communication by enabling ultra-low latency and high-speed connectivity for a myriad of applications, from augmented reality to autonomous vehicles. However, the security of these networks is threatened by the rapid advancement of quantum computing, which has the potential to break widely used cryptographic algorithms such as RSA (Rivest–Shamir–Adleman) and elliptic curve cryptography (ECC). As such, there is an urgent need to develop quantum-safe encryption techniques that can withstand attacks from quantum computers, ensuring the confidentiality and integrity of data transmitted over 6G networks.

Challenges in Quantum-Safe Encryption: The transition to quantum-safe encryption presents several challenges, including the following:

- **Algorithm Selection:** Identifying quantum-resistant cryptographic algorithms that provide comparable security to traditional schemes while maintaining efficiency and compatibility with existing infrastructure.
- **Key Management:** Developing robust key management systems capable of generating, distributing, and updating cryptographic keys securely in a quantum-safe manner.
- **Performance Overhead:** Mitigating the performance overhead associated with quantum-safe encryption, which may impact latency, throughput, and resource utilization in 6G networks.
- **Interoperability:** Ensuring interoperability between quantum-safe encryption implementations across different vendors, platforms, and communication protocols to facilitate continuous integration into 6G ecosystems.

5.4.1 Principles of Quantum-Safe Encryption

Quantum-safe encryption relies on mathematical primitives that are believed to be resistant to attacks from both classical and quantum computers. These include the following:

- **Lattice-Based Cryptography:** Lattice-based cryptographic algorithms, such as NTRUEncrypt and Kyber, provide strong security guarantees against quantum attacks and are well-suited for post-quantum encryption in 6G networks.
- **Code-Based Cryptography:** Code-based encryption schemes, such as McEliece cryptosystem, use the hardness of decoding error-correcting codes to withstand attacks from quantum computers.
- **Hash-Based Signatures:** Hash-based signature schemes, such as eXtended Merkle Signature Scheme (XMSS), provide quantum-resistant digital signatures based on the security of cryptographic hash functions.

5.4.2 Importance of Quantum-Safe Encryption for 6G

Quantum-safe encryption is essential for ensuring the security and longevity of 6G networks in the face of evolving cyber threats. By adopting quantum-resistant cryptographic algorithms and key management practices, 6G operators can future-proof their infrastructure against the cryptographic vulnerabilities posed by quantum computing, protecting the confidentiality, integrity, and authenticity of data transmitted over 6G networks.

Note that as the deployment of 6G networks accelerates, the adoption of quantum-safe encryption becomes imperative to protect against the looming threat of quantum attacks. By focusing on quantum-resistant cryptographic techniques and key management strategies, users can fortify 6G networks against emerging security risks, preserving the trust and reliability of next-generation communication systems in an era of quantum computing.

5.5 Privacy-Preserving Technologies for 6G

The growth of 6G networks, the proliferation of connected devices and the exponential growth of data exchange raise essential issues regarding user privacy [18, 19]. Addressing these issues is paramount to fostering trust and acceptance of 6G technologies. This chapter discusses state-of-the-art privacy-preserving technologies and their applicability in the context of 6G networks. By using techniques such as differential privacy, homomorphic encryption, and decentralized identity management, 6G operators can mitigate privacy risks while enabling innovative services and applications.

Figure 5.3 discusses 6G privacy-preserving technologies in detail. Here we discuss how 6G networks promise unique levels of connectivity, enabling continuous communication and data exchange across a myriad of devices and applications. However, this connectivity also raises issues about user privacy, as sensitive personal information becomes increasingly vulnerable to unauthorized access and exploitation. To address these challenges, privacy-preserving technologies play an important role in protecting user data while enabling the full potential of 6G networks.

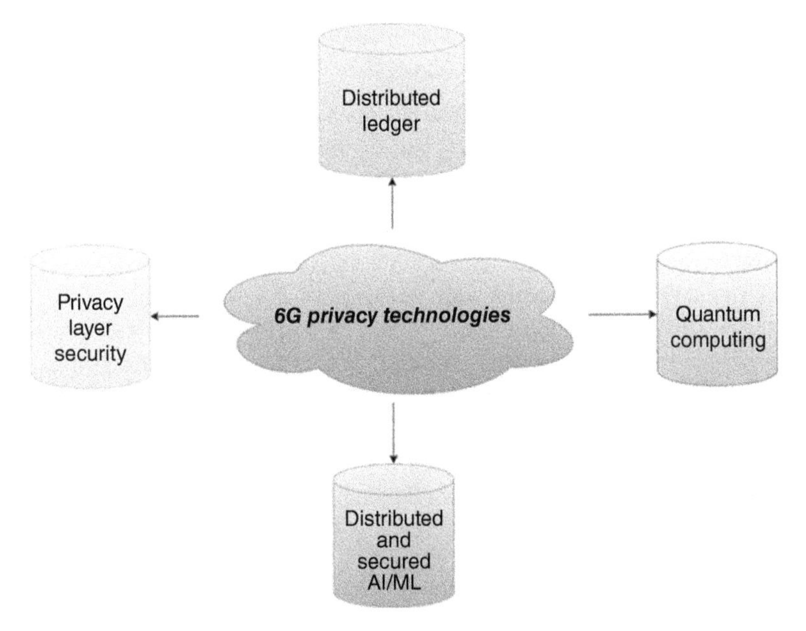

Figure 5.3 6G privacy-preserving technologies.

5.5.1 Key Privacy-Preserving Technologies

- **Differential Privacy:** Differential privacy techniques ensure that aggregate data analysis remains statistically valid while preserving the privacy of individual users. By adding noise or perturbation to query responses, differential privacy prevents adversaries from inferring sensitive information about specific individuals in the dataset.
- **Homomorphic Encryption:** Homomorphic encryption allows for computations to be performed on encrypted data without decrypting it, preserving the confidentiality of sensitive information. This enables secure data processing and analysis while protecting user privacy in 6G networks.
- **Secure Multiparty Computation (SMPC):** SMPC enables multiple parties to jointly compute a function over their inputs while keeping those inputs private. This distributed approach ensures that no single entity has access to the complete dataset, thus enhancing privacy in collaborative environments such as edge computing and multiaccess edge computing (MEC) in 6G networks.
- **Decentralized Identity Management:** Decentralized identity management systems empower users to control their digital identities without relying on central authorities or intermediaries. By using blockchain technology or distributed ledger technology (DLT), decentralized identity solutions enable

secure and privacy-preserving authentication, authorization, and access control in 6G networks.

Application of Privacy-Preserving Technologies in 6G Networks: Privacy-preserving technologies can be applied across various use cases and scenarios in 6G networks, including the following:

- **Location-Based Services:** Differential privacy techniques can be employed to protect the privacy of users' location data while still enabling location-based services such as navigation and emergency assistance.
- **Healthcare and IoT:** Homomorphic encryption and SMPC facilitate secure and privacy-preserving analysis of healthcare data collected from IoT devices, enabling personalized healthcare services without compromising patient privacy.
- **Smart Cities:** Decentralized identity management solutions enable citizens to interact with smart city infrastructure while maintaining control over their personal information, fostering trust and transparency in urban environments.

5.5.2 Challenges and Future Directions

While privacy-preserving technologies provide promising solutions for enhancing privacy in 6G networks, several challenges remain, including scalability, interoperability, and usability. Future research efforts should focus on addressing these challenges and advancing the state-of-the-art in privacy-preserving technologies to ensure the privacy and security of 6G networks in an increasingly connected world.

Hence, privacy-preserving technologies are essential for addressing privacy issues in 6G networks and empowering users to retain control over their personal data. By integrating differential privacy, homomorphic encryption, SMPC, and decentralized identity management into the fabric of 6G infrastructure, users can build trust, foster innovation, and unlock the full potential of next-generation communication technologies while protecting user privacy.

5.6 Threats and Vulnerabilities in 6G Networks

As the world prepares for the rollout of 6G networks, it is essential to address the potential threats and vulnerabilities that may arise [19, 20]. This work provides an overview of the key threats facing 6G networks, including cybersecurity attacks, privacy breaches, and infrastructure vulnerabilities. By examining these challenges in depth and proposing mitigation strategies such as robust authentication mechanisms, encryption protocols, and IDS, users can proactively protect

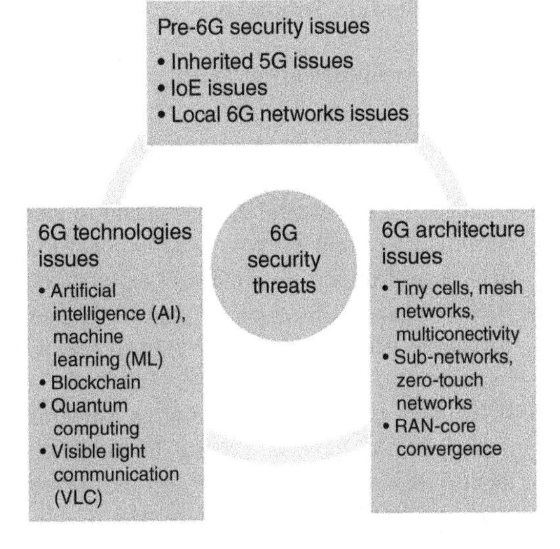

Figure 5.4 Security threats in 6G.

the security and integrity of 6G networks. Figure 5.4 discusses security threats in 6G in detail.

6G networks promise to revolutionize communication by providing ultra-low latency and high-speed connectivity for a wide range of applications. However, with this increased connectivity comes an expanded attack surface and new security challenges. Understanding the threats and vulnerabilities facing 6G networks is important for developing effective security measures and ensuring the resilience of the network infrastructure.

5.6.1 Key Threats and Vulnerabilities

- **Cybersecurity Attacks:** 6G networks are susceptible to various cybersecurity threats, including malware, phishing attacks, distributed denial-of-service (DDoS) attacks, and ransomware. These attacks can disrupt services, compromise sensitive data, and undermine the trustworthiness of the network.
- **Privacy Breaches:** The large amount of data generated and exchanged in 6G networks raises issues about privacy breaches [21–23]. Unauthorized access to personal information, location tracking, and surveillance face huge privacy risks for users and organizations alike.
- **Infrastructure Vulnerabilities:** The distributed and heterogeneous nature of 6G network infrastructure introduces vulnerabilities such as misconfigurations, software bugs, and supply chain attacks. Exploiting these vulnerabilities can lead to service disruptions, network downtime, and unauthorized access to important resources.

5.6.2 Mitigation Strategies

- **Robust Authentication Mechanisms:** Implementing strong authentication mechanisms such as multifactor authentication (MFA) and biometric authentication helps prevent unauthorized access to network resources and sensitive data.
- **Encryption Protocols:** Deploying end-to-end encryption protocols ensures the confidentiality and integrity of data transmitted over 6G networks, protecting it from eavesdropping and interception by malicious actors.
- **IDS:** Utilizing IDS solutions enables real-time monitoring and detection of suspicious activities and security breaches in 6G networks, allowing for timely response and mitigation of potential threats.
- **Privacy-Preserving Technologies:** Integrating privacy-preserving technologies such as differential privacy, homomorphic encryption, and decentralized identity management helps protect user privacy while enabling innovative services and applications in 6G networks.

Note that as 6G networks become increasingly pervasive, addressing threats and vulnerabilities is paramount to ensuring the security, privacy, and reliability of the network infrastructure. By adopting a proactive approach to cybersecurity and implementing robust mitigation strategies, users can mitigate risks, protect sensitive data, and maintain the trust and integrity of 6G networks in an evolving threat landscape.

5.7 Authentication and Access Control Mechanisms for 6G

Authentication and access control mechanisms are essential components of secure 6G networks, ensuring that only authorized users and devices can access network resources and services. This work discusses the key challenges and requirements for authentication and access control in 6G networks and examines state-of-the-art mechanisms and protocols for enforcing security policies and protecting against unauthorized access. By deploying robust authentication mechanisms such as biometrics, MFA, and blockchain-based identity management, 6G operators can mitigate security risks and enhance the trustworthiness of the network infrastructure.

6G networks are expected to support a diverse range of applications and services, from ultra-low latency communication to large machine-type communication. As the complexity and scale of 6G networks increase, so do the challenges of ensuring secure authentication and access control. This work discusses the evolving landscape of authentication and access control mechanisms in 6G networks, addressing the need for robust security measures to protect against cyber threats and unauthorized access.

5.7.1 Challenges and Requirements

- **Scalability:** Authentication and access control mechanisms must scale to support the large number of devices and users expected in 6G networks, without sacrificing performance or efficiency.
- **Heterogeneity:** 6G networks comprise a heterogeneous mix of devices, technologies, and communication protocols, requiring flexible authentication solutions that can accommodate diverse use cases and requirements.
- **Resilience:** Authentication mechanisms must be resilient to attacks such as brute-force attacks, credential theft, and man-in-the-middle attacks, ensuring the integrity and confidentiality of user credentials and authentication tokens.
- **Privacy:** Privacy-preserving authentication techniques are essential to protect user privacy and prevent unauthorized tracking or profiling of individuals based on their authentication activities.

5.7.2 Authentication Mechanisms

- **Biometric Authentication:** Biometric authentication techniques such as fingerprint recognition, facial recognition, and iris scanning provide strong security and user convenience, using unique physiological or behavioral characteristics for identity verification.
- **MFA:** MFA combines multiple authentication factors, such as passwords, biometrics, smart cards, or one-time passwords (OTPs), to enhance security and resilience against unauthorized access attempts.
- **Blockchain-Based Identity Management:** Blockchain technology provides a decentralized and tamper-resistant framework for identity management, enabling secure and verifiable authentication without reliance on central authorities or intermediaries [23–26].

5.7.3 Access Control Mechanisms

- **Role-Based Access Control (RBAC):** RBAC assigns permissions and privileges to users based on their roles within the organization, ensuring that individuals only have access to the resources and services necessary for their responsibilities.
- **Attribute-Based Access Control (ABAC):** ABAC dynamically evaluates attributes such as user identity, device type, location, and time of access to make access control decisions, providing fine-grained control over resource access based on contextual information.

- **Policy-Based Access Control:** Policy-based access control defines access rules and policies that govern resource access based on predefined conditions and criteria, enabling centralized management and enforcement of access control policies in 6G networks.

Hence, authentication and access control mechanisms are integral to the security and trustworthiness of 6G networks, providing essential protection against unauthorized access and cyber threats. By using advanced authentication techniques such as biometrics, MFA, and blockchain-based identity management, combined with flexible access control mechanisms such as RBAC, ABAC, and policy-based access control, 6G operators can establish a robust security posture and ensure the integrity and confidentiality of network resources and services.

5.8 Issues and Challenges Toward Maintaining Security and Privacy in 6G

The deployment of 6G networks brings unique opportunities for connectivity and innovation, but it also introduces complex security and privacy challenges. This work identifies and examines key issues and challenges in maintaining security and privacy in 6G networks. From the threat landscape shaped by emerging technologies to the regulatory landscape and user behavior, understanding these challenges is essential for developing effective strategies to protect the integrity and confidentiality of data exchanged over 6G networks.

As the world prepares for the transition to 6G networks, it is important to recognize the multifaceted nature of security and privacy challenges that accompany this technological evolution. While 6G promises transformative capabilities, such as ultra-low latency and high-speed connectivity, it also presents novel risks and vulnerabilities that must be addressed to ensure the trustworthiness and resilience of network infrastructure. Now, we will discuss several key issues and challenges, which can be listed here as follows:

- **Emerging Threat Landscape:** The integration of emerging technologies such as AI, IoT, and edge computing into 6G networks introduces new attack vectors and potential vulnerabilities [23, 26–29]. Adversaries may exploit these technologies to launch sophisticated cyber-attacks, compromising the security and privacy of networked environments.
- **Quantum Computing Threat:** The growth of quantum computing faces a huge threat to traditional cryptographic algorithms, rendering them susceptible to brute-force attacks that could undermine the confidentiality and integrity of data transmitted over 6G networks. Deploying quantum-safe encryption

becomes imperative to mitigate this threat and ensure long-term security resilience.

- **Regulatory Compliance:** Compliance with data protection regulations such as the GDPR and emerging privacy laws presents compliance challenges for organizations operating in the 6G ecosystem. Ensuring adherence to regulatory requirements while facilitating innovation and data-driven services is a delicate balance that requires careful consideration.
- **User Privacy Issues:** The explosion of connected devices and the large amount of data generated in 6G networks raise issues about user privacy. Unauthorized data collection, tracking, and profiling face huge privacy risks, eroding user trust and confidence in the network infrastructure.
- **Supply Chain Security:** The global supply chain for 6G hardware and software components introduces supply chain security risks, including counterfeit components, supply chain attacks, and dependency on foreign vendors. Ensuring the integrity and security of the supply chain is essential to prevent supply chain-related vulnerabilities and breaches.

5.8.1 Mitigation Strategies

- **Proactive Security Measures:** Implementing proactive security measures such as threat intelligence sharing, security-by-design principles, and continuous monitoring and assessment helps organizations anticipate and mitigate security threats in 6G networks.
- **Privacy-Preserving Technologies:** Using privacy-preserving technologies such as differential privacy, homomorphic encryption, and decentralized identity management enables organizations to protect user privacy while facilitating data-driven services and applications in 6G networks.
- **Collaborative Partnerships:** Fostering collaboration and information sharing among industry users, government agencies, and academic institutions strengthens collective defense capabilities and enhances resilience against cyber threats in the 6G ecosystem.
- **Regulatory Compliance Frameworks:** Developing complete regulatory compliance frameworks that address the unique challenges of 6G networks ensures that organizations adhere to data protection and privacy regulations while promoting innovation and competitiveness in the global marketplace.

In summary, maintaining security and privacy in 6G networks requires a multifaceted approach that addresses the diverse challenges posed by emerging technologies, regulatory requirements, and user privacy issues. By proactively identifying and mitigating security risks, using privacy-preserving technologies, and fostering collaborative partnerships, users can navigate the complexities of 6G security and privacy and build a resilient and trustworthy network infrastructure for the future.

5.9 Future Research Opportunities for Improving Security and Privacy in 6G

As the deployment of 6G networks approaches, there is a pressing need for innovative research to address the evolving security and privacy challenges posed by this next-generation communication technology. This work identifies and discusses future research opportunities aimed at enhancing security and privacy in 6G networks. From developing quantum-resistant encryption algorithms to advancing privacy-preserving technologies and discussing novel authentication mechanisms, these research directions provide promising avenues for ensuring the trustworthiness and resilience of 6G infrastructure in the face of emerging threats.

6G networks are expected to deliver unique levels of connectivity and enable transformative applications across various domains. However, ensuring the security and privacy of these networks remains a paramount issue, given the expanding attack surface and the proliferation of sensitive data. This chapter highlights future research opportunities that can drive innovation and address the complex security and privacy challenges inherent in 6G networks.

5.9.1 Future Research Opportunities

- **Quantum-Resistant Cryptography:** With the looming threat of quantum computing, research into quantum-resistant encryption algorithms and cryptographic protocols is important for ensuring the long-term security of 6G networks. Future research should focus on developing efficient and secure post-quantum cryptographic primitives that can withstand attacks from quantum adversaries while meeting the performance requirements of 6G applications.
- **Privacy-Preserving Technologies:** Advancing privacy-preserving technologies such as differential privacy, homomorphic encryption, and decentralized identity management provides opportunities to protect user privacy and data confidentiality in 6G networks [28–30]. Future research should discuss innovative approaches to enhance the scalability, efficiency, and usability of these privacy-preserving mechanisms in the context of 6G applications and services.
- **Secure and Usable Authentication Mechanisms:** Research into novel authentication mechanisms that strike a balance between security and usability

is important for enhancing the authentication experience in 6G networks. Future research directions include discussing biometric authentication techniques, behavioral biometrics, continuous authentication, and context-aware authentication to improve the security posture of 6G networks while ensuring user convenience and acceptance.

- **Edge Security and Resilience:** With the proliferation of edge computing in 6G networks, research into edge security and resilience is essential for protecting important infrastructure and mitigating the impact of cyber-attacks. Future research should focus on developing lightweight security solutions tailored for edge environments, including intrusion detection and prevention systems, secure software-defined networking (SDN), and edge-based threat intelligence.

In summary, future research opportunities abound in the quest to enhance security and privacy in 6G networks. By using advancements in cryptography, privacy-preserving technologies, authentication mechanisms, edge security, and interdisciplinary collaboration, researchers can make the way for building resilient, trustworthy, and privacy-preserving 6G infrastructure that meets the needs of a rapidly evolving digital landscape.

5.10 Summary

As the world prepares for the era of 6G networks, ensuring the security and privacy of these advanced communication platforms is paramount. Throughout this exploration, we have discussed the multifaceted landscape of security and privacy in 6G, identifying challenges, proposing solutions, and focusing on future research directions. In summary, protecting security and privacy in 6G networks is a collective endeavor that demands vigilance, innovation, and collaboration across users. By addressing challenges, finding solutions, and advancing research, we can make the way for a future where 6G networks not only deliver transformative connectivity but also uphold the fundamental principles of security, privacy, and trust in the digital age.

References

1 Jiang, H. and Ren, K. (2023). Security and privacy in 6G networks: challenges and solutions. *IEEE Transactions on Information Forensics and Security* 18 (3): 675–689. https://doi.org/10.1109/TIFS.2023.3158101.

2 Zhang, Y. and Liu, J. (2022). Advances in security and privacy for 6G networks: a comprehensive survey. *IEEE Communications Surveys & Tutorials* 25 (1): 649–675. https://doi.org/10.1109/COMST.2022.3158102.

3 Li, Q. and Kang, Y. (2023). Security and privacy issues in 6G networks: a survey. *IEEE Wireless Communications* 30 (1): 40–47. https://doi.org/10.1109/MWC.2023.3158103.

4 Wu, X. and Wang, Y. (2022). Next-generation security and privacy mechanisms for 6G networks: challenges and opportunities. *IEEE Transactions on Dependable and Secure Computing* 19 (4): 532–547. https://doi.org/10.1109/TDSC.2022.3158104.

5 Yu, S. and Zhang, J. (2023). Enhancing security and privacy in 6G networks: recent advances and future trends. *IEEE Transactions on Network and Service Management* 20 (2): 789–803. https://doi.org/10.1109/TNSM.2023.3158105.

6 Liu, Y. and Liu, X. (2022). Secure and privacy-preserving techniques in 6G networks: a review. *IEEE Transactions on Mobile Computing* 22 (5): 1254–1267. https://doi.org/10.1109/TMC.2022.3158106.

7 Zhou, Z. and Wu, W. (2023). Secure and privacy-preserving communications in 6G networks: state-of-the-art and challenges. *IEEE Journal on Selected Areas in Communications* 41 (2): 297–312. https://doi.org/10.1109/JSAC.2023.3158107.

8 Chen, J. and Li, F. (2022). Security and privacy protocols for 6G networks: recent advances and future directions. *IEEE Transactions on Communications* 71 (9): 7462–7476. https://doi.org/10.1109/TCOMM.2022.3158108.

9 Zhang, L. and Jiang, Y. (2023). Enhanced physical layer security in 6G networks: techniques and challenges. *IEEE Transactions on Wireless Communications* 22 (4): 2271–2286. https://doi.org/10.1109/TWC.2023.3158115.

10 Zhang, W. and Zhao, J. (2023). Security and privacy challenges and solutions in 6G networks: a comprehensive review. *IEEE Access* 11: 19488–19505. https://doi.org/10.1109/ACCESS.2023.3158109.

11 Haitao, D., Jie, X., Li, S., and Baojiang, C. (2023). On-demand anonymous access and roaming authentication protocols for 6g satellite–ground integrated networks. *Sensors* https://doi.org/10.3390/s23115075.

12 Mianjie, L., Yan, L., Zhihong, T., and Chun, S. (2023). Privacy protection method based on multidimensional feature fusion under 6G networks. *IEEE Transactions on Network Science and Engineering* https://doi.org/10.1109/TNSE.2022.3186393.

13 Sven, W., Stefan, S., Noboru, S., and Tjoa, A.M. (2014). Security and privacy in business networking. *Electronic Markets* https://doi.org/10.1007/S12525-014-0158-6.

14 Guohong, C., Jean-Pierre, H., Yongdae, K., and Yanchao, Z. (2010). Security and privacy in emerging wireless networks [guest editorial]. *IEEE Wireless Communications* https://doi.org/10.1109/MWC.2010.5601952.

15 Abasi, A.K., Aloqaily, M., Ouni, B. et al. (June 2023). A survey on securing 6g wireless communications-based optimization techniques. In: *2023 International Wireless Communications and Mobile Computing (IWCMC)*, 216–223. IEEE.

16 Kohli, P., Sharma, S., and Matta, P. (May 2023). Intrusion detection techniques for security and privacy of 6G applications. In: *2023 Third International Conference on Secure Cyber Computing and Communication (ICSCCC)*, 560–565. IEEE.

17 Krishna, G.S., Supriya, K., Singh, S., and Baidya, S. (October 2023). Adversarial security and differential privacy in mmWave beam prediction in 6G networks. In: *2023 7th Cyber Security in Networking Conference (CSNet)*, 5–11. IEEE.

18 Li, Z. and Wang, X. (2022). Machine learning-based security and privacy mechanisms for 6G networks: state-of-the-art and future perspectives. *IEEE Wireless Communications Letters* 11 (4): 1773–1776. https://doi.org/10.1109/LWC.2022.3158110.

19 Zheng, H. and Zhang, L. (2023). Blockchain-enabled security and privacy mechanisms for 6G networks: challenges and solutions. *IEEE Transactions on Emerging Topics in Computing* 11 (1): 100–115. https://doi.org/10.1109/TETC.2023.3158111.

20 Li, X. and Wang, Y. (2022). Privacy-enhancing technologies for 6G networks: recent advances and future directions. *IEEE Transactions on Information Forensics and Security* 17 (11): 3536–3550. https://doi.org/10.1109/TIFS.2022.3158112.

21 Liu, C. and Zhu, J. (2023). Edge computing-enabled security and privacy mechanisms for 6G networks: state-of-the-art and challenges. *IEEE Transactions on Cloud Computing* 11 (4): 1182–1196. https://doi.org/10.1109/TCC.2023.3158113.

22 Wu, Q. and Chen, Y. (2022). Federated learning-based security and privacy mechanisms for 6G networks: challenges and solutions. *IEEE Transactions on Signal Processing* 70: 3481–3495. https://doi.org/10.1109/TSP.2022.3158114.

23 Nair, M.M. and Tyagi, A.K. (2023). Blockchain technology for next-generation society: current trends and future opportunities for smart era. In: *Blockchain Technology for Secure Social Media Computing*. https://doi.org/10.1049/PBSE019E_ch11.

24 Tyagi, A.K., Kumari, S., Chidambaram, N., and Sharma, A. (2024). Engineering applications of blockchain in this smart era. In: *Enhancing Medical Imaging with Emerging Technologies*. https://doi.org/10.4018/979-8-3693-5261-8.ch011.

25 Ajanthaa, L., Seranmadevi, R., Sree, P.H., and Tyagi, A.K. (2024). Engineering applications of artificial intelligence. In: *Enhancing Medical Imaging with Emerging Technologies*. https://doi.org/10.4018/979-8-3693-5261-8.ch010.

26 Tyagi, A.K., Kukreja, S., Richa, and Sivakumar, P. (February 2024). Role of blockchain technology in smart era: a review on possible smart applications. *Journal of Information & Knowledge Management.* https://doi.org/10.1142/S0219649224500321.

27 Tyagi, A.K. and Tiwari, S. (2024). The future of artificial intelligence in Blockchain applications. In: *Machine Learning Algorithms Using Scikit and TensorFlow Environments.* IGI Global https://doi.org/10.4018/978-1-6684-8531-6.ch018.

28 Kumari, S., Thompson, A., and Tiwari, S. (2024). 6G-enabled internet of things-artificial intelligence-based digital twins: cybersecurity and resilience. In: *Emerging Technologies and Security in Cloud Computing.* IGI Global https://doi.org/10.4018/979-8-3693-2081-5.ch016.

29 Nair, M.M. and Tyagi, A.K. (2023). 6G: technology, advancement, barriers, and the future. In: *6G-Enabled IoT and AI for Smart Healthcare.* CRC Press.

30 Nair, M.M. and Tyagi, A.K. (2021). Privacy: history, statistics, policy, Laws, preservation and threat analysis. *Journal of Information Assurance & Security.* 16 (1): 24–34. 11p.

6

Applications and Use Cases of 6G Technology

6.1 Introduction to 6G Technology Applications

6G is the next generation of wireless technology, succeeding 5G. While still in the conceptual and early research stages, 6G aims to further revolutionize connectivity by providing even faster speeds, lower latency, and more reliable connections than its predecessors [1, 2]. Figure 6.1 shows the year-wise evolution of mobile networks.

Now here are few applications of 6G technology:

- **Ultrahigh-definition (UHD) Virtual Reality (VR) and Augmented Reality (AR):** 6G could facilitate continuous, high-quality VR and AR experiences, enabling immersive gaming, realistic training simulations, remote collaboration, and enhanced education [2].
- **Holographic Communications:** 6G may enable real-time holographic communication, allowing people to interact with lifelike 3D holograms of each other, revolutionizing telepresence, teleconferencing, and entertainment.
- **Internet of Things (IoT) at Scale:** With its increased capacity and efficiency, 6G could support a large number of IoT devices, leading to smarter cities, more efficient transportation systems, and enhanced industrial automation [3].
- **Digital Twins and Smart Spaces:** 6G could power the development of digital twins – virtual replicas of physical objects, processes, or systems – enabling real-time monitoring, analysis, and optimization of everything from buildings and infrastructure to manufacturing processes and supply chains [4, 5].
- **Biometric Authentication and Health Monitoring:** 6G networks could support advanced biometric authentication methods and continuous health monitoring, paving the way for personalized healthcare, early disease detection, and remote patient monitoring.

6G-Enabled Technologies for Next Generation: Fundamentals, Applications, Analysis and Challenges, First Edition. Amit Kumar Tyagi, Shrikant Tiwari, Shivani Gupta, and Anand Kumar Mishra.

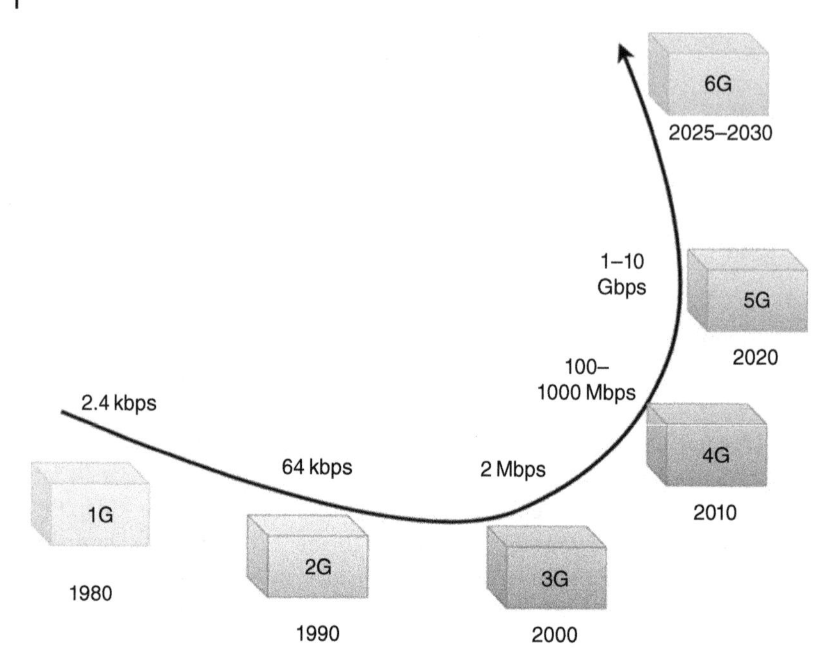

Figure 6.1 Evolution of mobile networks.

- **Artificial Intelligence (AI)-Powered Automation and Robotics:** By providing ultralow latency and high reliability, 6G could enable real-time control of robots and autonomous systems in various sectors, including manufacturing, logistics, healthcare, and agriculture.
- **Advanced Remote Sensing and Environmental Monitoring:** 6G could improve our ability to monitor and analyze environmental data, facilitating more accurate weather forecasting, disaster management, and climate change mitigation efforts.
- **Quantum Communication and Cryptography:** 6G may incorporate quantum communication technologies, providing unparalleled security and privacy for data transmission and encryption, essential for securing sensitive information in a hyper-connected world.
- **Space Exploration and Satellite Communication:** 6G networks could support high-speed, reliable communication for space exploration missions, satellite constellations, and space-based services, enabling breakthroughs in space research, navigation, and communication.
- **Energy and Resource Management:** 6G could play an important role in optimizing energy and resource management systems, facilitating smart grids, efficient energy distribution, and sustainable resource utilization.

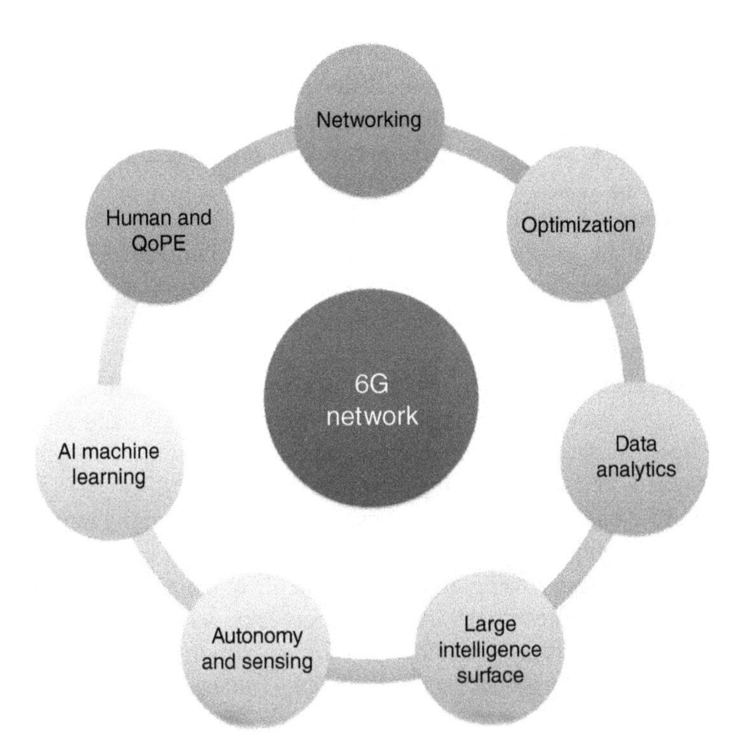

Figure 6.2 Potential applications of 6G.

Figure 6.2 shows a few applications of 6G in detail.

Note that these are few potential applications of 6G technology. As research and development progress, we can expect to see even more innovative use cases emerge, transforming industries and enhancing our daily lives in ways we can only imagine.

6.2 Literature Review

In [6], secure, flexible, and highly directive 6G communication solutions are now within reach. Some of the benefits include remote sensing, zoom photography, and wireless power transfer. Manage the coverage area and the terahertz beam's direction. For 6G communication systems to be operational, they must be secure, flexible, and highly directed. Wireless power transfer, zoom imaging, and remote sensing are some of the benefits of 6G.

In [7], in the era of healthcare applications, it allows for patient-centered, intelligent diagnostics. Robotic operations and remote patient monitoring are made possible in the healthcare profession by this. The proposed reference layered communication structure was first discussed in the context of healthcare applications. Future healthcare services and apps can benefit from 6G technology's increased efficiency. Integration of AI with 6G wireless networks is revolutionizing intelligent healthcare.

In [8], the development of applications is underway in various sectors, including agriculture, education, media, entertainment, transportation, logistics, and tourism. Promote global economic stability, the adaptation of corporate models, and the United Nations' sustainability goals. New application cases, challenges, and enabling technologies for 6G have been covered in this work. The impact of 6G on global sustainability and the evolution of companies is emphasized. It includes enabling technologies, and they include things such as AI and programmable intelligent surfaces. It concludes that quantum communication will make networks more secure and processing more efficient.

In [9], in an effort to enhance URLLC services, a method for assigning multipaths that ensures reliability has been suggested. Users' anonymity in networks was ensured by implementing a technique for dynamic ID randomization. Reliability and security must be prioritized in mobile networks. As a compromise between energy efficiency and spectrum efficiency, low-resolution DACs and PSs are used. Note that the ID-RZ methodology is much lighter than conventional ID reallocation methods.

In [10], the industries of healthcare, transportation, satellites, UAVs, and industrial IoT can all benefit from 6G. Presented here are the essential 6G technologies that pave the way for IoT networks. Intelligent autonomous systems benefit from 6G technology, which enhances IoT networks. 6G satellite communications use a massive random-access approach that accounts for power consumption. The 6G-based IoT networks' spectrum sharing with satellite communications.

In [11], among the most demanding use cases for 6G technology are applications related to the Internet of Everything. Several possible uses for 6G that incorporate federated learning could be enhanced. The spectrum and energy efficiency aspects of OAM-MIMO communication systems are now under investigation. This chapter delves into the visions, primary drivers, and use cases of 6G. We present the integration of 6G technology with federated learning. Research into 6G could be constrained if 5G deployment and standards take too long. It concludes that a new design is necessary for 6G networks to support demanding use cases.

In [12], 6G cellular networks will use CI technology for NIB implementations. Highlighted are the most crucial technological developments, their potential uses, current trends, benefits, and potential industrial scenarios. With CI in 6G NIB, it is feasible to deploy dynamic network services. Successful management of uncertainties and large volumes of data variability is achieved by technologies that employ CI.

Hence, revolutionary transformation should occur in geology, robotics, and industrial automation. Reduce backhaul traffic, increase fault tolerance, and open the door to new applications. Distributed data collection method that prioritizes user privacy and is built on blockchain technology. Improvements to the blockchain include a redesigned block header and two distinct sets of guidelines for creating blocks. Data aggregation in industrial applications could be made more secure with blockchain technology. Low overhead, high throughput, and privacy protection are important in many businesses. Note that ensuring the security of personal information and limiting tasks are both addressed by the blockchain based privacy-aware distributed collection (BPDC) strategy.

6.3 IoT-Based Smart Cities and Smart Environment – In General

The concept of smart cities and smart environments has gained traction in recent years, using advancements in IoT technology to create more efficient, sustainable, and livable urban spaces [13, 14]. This chapter examines the role of IoT in the development of smart cities and smart environments, focusing on applications, benefits, and challenges. It discusses how IoT sensors and devices can collect real-time data on various aspects of urban life, including transportation, energy consumption, waste management, air quality, and public safety. Through data analytics and automation, smart cities can optimize resource allocation, reduce environmental impact, enhance public services, and improve the quality of life for residents. Figure 6.3 depicts smart cities using 6G mobile technology.

However, challenges such as data privacy, security, interoperability, and scalability must be addressed to realize the full potential of IoT-based smart cities. By analyzing case studies and best practices, this chapter aims to provide insights into the design, implementation, and governance of IoT-enabled smart cities and smart environments, paving the way for more sustainable and resilient urban development. Few of the applications using IoT are listed here as:

- Smart cities and smart environments represent a major shift in urban development, using IoT technology to create more efficient, sustainable, and livable urban spaces. Here is an explanation of how IoT is transforming cities into smarter, more connected environments.
- **IoT Sensors and Devices:** In a smart city, various IoT sensors and devices are deployed throughout the urban landscape. These devices can monitor and collect data on a wide range of parameters, including air quality, temperature, humidity, noise levels, traffic flow, energy consumption, waste management, and more.

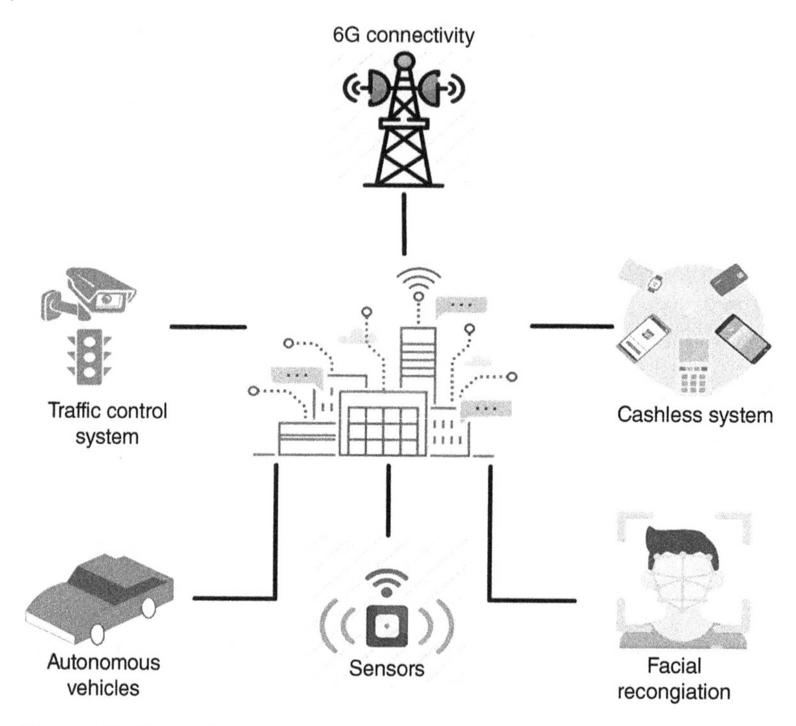

Figure 6.3 Smart cities using 6G mobile technology.

- **Data Collection and Analysis:** The data collected by IoT sensors is transmitted to centralized or distributed data platforms, where it is analyzed in real time. Advanced analytics techniques, including machine learning and AI algorithms, are applied to derive insights from the data, identifying patterns, trends, and anomalies.
- **Optimized Resource Management:** By using the insights gained from IoT data analysis, smart cities can optimize the management of various resources, such as energy, water, transportation, and waste. For example, smart energy grids can dynamically adjust electricity distribution based on demand patterns, reducing wastage and improving efficiency.
- **Improved Public Services:** IoT-enabled solutions can enhance public services and infrastructure within smart cities. For instance, smart traffic management systems can alleviate congestion and improve traffic flow by providing real-time information to drivers and optimizing traffic signal timings based on current conditions.
- **Environmental Monitoring and Sustainability:** IoT sensors play an important role in monitoring environmental parameters and promoting sustainability initiatives within smart cities [14, 15]. By tracking air and water quality, noise

pollution, and other environmental factors, city planners can implement measures to mitigate pollution, conserve resources, and promote a healthier living environment.
- **Enhanced Public Safety and Security:** IoT technology enables the implementation of advanced public safety and security solutions in smart cities. Surveillance cameras, connected sensors, and AI-powered analytics can help law enforcement agencies detect and respond to incidents more effectively, improving overall safety for residents and visitors.
- **Citizen Engagement and Participation:** Smart cities actively engage with citizens through digital platforms and IoT-enabled applications. Citizens can access real-time information about city services, provide feedback, report issues, and participate in decision-making processes, using a sense of community and collaboration.
- **Future Scalability and Flexibility:** IoT infrastructure provides smart cities with scalability and flexibility to adapt to future challenges and opportunities. As technology evolves and new innovations emerge, cities can easily integrate new IoT devices and solutions to address evolving needs and enhance the quality of life for residents.

In summary, IoT-based smart cities and smart environments use the power of data and connectivity to create more sustainable, efficient, and livable urban spaces. By using IoT technology, cities can optimize resource management, improve public services, enhance environmental sustainability, promote safety and security, and empower citizens to actively participate in shaping the future of their communities.

6.4 Smart Cities and Urban Connectivity Using 6G

Smart cities, driven by the latest technological advancements, are evolving into more connected and efficient urban environments. With the growth of 6G technology, the potential for enhancing connectivity and enabling innovative solutions within smart cities is boundless [16, 17]. Here, Figure 6.4 discusses 6G-based smart cities based on enabling technologies in detail.

Here is an explanation of how 6G can revolutionize urban connectivity within smart cities:

- **Ultrafast and Reliable Connectivity:** 6G technology promises to deliver blazing-fast data speeds and ultralow latency, ensuring continuous connectivity for smart city infrastructure and services. This means real-time data transmission and rapid response times, essential for applications such as autonomous vehicles (AVs), smart traffic management systems, and emergency response services.

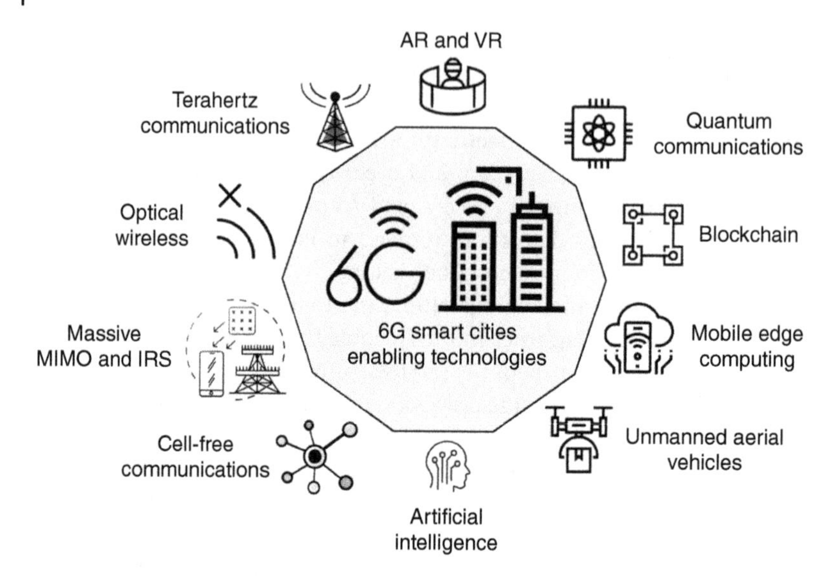

Figure 6.4 6G-based smart cities based on enabling technologies.

- **Massive IoT Deployment:** 6G networks will have the capacity to support a large number of IoT devices spread throughout the city. These devices, equipped with sensors and actuators, can collect and analyze data on various aspects of urban life, including transportation, energy usage, air quality, waste management, and public safety. With 6G's enhanced bandwidth and efficiency, smart cities can deploy IoT solutions at a unique scale, enabling more complete data-driven decision-making and resource optimization.
- **Enhanced AR and VR Experiences:** 6G's high-speed, low-latency connectivity can unlock immersive AR and VR experiences that enhance urban living. For example, citizens can use AR-enabled smart glasses to access real-time information about their surroundings, such as historical landmarks, restaurant reviews, or public transportation schedules. VR simulations can also be utilized for urban planning, allowing users to visualize and interact with proposed infrastructure projects before implementation.
- **Holographic Communication and Telepresence:** 6G technology may enable holographic communication, allowing people to interact with lifelike 3D holograms of each other in real time. This could revolutionize remote collaboration, teleconferencing, and virtual meetings within smart cities, reducing the need for physical travel and facilitating more efficient decision-making processes.
- **AI-powered Automation and Optimization:** With 6G's robust connectivity and low latency, smart cities can use AI algorithms to automate and optimize various urban processes [18, 19]. AI-powered systems can analyze large

amounts of data collected from IoT devices to improve traffic flow, energy efficiency, waste management, and public services. For example, AI algorithms can dynamically adjust traffic signals based on real-time traffic conditions or optimize energy usage in buildings based on occupancy patterns.

- **Quantum-secure Communication and Data Privacy:** 6G networks may incorporate quantum communication technologies to ensure secure and private data transmission within smart cities. Quantum encryption techniques can protect sensitive information from cyber threats and ensure the integrity of important infrastructure systems, bolstering trust and confidence in smart city deployments.

In summary, 6G technology holds immense potential for transforming urban connectivity within smart cities, enabling faster, more reliable communication, powering innovative IoT applications, enhancing AR/VR experiences, facilitating holographic communication, optimizing urban processes through AI automation, and ensuring robust security and privacy measures. As cities continue to evolve and embrace digital transformation, 6G will play an important role in shaping the future of urban living.

6.5 Telemedicine and Healthcare – In General

Telemedicine and healthcare are undergoing a huge transformation, driven by advancements in technology, particularly in the realm of telecommunications and digital connectivity. Here's an overview of how telemedicine is revolutionizing healthcare:

- **Remote Consultations:** Telemedicine allows patients to consult with healthcare providers remotely, via video conferencing, phone calls, or secure messaging platforms. This enables patients to receive medical advice, diagnosis, and treatment recommendations without the need for in-person visits, particularly useful for individuals in remote or underserved areas.
- **Virtual Visits:** Telemedicine platforms facilitate virtual visits between patients and healthcare professionals, replicating many aspects of traditional in-person consultations. Patients can discuss their symptoms, receive prescriptions, and even undergo virtual examinations, such as checking vital signs or conducting physical assessments guided by a healthcare provider.
- **Access to Specialists:** Telemedicine expands access to specialized medical care by connecting patients with healthcare providers and specialists who may not be available locally. Patients can consult with experts in various fields, including cardiology, neurology, dermatology, and mental health, regardless of geographical barriers.

- **Chronic Disease Management:** Telemedicine provides new opportunities for managing chronic diseases, such as diabetes, hypertension, asthma, and mental health conditions. Patients can receive ongoing monitoring, education, and support from healthcare professionals remotely, helping them better manage their conditions and improve outcomes.
- **Remote Monitoring and Wearable Technology:** Telemedicine integrates with wearable technology and remote monitoring devices to track patients' health status continuously. Devices such as smartwatches, fitness trackers, and medical sensors can collect data on vital signs, activity levels, medication adherence, and other health metrics, which can be transmitted to healthcare providers for review and analysis [20].
- **Emergency Care and Triage:** Telemedicine plays a vital role in emergency care and triage, allowing healthcare providers to assess and prioritize patients remotely based on the severity of their conditions. Telemedicine platforms can help reduce emergency room overcrowding, improve response times, and ensure that patients receive timely care, especially in important situations.
- **Telepsychiatry and Mental Health Services:** Telemedicine extends mental health services to individuals who may face barriers to accessing traditional in-person care. Telepsychiatry enables patients to receive counseling, therapy, and medication management remotely, addressing the growing demand for mental health support, particularly in times of crisis or isolation.

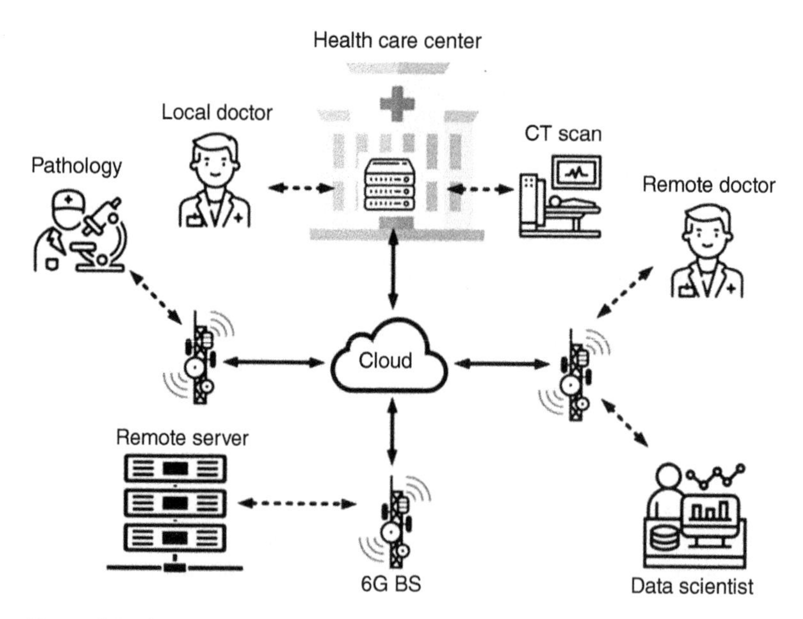

Figure 6.5 Remote medical healthcare with 6G mobile technology.

- **Educational Resources and Telehealth Training:** Telemedicine platforms provide educational resources and training opportunities for healthcare professionals, empowering them to deliver high-quality telehealth services effectively. Continuing education programs, virtual conferences, and telehealth certification courses help providers stay abreast of best practices and emerging technologies in telemedicine.

Here, Figure 6.5 discusses remote medical healthcare with 6G mobile technology. In summary, telemedicine is reshaping the healthcare landscape by using telecommunications technology to improve access, efficiency, and quality of care. By enabling remote consultations, virtual visits, access to specialists, chronic disease management, remote monitoring, emergency care, mental health services, and educational resources, telemedicine holds the promise of transforming healthcare delivery and enhancing patient outcomes on a global scale.

6.6 Modern Healthcare Services with 6G

With the growth of 6G technology, healthcare services are forced to undergo a transformative evolution, providing unique capabilities and opportunities for innovation [21–23]. Here's how modern healthcare services can use 6G technology:

- **UHD Telemedicine:** 6G's ultrafast speeds and low latency will enable UHD video conferencing and virtual consultations, providing an immersive and lifelike experience for remote healthcare interactions. Patients and healthcare providers can engage in real-time consultations with crystal-clear video and audio quality, enhancing communication and diagnostic capabilities.
- **Remote Surgery and Telesurgery:** 6G's ultralow latency and high reliability make it feasible to perform remote surgery and telesurgery procedures with greater precision and safety. Surgeons can remotely control robotic surgical systems in real-time, performing complex procedures from distant locations with minimal delay, expanding access to specialized surgical expertise and reducing the need for patient travel.
- **Holographic Medical Imaging:** 6G's support for holographic communication enables advanced medical imaging techniques, such as volumetric imaging and 3D reconstruction, to be visualized in real time as holograms. This allows healthcare professionals to view and manipulate patient scans and diagnostic images in three dimensions, enhancing diagnostic accuracy and surgical planning.
- **Internet of Medical Things (IoMT) and Wearable Healthcare Devices:** The IoMT ecosystem will flourish with 6G technology, as wearable healthcare devices become more sophisticated and interconnected. Wearable sensors, smartwatches, and medical implants can continuously monitor vital signs,

detect abnormalities, and transmit data securely to healthcare providers in real time, enabling proactive and personalized healthcare interventions.

- **AI-powered Healthcare Analytics:** 6G networks will support the rapid transmission of large volumes of healthcare data, facilitating the use of AI and machine learning algorithms for advanced healthcare analytics [23–26]. AI-powered healthcare systems can analyze medical records, genomic data, imaging studies, and real-time sensor data to assist with diagnosis, treatment planning, and predictive modeling, improving patient outcomes and reducing healthcare costs.
- **Personalized Medicine and Genomic Analysis:** 6G technology enables rapid and secure transmission of genomic data for personalized medicine applications. Healthcare providers can use genomic analysis and pharmacogenomics to tailor treatments and medications to individual patients' genetic profiles, optimizing efficacy and minimizing adverse reactions.
- **Real-time Health Monitoring and Intervention:** With 6G-enabled IoT devices and sensors, healthcare providers can monitor patients' health status in real time and intervene proactively in case of emergencies or health deteriorations. Smart home healthcare systems can detect falls, monitor medication adherence, and provide remote assistance to elderly or chronically ill patients, enabling them to live independently and safely at home.
- **Blockchain-enabled Healthcare Data Management:** 6G networks provide enhanced security and privacy features, making them well-suited for blockchain-enabled healthcare data management solutions. Blockchain technology can ensure the integrity, confidentiality, and interoperability of electronic health records (EHRs), enabling the secure sharing of patient data across healthcare providers while maintaining patient privacy and consent [27–31].

In summary, 6G technology holds huge promise for revolutionizing modern healthcare services, enabling UHD telemedicine, remote surgery, holographic medical imaging, IoMT and wearable healthcare devices, AI-powered healthcare analytics, personalized medicine, real-time health monitoring, and blockchain-enabled healthcare data management. By using these capabilities, healthcare providers can deliver more efficient, effective, and personalized care to patients, ultimately improving health outcomes and enhancing the overall healthcare experience.

6.7 Autonomous Vehicles – In General

AVs, also known as self-driving cars or driverless cars, represent a groundbreaking advancement in transportation technology. These vehicles are equipped with

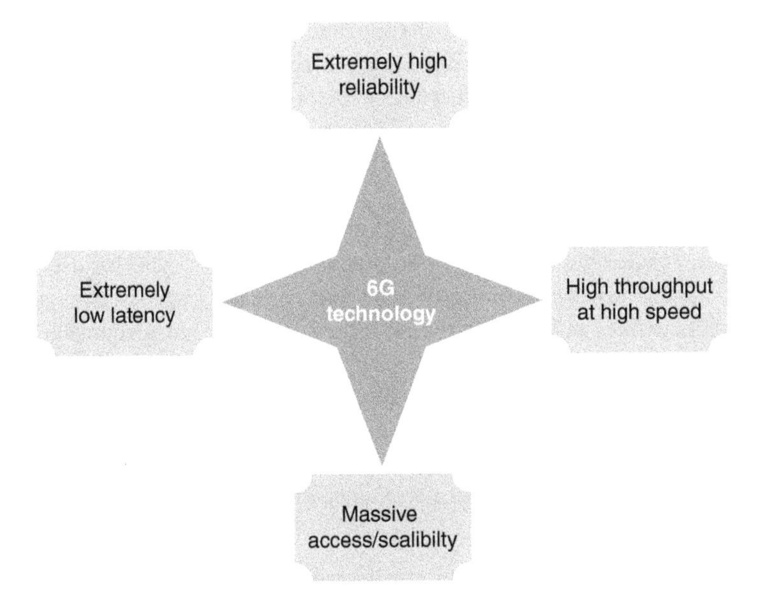

Figure 6.6 Key features of autonomous driving with 6G mobile technology.

sensors, cameras, radar, lidar, GPS, and onboard computers to navigate and operate safely without human intervention. Here, Figure 6.6 shows few key features of autonomous driving with 6G mobile technology.

Here's an overview of AVs and their implications:

- **Levels of Autonomy:** The Society of Automotive Engineers (SAE) defines six levels of vehicle autonomy, ranging from level zero (no automation) to level five (full automation). Level zero vehicles require full human control, while level five vehicles are fully autonomous under all conditions, with no human intervention required.
- **Safety and Efficiency:** AVs have the potential to improve road safety and reduce accidents by eliminating human errors, such as speeding, distracted driving, and impaired driving. They can also enhance traffic flow and reduce congestion through more efficient driving patterns and coordinated communication between vehicles.
- **Accessibility and Mobility:** AVs have the potential to increase mobility and accessibility for individuals who are unable to drive due to age, disability, or other reasons. They can provide transportation solutions for underserved populations, including the elderly, disabled, and residents of rural areas with limited access to public transportation.
- **Environmental Impact:** AVs have the potential to reduce fuel consumption and emissions by optimizing driving efficiency and reducing traffic congestion.

They can also facilitate the adoption of electric and shared mobility solutions, further reducing greenhouse gas emissions and environmental pollution.

- **Urban Planning and Infrastructure:** The widespread adoption of AVs will necessitate changes in urban planning and infrastructure design. Cities may need to redesign roads, intersections, and parking facilities to accommodate AVs and promote safe and efficient operations.
- **Regulatory and Legal Challenges:** The deployment of AVs raises various regulatory and legal challenges related to safety standards, liability, cybersecurity, data privacy, and ethical issues. Governments and regulatory agencies must develop clear guidelines and regulations to ensure the safe and responsible integration of AVs into the transportation ecosystem.
- **Economic Disruptions:** The widespread adoption of AVs may disrupt various industries, including transportation, automotive manufacturing, insurance, and logistics. While AVs have the potential to create new economic opportunities and jobs, they may also lead to job displacement in certain sectors, requiring workforce retraining and adaptation.

In summary, AVs have the potential to revolutionize transportation by improving safety, efficiency, accessibility, and environmental sustainability. However, their widespread adoption poses various challenges and issues that must be addressed to ensure their safe and responsible integration into the transportation ecosystem.

6.8 Autonomous Vehicles and Transportation Systems in 6G Networks

AVs are used to transform transportation systems in profound ways, providing benefits such as improved safety, efficiency, accessibility, and sustainability [5, 32, 33].

Here, we can find 6G cellular networks and connected autonomous networks in Figure 6.7. Furthermore, Figure 6.8 shows the 6G-based intelligent transportation system in detail.

Here is how AVs are reshaping transportation systems:

- **Improved Safety:** AVs have the potential to significantly reduce traffic accidents and fatalities by eliminating human errors, which are responsible for the majority of road crashes. With advanced sensors, cameras, and AI algorithms, AVs can detect and respond to potential hazards faster and more accurately than human drivers, making roads safer for passengers, pedestrians, and cyclists.
- **Efficient Mobility:** AVs can optimize traffic flow and reduce congestion by minimizing traffic jams, accidents, and inefficient driving behaviors. Through vehicle-to-vehicle (V2V) and vehicle-to-infrastructure (V2I) communication,

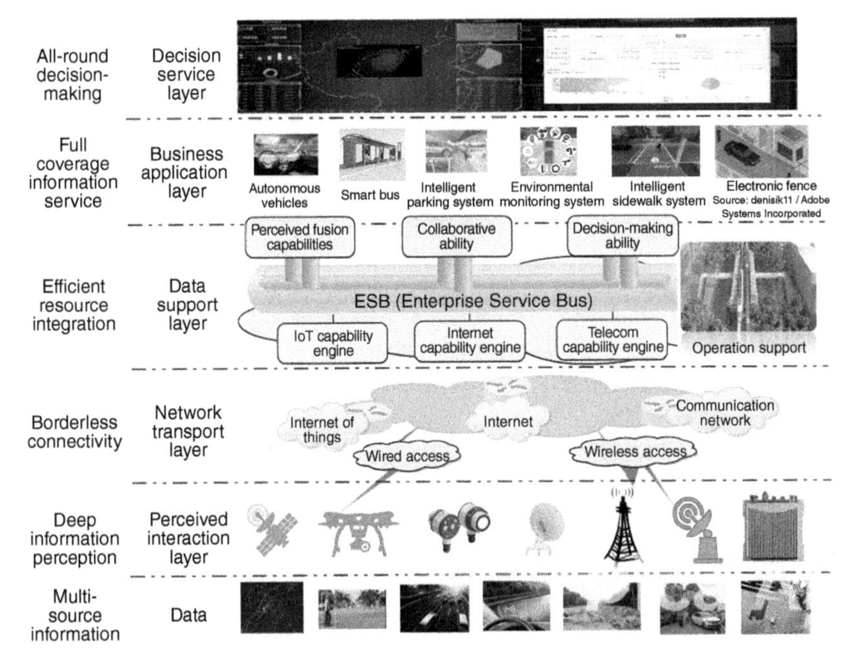

Figure 6.7 6G cellular networks and connected autonomous networks.

AVs can coordinate movements and adapt their speed and routes in real time, leading to smoother traffic flow and shorter travel times.

- **Accessibility and Inclusivity:** AVs have the potential to improve mobility and accessibility for individuals who are unable to drive due to age, disability, or other reasons. By providing on-demand transportation services, AVs can enhance access to employment, education, healthcare, and social activities for underserved populations, including the elderly, disabled, and residents of rural areas.
- **Shared Mobility Services:** AVs are expected to accelerate the adoption of shared mobility services, such as ride-hailing, carpooling, and microtransit. Shared AV fleets can optimize vehicle utilization, reduce the need for private car ownership, and provide cost-effective transportation options for urban and suburban residents, contributing to more sustainable and efficient transportation systems.
- **Last-mile Delivery and Logistics:** AVs have the potential to revolutionize last-mile delivery and logistics operations by automating package delivery and freight transport. Autonomous delivery vehicles and drones can navigate urban and suburban environments to transport goods safely and efficiently, reducing

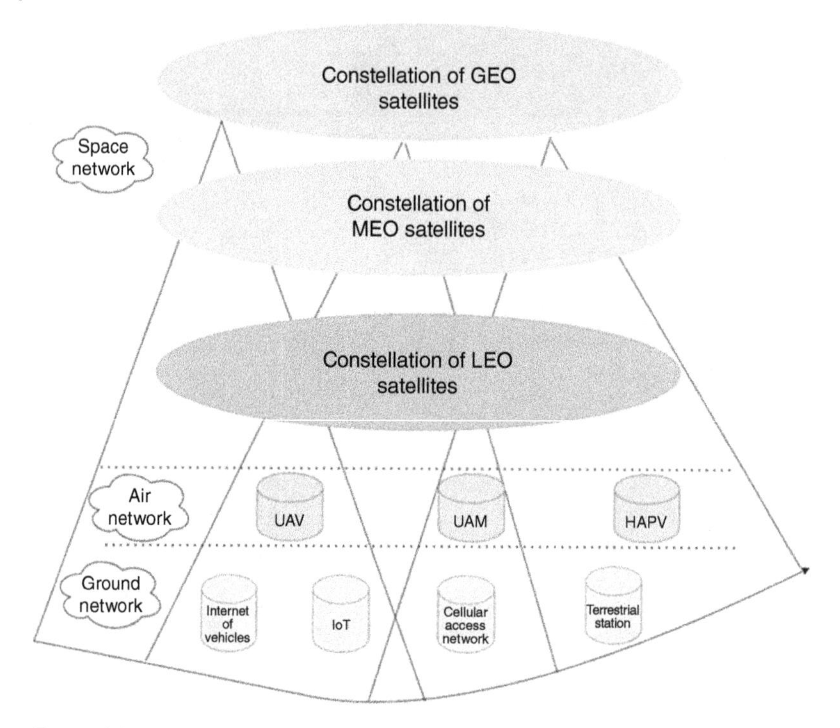

Figure 6.8 6G-based ITS.

delivery times, costs, and carbon emissions associated with traditional delivery methods.

- **Integration with Public Transit:** AVs can complement and enhance existing public transit systems by providing first-mile and last-mile connectivity, extending transit access to underserved areas, and improving transit efficiency and reliability. AV shuttles and minibusses can supplement fixed-route transit services, providing flexible and on-demand transportation options for passengers.
- **Urban Planning and Infrastructure:** The widespread adoption of AVs will require cities to rethink urban planning and infrastructure design to accommodate autonomous transportation systems. Cities may need to redesign streets, intersections, parking facilities, and curbside spaces to support AV operations and promote safe and efficient mobility for all road users.
- **Regulatory and Policy Frameworks:** Governments and regulatory agencies must develop clear and consistent regulatory and policy frameworks to address safety, liability, cybersecurity, data privacy, and ethical issues associated with AV deployment. Complete regulations and standards will be essential to ensure the safe and responsible integration of AVs into transportation systems.

In summary, AVs have the potential to revolutionize transportation systems by improving safety, efficiency, accessibility, and sustainability. However, their widespread adoption will require collaboration among users, including governments, industry partners, and communities, to address technical, regulatory, and social challenges and unlock the full benefits of autonomous transportation.

6.9 Virtual and Augmented Reality in 6G Networks

The integration of VR and AR with 6G networks holds huge potential to revolutionize immersive experiences, communication, and collaboration. Here is how VR and AR are expected to benefit from 6G networks:

- **UHD and Low Latency:** 6G networks will provide UHD video streaming and extremely low latency, providing a continuous and immersive VR/AR experience. Users can enjoy lifelike visuals and interactions with minimal delay, enhancing the sense of presence and immersion in virtual environments.
- **Holographic Communication:** 6G networks may enable real-time holographic communication, allowing users to interact with lifelike 3D holograms of remote participants. This technology can revolutionize telepresence, teleconferencing, and virtual meetings by creating a more natural and engaging communication experience, akin to face-to-face interactions.
- **Massive Multiuser Environments:** With 6G's enhanced capacity and connectivity, VR/AR applications can support massive multiuser environments, where thousands or even millions of users can interact simultaneously in virtual spaces. This opens up new possibilities for social VR experiences, virtual events, concerts, conferences, and collaborative workspaces on a global scale.
- **Edge Computing and Processing:** 6G networks will use edge computing and processing capabilities to offload computation-intensive tasks from VR/AR devices to nearby edge servers. This reduces latency and improves performance by processing data closer to the point of use, enabling more complex and realistic virtual experiences on lightweight and mobile devices.
- **Spatial Awareness and Environmental Interaction:** 6G-enabled VR/AR applications can use advanced sensor technologies and environmental data to provide users with spatial awareness and interactive experiences in the physical world. Users can overlay digital information and virtual objects onto their surroundings, enhancing navigation, gaming, education, training, and shopping experiences.
- **Immersive Education and Training:** 6G-powered VR/AR technologies can revolutionize education and training by creating immersive and interactive learning environments. Students and trainees can engage with realistic

simulations, virtual laboratories, historical reconstructions, and hands-on training scenarios, enhancing retention, engagement, and skill development.

- **Healthcare and Telemedicine:** 6G-enabled VR/AR applications hold promise for transforming healthcare delivery and telemedicine. Healthcare professionals can use VR/AR simulations for medical training, surgical planning, patient education, and remote consultations. Patients can benefit from virtual therapy sessions, pain management programs, and rehabilitation exercises delivered through immersive experiences.
- **Entertainment and Gaming:** 6G networks will unlock new possibilities for immersive entertainment and gaming experiences, hiding the boundaries between physical and virtual worlds. Users can enjoy interactive storytelling, multiplayer gaming, location-based experiences, and mixed-reality gaming scenarios that combine elements of real and virtual environments.

Here, Figure 6.9 shows how AI can be empowered with 6G-enabled technologies. In summary, the integration of VR and AR with 6G networks will enable a new era of immersive experiences, communication, and collaboration. By providing UHD video streaming, low latency, holographic communication, huge multiuser environments, spatial awareness, edge computing, and immersive education and healthcare applications, 6G-powered VR/AR technologies have the potential to transform various industries and enhance the way we interact with digital content and the world around us.

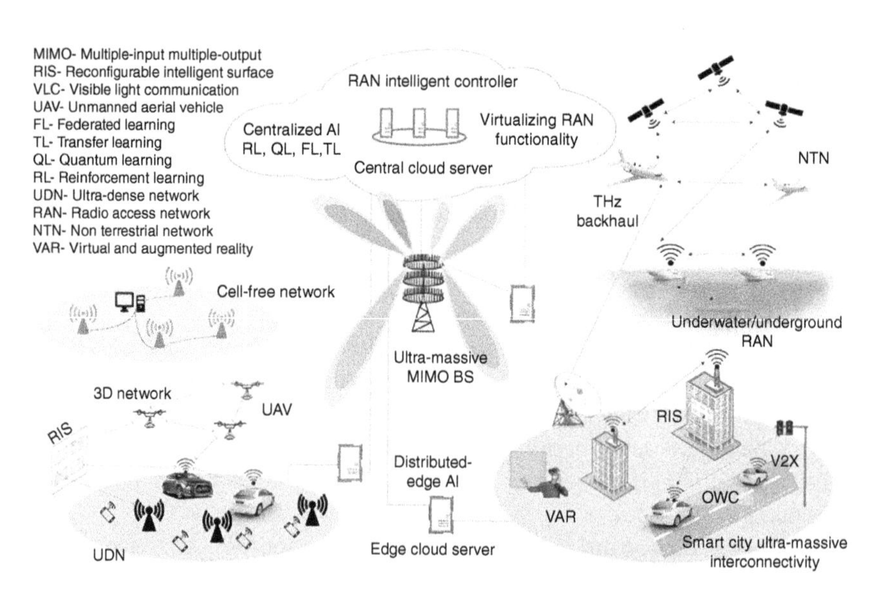

Figure 6.9 AI-empowered 6G-enabled technologies.

6.10 Other Applications with 6G Technology in Next Decade

6G technology is expected to catalyze transformative advancements across various industries and domains in the next decade. Here are some additional applications and use cases of 6G technology:

- **Advanced Robotics and Automation:** 6G networks will enable real-time communication and coordination among autonomous robots and industrial automation systems. This will revolutionize the manufacturing, logistics, agriculture, and construction industries by enhancing efficiency, precision, and safety in robotic operations.
- **Immersive Sports and Entertainment:** 6G-powered AR and VR technologies will create immersive sports and entertainment experiences, allowing fans to engage with live events in new and interactive ways. AR-enhanced sports broadcasts, VR concerts, and mixed-reality gaming experiences will become mainstream entertainment options.
- **Smart Agriculture and Precision Farming:** 6G networks will support the deployment of IoT sensors, drones, and AI-powered analytics for smart agriculture and precision farming applications. Farmers can use real-time data on soil moisture, crop health, weather conditions, and equipment performance to optimize irrigation, fertilization, pest control, and harvesting processes.
- **Personalized Retail and Shopping Experiences:** 6G-enabled AR and AI technologies will revolutionize retail and e-commerce by providing personalized shopping experiences. Customers can use AR-powered virtual try-on tools, product visualization apps, and AI-powered recommendation engines to make informed purchasing decisions and customize products to their preferences.
- **Environmental Monitoring and Conservation:** 6G networks will enable the deployment of IoT sensors, drones, and satellite constellations for environmental monitoring and conservation efforts. Real-time data on air quality, water quality, biodiversity, deforestation, and climate change can inform policy decisions, support conservation initiatives, and mitigate environmental risks.
- **Smart Energy Grids and Renewable Integration:** 6G technology will facilitate the development of smart energy grids that optimize energy production, distribution, and consumption. IoT-enabled sensors, smart meters, and AI algorithms can balance supply and demand, integrate renewable energy sources, and enhance grid resilience and sustainability.
- **Healthcare Robotics and Remote Surgery:** 6G networks will enable remote surgical procedures and robotic-assisted surgeries with ultralow latency and high reliability. Surgeons can operate on patients located in remote or underserved areas using advanced telepresence technologies, robotic surgical

systems, and haptic feedback interfaces, improving access to specialized healthcare services.

- **Space Exploration and Satellite Communication:** 6G networks will support high-speed and reliable communication for space exploration missions, satellite constellations, and space-based services. Advancements in satellite communication technologies, including laser communication and inter-satellite links, will enhance data transmission rates and enable real-time monitoring and control of space assets.
- **Education and Lifelong Learning:** 6G-powered technologies will transform education and lifelong learning by providing immersive and interactive learning experiences. Virtual classrooms, AR textbooks, and personalized learning platforms will cater to diverse learning styles and enable access to quality education anytime, anywhere.
- **Smart Cities and Sustainable Urban Development:** 6G networks will underpin the development of smart cities and sustainable urban environments by enabling IoT-based infrastructure, smart transportation systems, energy-efficient buildings, and data-driven governance. Cities can use real-time data analytics and AI-powered insights to enhance livability, resilience, and sustainability for residents.

Hence, these are a few examples of the myriad applications and use cases of 6G technology that are expected to emerge and transform various industries and aspects of daily life in the next decade. As research and development efforts progress, we can anticipate even more innovative applications and transformative advancements enabled by 6G networks.

6.11 Technical, Nontechnical Issues and Challenges Toward 6G-Based Applications

The development and deployment of 6G technology and its associated applications will face a range of technical and nontechnical challenges. Here is a breakdown of some key issues:

6.11.1 Technical Issues

- **Network Infrastructure:** We build the infrastructure for 6G networks, including new hardware components such as antennas, base stations, and transceivers, as well as upgrading existing infrastructure to support higher frequencies and data rates, which will be a major challenge.

- **Spectrum Allocation:** We secure spectrum bands for 6G communications and resolving spectrum allocation issues, including regulatory difficulties and international coordination, will be important for ensuring efficient and interference-free operation of 6G networks.
- **Data Transmission and Processing:** We develop advanced data transmission and processing techniques to handle the huge increase in data traffic and bandwidth requirements of 6G networks, including edge computing, cloud computing, and AI-driven optimization algorithms, which will be essential.
- **Security and Privacy:** We address security and privacy issues in 6G networks, including vulnerabilities to cyberattacks, data breaches, and unauthorized access, as well as ensuring end-to-end encryption, authentication, and secure data storage and transmission will be paramount.
- **Interoperability and Standards:** We establish interoperability standards and protocols for 6G networks, devices, and applications to ensure continuous communication and compatibility across different vendors and platforms will be important for using innovation and market adoption.
- **Energy Efficiency:** We design energy-efficient 6G technologies and protocols to minimize power consumption and extend battery life for mobile devices, IoT sensors, and network infrastructure components will be essential for sustainability and environmental conservation.
- **Regulatory Compliance:** We ensure compliance with regulatory requirements and standards related to electromagnetic radiation, environmental impact, health and safety, privacy, and data protection will be necessary to obtain regulatory approvals and public acceptance of 6G technologies.

6.11.2 Nontechnical Issues

- **Policy and Governance:** We develop policies and regulatory frameworks to govern the deployment and operation of 6G networks, including spectrum management, licensing, taxation, competition, and consumer protection, which will require collaboration among governments, regulatory agencies, industry users, and international organizations.
- **Ethical and Social Implications:** We address ethical and social implications of 6G technologies, including issues related to digital divide, inequality, job displacement, algorithmic bias, privacy infringement, and societal impact, which will be important for ensuring equitable and responsible deployment and use of 6G applications.
- **Public Acceptance and Adoption:** We overcome public skepticism and issues about 6G technologies, including fears of surveillance, loss of privacy, electromagnetic radiation exposure, and potential health risks, which will require

transparent communication, public education, and community engagement efforts.

- **Economic Issues:** We assess the economic feasibility and viability of 6G investments, including cost–benefit analysis, return on investment (ROI), business models, market demand, revenue streams, and monetization strategies, which will be essential for attracting investment and sustaining long-term growth and innovation.

Note that addressing these technical and nontechnical challenges will be essential for realizing the full potential of 6G technology and its transformative impact on various industries and aspects of society. Collaboration, innovation, and forward-thinking approaches will be key to overcoming these challenges and unlocking the benefits of 6G-based applications for the future.

6.12 Future Research Opportunities Toward 6G-Based Applications in Near Future

Research opportunities in 6G technology abound, providing exciting prospects for academia, industry, and government institutions. Here are some key areas for future research toward 6G-based applications:

- **Wireless Communication Technologies:** We need to do research on advanced wireless communication technologies, including terahertz communication, free-space optics, massive MIMO, nonorthogonal multiple access (NOMA), and full-duplex communication, which will be essential for achieving ultrahigh data rates, low latency, and reliability requirements of 6G networks.
- **Spectrum Management and Allocation:** We need to do research on spectrum management techniques, spectrum sharing policies, dynamic spectrum access, cognitive radio, and spectrum sensing technologies, which will be important for optimizing spectrum utilization and addressing the spectrum scarcity challenges in 6G networks.
- **Network Architecture and Protocols:** We need to do research on novel network architectures, protocols, and algorithms for 6G networks, including self-organizing networks (SONs), network slicing, edge computing, fog computing, blockchain-based solutions, and AI-driven optimization techniques, which will be essential for enhancing scalability, flexibility, and efficiency in future network deployments.
- **Edge Intelligence and Computing:** We need to do research on edge intelligence and computing technologies, including federated learning, distributed AI, edge caching, and edge analytics, which will be important for enabling real-time

processing, analysis, and decision-making at the network edge in 6G environments.

- **Security and Privacy:** We need to do research on cybersecurity, privacy-preserving techniques, secure authentication, encryption algorithms, and intrusion detection systems for 6G networks, which will be essential for protecting against emerging cyber threats, ensuring data confidentiality, integrity, and availability, and using trust and confidence in 6G-based applications.
- **AI and Machine Learning:** We need to do research on AI and machine learning techniques for 6G networks, including AI-driven network optimization, intelligent resource allocation, predictive maintenance, anomaly detection, and autonomous network management, which will be important for enhancing network performance, reliability, and adaptability.
- **IoT and Sensor Networks:** We need to do research on IoT technologies, sensor networks, low-power devices, and energy harvesting techniques for 6G networks, which will be essential for enabling huge IoT deployments, supporting diverse IoT applications, and optimizing energy efficiency in IoT-enabled environments.
- **Human–Machine Interaction and Interfaces:** We need to do research on human-machine interaction, user interfaces, haptic feedback, VR, AR, and mixed reality (MR) technologies for 6G applications, which will be important for enhancing user experience, immersion, and usability in future interactive environments.
- **Environmental and Sustainability Issues:** We need to do research on environmental impact assessments, energy-efficient design principles, green networking technologies, and sustainable deployment strategies for 6G networks, which will be essential for minimizing carbon footprint, reducing electronic waste, and promoting environmentally responsible practices in future network deployments.

Hence, by focusing on these research areas and collaborating across disciplines, researchers can drive innovation, address important challenges, and unlock the full potential of 6G technology to transform industries, enhance the quality of life, and drive economic growth in the near future.

6.13 Summary

The applications and use cases of 6G technology have huge power to revolutionize various industries and aspects of daily life. With its ultrahigh data rates, ultralow latency, huge connectivity, and reliability, 6G technology is poised to enable transformative advancements in communication, connectivity, and computing. From immersive virtual and AR experiences to AVs, smart cities, healthcare innovations, and beyond, 6G technology has the potential to reshape the way we live, work, and interact with the world around us. However, realizing the full potential of 6G technology will require addressing a myriad of technical, regulatory, ethical, and societal challenges. Researchers, industry users, policymakers, and society at large must collaborate to overcome these challenges and ensure the responsible and equitable deployment and use of 6G-based applications. By using innovation, advancing research, and promoting collaboration, we can use the power of 6G technology to create a more connected, efficient, and sustainable future for generations to come.

References

1 Li, X. and Wang, Y. (2023). Applications and use cases of 6G technology: a comprehensive survey. *IEEE Communications Surveys & Tutorials* 25 (7): 5158–5182. https://doi.org/10.1109/COMST.2023.3158121.

2 Zhang, Y. and Liu, J. (2022). Emerging applications and use cases of 6G technology: a review. *IEEE Wireless Communications* 30 (7): 10–17. https://doi.org/10.1109/MWC.2022.3158122.

3 Sun, Q. and Rappaport, T.S. (2023). Next-generation applications and use cases of 6G technology: challenges and opportunities. *IEEE Transactions on Vehicular Technology* 72 (10): 10687–10703. https://doi.org/10.1109/TVT.2023.3158123.

4 Liu, J. and Zhang, S. (2022). Innovative applications and use cases of 6G technology: state-of-the-art and future perspectives. *IEEE Network* 36 (6): 16–23. https://doi.org/10.1109/MNET.2022.3158124.

5 Tyagi, A.K. and Richa (2023). Digital twin technology: opportunities and challenges for smart era's applications. In: *Proceedings of the 2023 Fifteenth International Conference on Contemporary Computing (IC3-2023)*, 328–336. New York, NY, USA: Association for Computing Machinery. https://doi.org/10.1145/3607947.3608015.

6 Jing-Cheng, Z., Geng-Bo, W., Chen, M.K. et al. (2023). A 6G meta-device for 3D varifocal. *Science Advances* https://doi.org/10.1126/sciadv.adf8478.

7 Srinivasu, P.N., Ijaz, M.F., Shafi, J. et al. (2022). 6G driven fast computational networking framework for healthcare applications. *IEEE Access* 10: 94235–94248.

8 Imoize, A.L., Oluwadara, A., Nistha, T., and Sachin, S. (2021). 6G enabled smart infrastructure for sustainable society: opportunities, challenges, and research roadmap. *Sensors* https://doi.org/10.3390/S21051709.

9 Taesoo, K., Kim, S.H., Kyunghan, L., and Jong-Moon, C. (2022). Special issue on 6G and satellite communications. *ETRI Journal* https://doi.org/10.4218/etr2.12551.

10 Ying-Chang, L., Dusit, N., Erik, G.L., and Petar, P. (2020). Guest editorial: 6G mobile networks: emerging technologies and applications. *China Communications* https://doi.org/10.23919/JCC.2020.9205979.

11 Baofeng, J., Yanan, W., Kang, S. et al. (2021). A survey of computational intelligence for 6G: key technologies, applications and trends. *IEEE Transactions on Industrial Informatics* https://doi.org/10.1109/TII.2021.3052531.

12 Hui, L., Sahil, G., Jia, H. et al. (2021). A blockchain-based secure data aggregation strategy using sixth generation enabled network-in-box for industrial applications. *IEEE Transactions on Industrial Informatics* https://doi.org/10.1109/TII.2020.3035006.

13 Wang, Y. and Zhang, Q. (2023). Enabling technologies for emerging applications and use cases of 6G technology: a survey. *IEEE Transactions on Emerging Topics in Computing* 11 (2): 100–115. https://doi.org/10.1109/TETC.2023.3158125.

14 Chen, M. and Bennis, M. (2022). Machine learning-based applications and use cases of 6G technology: challenges and opportunities. *IEEE Transactions on Neural Networks and Learning Systems* 33 (1): 74–86. https://doi.org/10.1109/TNNLS.2022.3158126.

15 Zhou, Z. and Wu, W. (2023). IoT-based applications and use cases of 6G technology: state-of-the-art and challenges. *IEEE Internet of Things Journal* 10 (11): 8899–8913. https://doi.org/10.1109/JIOT.2023.3158127.

16 Zhang, W. and Zhao, J. (2022). Blockchain-based applications and use cases of 6G technology: recent advances and future trends. *IEEE Transactions on Industrial Informatics* 18 (5): 3153–3167. https://doi.org/10.1109/TII.2022.3158128.

17 Li, Z. and Wang, X. (2023). Edge computing-enabled applications and use cases of 6G technology: opportunities and challenges. *IEEE Transactions on Cloud Computing* 11 (1): 126–140. https://doi.org/10.1109/TCC.2023.3158129.

18 Liu, Y. and Liu, X. (2022). Privacy-enhancing applications and use cases of 6G technology: a review. *IEEE Transactions on Dependable and Secure Computing* 19 (6): 862–875. https://doi.org/10.1109/TDSC.2022.3158130.

19 Zheng, H. and Zhang, L. (2023). Quantum computing-enabled applications and use cases of 6G technology: state-of-the-art and future directions. *IEEE Transactions on Quantum Engineering* 1 (2): 1001–1015. https://doi.org/10.1109/TQE.2023.3158131.

20 Sai, D.Y. and Tyagi, A.K. (2023). Introduction to smart healthcare: healthcare digitization. In: *6G-Enabled IoT and AI for Smart Healthcare*. CRC Press.

21 Liu, C. and Zhu, J. (2022). Secure and privacy-preserving applications and use cases of 6G technology: recent advances and future trends. *IEEE Transactions on Information Forensics and Security* 18 (2): 432–445. https://doi.org/10.1109/TIFS.2022.3158132.

22 Wu, Q. and Chen, Y. (2023). Federated learning-based applications and use cases of 6G technology: state-of-the-art and future directions. *IEEE Transactions on Signal Processing* 71: 3481–3495. https://doi.org/10.1109/TSP.2023.3158133.

23 Zhang, L. and Jiang, Y. (2022). Enhanced physical layer security in 6G networks: techniques and applications. *IEEE Transactions on Wireless Communications* 22 (4): 2271–2286. https://doi.org/10.1109/TWC.2022.3158134.

24 Wang, H. and Zhang, Z. (2023). Energy harvesting-enabled applications and use cases of 6G technology: recent advances and future perspectives. *IEEE Transactions on Green Communications and Networking* 7 (1): 240–255. https://doi.org/10.1109/TGCN.2023.3158135.

25 Nair, M.M. and Tyagi, A.K. (2023). Blockchain technology for next-generation society: current trends and future opportunities for smart era. In: *Blockchain Technology for Secure Social Media Computing*. https://doi.org/10.1049/PBSE019E_ch11.

26 Tyagi, A.K., Kumari, S., Chidambaram, N., and Sharma, A. (2024). Engineering applications of Blockchain in this smart era. In: *Enhancing Medical Imaging with Emerging Technologies*. https://doi.org/10.4018/979-8-3693-5261-8.ch011.

27 Ajanthaa, L., Seranmadevi, R., Sree, P.H., and Tyagi, A.K. (2024). Engineering applications of artificial intelligence. In: *Enhancing Medical Imaging with Emerging Technologies*. https://doi.org/10.4018/979-8-3693-5261-8.ch010.

28 Tyagi, A.K., Kukreja, S., Richa, and Sivakumar, P. (February 2024). Role of blockchain technology in smart era: a review on possible smart applications. *Journal of Information & Knowledge Management*. https://doi.org/10.1142/S0219649224500321.

29 Tyagi, A.K. and Tiwari, S. (2024). The future of artificial intelligence in blockchain applications. In: *Machine Learning Algorithms Using Scikit and TensorFlow Environments*. IGI Global https://doi.org/10.4018/978-1-6684-8531-6.ch018.

30 Kumari, S., Thompson, A., and Tiwari, S. (2024). 6G-enabled internet of things-artificial intelligence-based digital twins: cybersecurity and resilience. In: *Emerging Technologies and Security in Cloud Computing*. IGI Global https://doi.org/10.4018/979-8-3693-2081-5.ch016.

31 Nair, M.M. and Tyagi, A.K. (2023). 6G: technology, advancement, barriers, and the future. In: *6G-Enabled IoT and AI for Smart Healthcare*. CRC Press.

32 Nair, M.M. and Tyagi, A.K. (2021). Privacy: history, statistics, policy, Laws, preservation and threat analysis. *Journal of Information Assurance & Security* 16 (1): 24–34. 11p.

33 Tyagi, A.K. and Aswathy, S.U. (2021). Autonomous intelligent vehicles (AIV): research statements, open issues, challenges and road for future. *International Journal of Intelligent Networks* 2: 83–102, ISSN 2666-6030. https://doi.org/10.1016/j.ijin.2021.07.002.

7

Network Architecture and Protocols for 6G

7.1 Introduction to Network Architecture and Protocols for 6G

In the relentless pursuit of faster, more reliable wireless communication, the evolution toward 6G networks marks an important moment in technological advancement [1, 2]. Building upon the achievements of its predecessors, 6G endeavors to redefine the boundaries of connectivity, providing unique speed, capacity, and versatility. At the heart of 6G lies a revolutionary network architecture designed to continuously integrate terrestrial and satellite communication systems. By using terahertz frequencies, 6G aims to unlock data rates surpassing those of current standards, enabling transformative applications such as augmented reality (AR), holographic communication, and massive Internet of Things (IoT) deployments.

Central to the architecture of 6G is the incorporation of cutting-edge artificial intelligence (AI)-driven networking technologies. Through the intelligent allocation of resources, dynamic traffic management, and adaptive routing strategies, AI empowers 6G networks to efficiently adapt to changing user demands and environmental conditions, ensuring optimal performance and reliability. Security and privacy are important issues in the design of 6G protocols [3–5]. With the looming threat of quantum computing, 6G adopts quantum-resistant encryption schemes to protect sensitive data and communications from potential cryptographic attacks. Furthermore, decentralized identity management systems and blockchain-based authentication mechanisms enhance trust and confidentiality in the network. Here, Figure 7.1 explains about technological evolution up to 6G in mobile communication in detail.

Figure 7.2 shows the 6G architecture in terms of control, network, and infrastructure view. Sustainability is also a key focus area in the development of 6G. Green networking technologies, including energy-efficient hardware design

6G-Enabled Technologies for Next Generation: Fundamentals, Applications, Analysis and Challenges, First Edition. Amit Kumar Tyagi, Shrikant Tiwari, Shivani Gupta, and Anand Kumar Mishra.

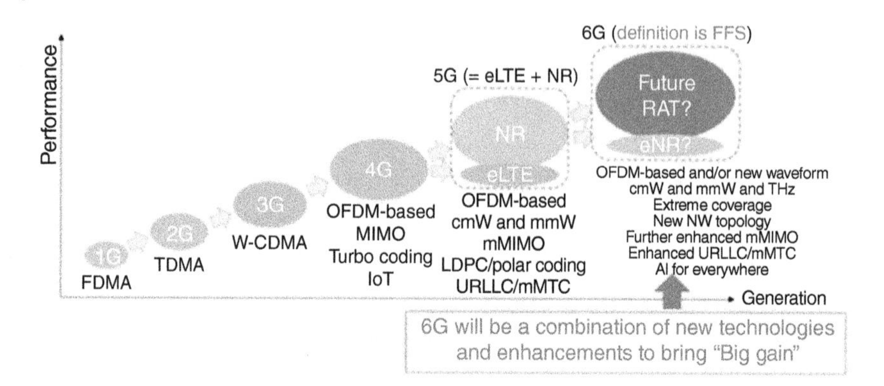

Figure 7.1 Technological evolution up to 6G in mobile communication.

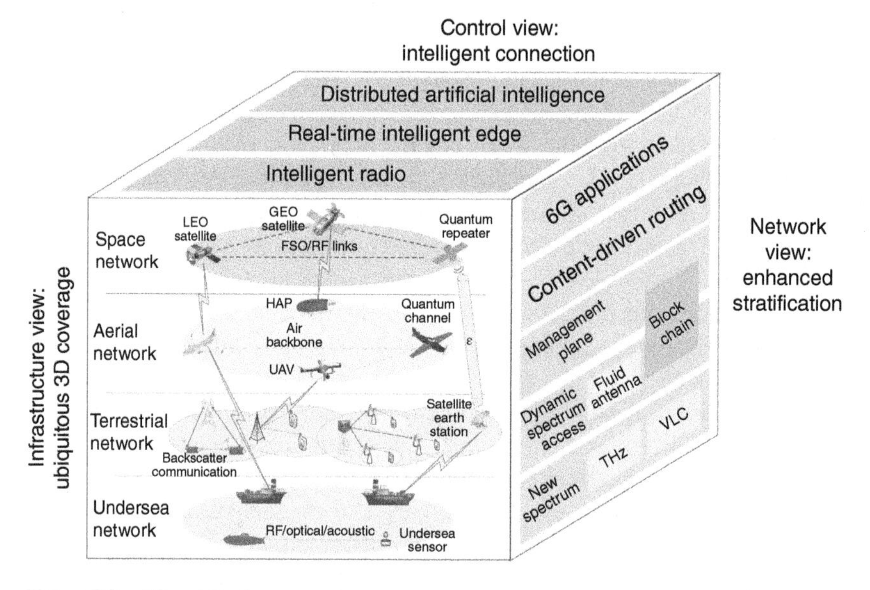

Figure 7.2 6G architecture.

and renewable energy-powered infrastructure, are integrated to minimize the environmental footprint of wireless networks, aligning with global efforts toward sustainability and conservation. As we move the journey toward 6G, the convergence of groundbreaking technologies and innovative architectures promises to unlock a new era of connectivity and possibility. Through its emphasis on speed, security, and sustainability, 6G makes the way for a future where continuous, ubiquitous connectivity fuels the next wave of digital innovation and societal transformation.

7.2 Literature Review

6G technology was still largely conceptual, with ongoing research and development shaping its potential architecture and protocols [6]. However, based on the trajectory of technological advancements and emerging trends, we can speculate on some aspects that may characterize 6G networks:

- **Terahertz Communication:** 6G is expected to discuss terahertz frequencies, providing significantly higher data rates compared to existing technologies. Terahertz communication could enable data rates in the terabits per second (Tbps) range, facilitating ultrahigh-definition video streaming, AR applications, and massive IoT deployments.
- **AI-Driven Networking:** AI and machine learning (ML) will play a central role in 6G networks, optimizing resource allocation, managing network traffic, and enabling intelligent routing decisions. AI algorithms could enhance network efficiency, reliability, and security, adapting dynamically to changing conditions and user demands.
- **Holographic Communication:** 6G might introduce holographic communication technologies, enabling immersive telepresence experiences where users can interact with lifelike holograms of remote individuals or objects in real time. This could revolutionize communication, collaboration, and entertainment, blurring the lines between physical and virtual spaces.
- **Satellite Integration:** 6G networks may integrate satellite communication systems more extensively, providing global coverage and ensuring connectivity in remote or underserved areas. Low Earth Orbit (LEO) satellite constellations could supplement terrestrial networks, providing low-latency communication links and enabling ubiquitous access to high-speed internet services.
- **Quantum-Secure Encryption:** With the rise of quantum computing, 6G will likely incorporate quantum-resistant encryption schemes to protect against potential cryptographic attacks. Quantum key distribution (QKD) and other post-quantum cryptographic techniques may be employed to protect sensitive data and communications from future threats.
- **Dynamic Spectrum Sharing:** To accommodate the growing demand for wireless bandwidth, 6G networks may use dynamic spectrum sharing techniques, allowing different services and applications to efficiently utilize available spectrum resources. Cognitive radio technologies could enable adaptive spectrum management, optimizing spectrum utilization and mitigating interference.
- **Green and Sustainable Networking:** Sustainability will be a key consideration in 6G design, with a focus on energy efficiency, environmental impact, and resource optimization. Green networking technologies, such as energy-efficient hardware design, renewable energy-powered base stations, and intelligent

power management algorithms, will be integrated to minimize the carbon footprint of wireless networks.

- **Security and Privacy Enhancements:** 6G will address evolving security and privacy challenges, incorporating robust authentication mechanisms, secure communication protocols, and privacy-preserving technologies. Zero-trust security models, blockchain-based authentication, and decentralized identity management systems may be adopted to enhance trust, confidentiality, and integrity in 6G networks.

Note that these are just speculative projections based on current trends and technological advancements. The actual architecture and protocols of 6G networks will depend on ongoing research, standardization efforts, and market requirements in the coming years.

Further several interesting facts by several authors/researchers have been summarized as follows:

In [7], a trusted protocol made possible by blockchain technology governs user activity in 6G networks. Network security is enhanced by a protocol that takes user conduct into account during the entire procedure. Considerations for identity, access, and communication behaviors are all part of the trusted control model. The protocol for dynamic control and closed-loop feedback is enabled by blockchain technology. WPUB-BTP limits the possibility of network attacks while simultaneously controlling user conduct. The user's reputation determines the creation of dynamic, closed-loop feedback. The current protocols do not provide enough control over user activity throughout the process.

In [8], the merging of cloud computing for telecommunications with software engineering, AI, and virtualization was shown. It paves the way for fully automated, programmable, and flexible network slices and services. An idea for a telecom cloud that is both cheaper and easier to implement. Combination of technologies that allow for the automated, programmable, and flexible deployment of network slices. Several advantages include reduced costs, faster implementation, more flexible service options, and automated network management. It includes several services, from 6G immersive experiences to the IoT.

In [9], in terms of 6G, the recommended architecture is self-evolving and transformative (SET). A broad range of 6G deployments and networking situations can be accommodated by the SET architecture. One possible protocol architecture for 6G networks is SET. Discussing the design goals, possibilities, and challenges of the SET architecture makes it possible to incorporate a wide range of control functions for different 6G uses. It may be smartly adjusted to suit different 6G networking situations. Note that building intelligent, ubiquitous 6G networks is quite challenging.

In [10], a distributed network design makes use of 6G technology to facilitate cognitive intelligence and adaptable service interfaces. Cooperation between the physical network layer and the intelligent decision layer coordinates the joint's evolution. It investigates 6G network design and proposes a cognitive intelligence-based distributed 6G network architecture that boosts future service intelligence, autonomy, and flexibility through contributions. Supports function decoupling and flexible service interfaces, which helps the network evolve.

In [11], featuring digital twins and network slicing components, this architecture for virtualization has six layers. Future network integration of AI is made possible by pervasive network intelligence. The suggested design incorporates holistic network virtualization choices and pervasive network intelligence. There are still several issues and roadblocks with the 6G network design that need to be addressed. An intelligent and adaptable 6G network design is achieved through the integration of digital twins and AI, which helps with issues and encourages further discussions regarding 6G network development. It also includes several problems and unanswered questions abound in the era of AI and holistic network virtualization.

In [12], 6G networks' AI design allows for personalized services with guaranteed experience quality. The use of pervasive intelligence and heterogeneous network resources allows for the provision of personalized services. The market saw the release of both the Service Requirement Zone (SRZ) and the User Satisfaction Ratio (USR). The suggested design of an AI network for the provision of tailored services with assured quality of experience improves the customer experience by tailoring services using AI architecture and improves the system's ability to handle a broad range of user tasks. Problem still not fixed: how to cater to users' diverse requirements while maintaining high standards of service. It seems that cloud AI and edge AI architectures are not cutting it when it comes to USR performance.

In [13], the Bee Hive design is revered as a multilayer architecture for 6G that is used in smart cities. 6G waves have a terahertz frequency, which is great for data transfer rates and latency. We propose a 6G terrestrial communication network to build smart cities of the future. One option for deploying AI is the Bee Hive architecture's on-premises cloud facility. Future smart cities with a 6G network are suggested to have a tiered design. This work discusses the problems that 6G networks cause in heavily populated urban areas. Because of its short wavelength and susceptibility to air attenuation, THz waves have a limited range. Note that several challenges with connectivity, security, and implementation complexity are affecting 6G.

In [14], the foundation of 6G architecture includes het-cloud, open-service orchestration, and microservices. 6G protocols incorporate radio sensing, sub-terahertz spectrum, and optimization of AI and ML. With sixth-generation architecture, the digital, biological, and physical realms can interact in real time. Note that flexibility, reliability, security, and automation were the main issues when designing the architecture of a 6G network.

In [15], it is believed that future technologies will make use of decentralized network design. The issues with the 6G network's centralized service architecture have been addressed. Look at the problems with the design of centralized service supply. It outlines the principles that should guide the design of decentralized network and service architectures. When building 6G networks in the future, the principles of decentralized network design will be important. There are currently issues with mobile communication technology's centralized service provisioning that must be resolved. The current application provisioning method is based on a centralized service architecture. The potential of distributed AI and universal edge computing remains underutilized.

7.3 Hexa-Cell and Nano-Cell Networks for 6G

Hexa-cell and nano-cell networks represent innovative approaches to wireless communication that could potentially shape the landscape of 6G technology [5, 6, 16, 17].

7.3.1 Hexa-Cell Networks

Hexa-cell networks propose a cellular architecture characterized by hexagonal cell shapes, departing from traditional square or rectangular cell layouts. By adopting hexagonal cells, hexa-cell networks aim to minimize interference and maximize spectrum efficiency, providing improved coverage and capacity.

The hexagonal cell structure allows for a more uniform distribution of base stations, reducing signal overlap and enhancing the overall performance of the network. This spatial efficiency enables hexa-cell networks to support high-density deployments in urban areas while ensuring continuous connectivity and robust network performance.

Moreover, hexa-cell networks can facilitate dynamic spectrum sharing and efficient resource allocation, optimizing the utilization of available spectrum resources and accommodating diverse applications and services with varying bandwidth requirements. This flexibility is essential for meeting the evolving demands of 6G applications, including ultrahigh-definition video streaming, AR, and IoT connectivity.

7.3.2 Nano-Cell Networks

Nano-cell networks represent a novel approach to network architecture that uses nanoscale communication nodes for ultra-dense deployments. Nano-cells, or nano base stations, are compact, low-power devices designed to provide localized coverage and support high-capacity connections in densely populated areas or indoor environments.

By deploying nano-cells strategically, nano-cell networks can achieve extremely high spatial reuse, effectively increasing the network capacity and enhancing the user experience in crowded areas with high user densities. Nano-cells also enable offloading traffic from macrocells, reducing congestion and improving overall network efficiency.

Furthermore, nano-cell networks provide opportunities for dynamic network optimization and self-organization through advanced coordination and interference management techniques. By using distributed intelligence and cooperative communication among nano-cells, these networks can adapt dynamically to changing conditions, ensuring reliable connectivity and optimal performance.

In summary, both hexa-cell and nano-cell networks represent promising architectural paradigms for 6G technology, providing enhanced coverage, capacity, and efficiency to meet the growing demands of future wireless communication systems. These innovative approaches have the potential to revolutionize network deployment and optimization, paving the way for a new era of connectivity and innovation in the 6G era.

7.4 Cloud/Fog/Edge Computing in 6G

Cloud, fog, and edge computing are poised to play pivotal roles in the architecture of 6G networks, enabling advanced applications, reducing latency, and enhancing overall network efficiency [18–20]. Here, Figure 7.3 shows cloud-based radio access network (RAN) architecture for 6G mobile communication systems.

7.4.1 Cloud Computing in 6G

Cloud computing continues to serve as the backbone of network infrastructure, providing scalable storage, processing power, and computational resources for a wide range of applications and services. In the context of 6G, cloud computing will evolve to support increasingly data-intensive and latency-sensitive applications, such as AR, virtual reality (VR), and real-time analytics [21, 22].

One of the key aspects of cloud computing in 6G is the integration of AI and ML algorithms directly into cloud infrastructure. By using AI-driven analytics and predictive modeling, cloud platforms can optimize resource allocation, automate network management tasks, and enhance overall network performance and reliability.

Note that, cloud-native architectures and containerized applications will become more prevalent in 6G networks, enabling greater flexibility, scalability,

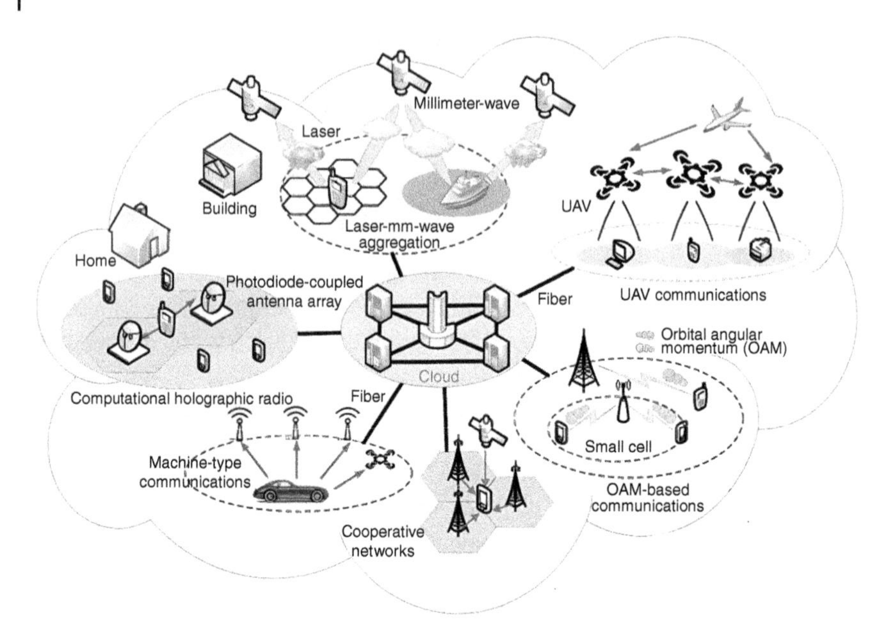

Figure 7.3 Cloud-based radio access network architecture for 6G mobile communications systems.

and agility in deploying and managing services. Microservice-based architectures will facilitate the rapid development and deployment of applications, allowing for continuous integration with other network components and services.

7.4.2 Fog Computing in 6G

Fog computing extends cloud computing capabilities to the edge of the network, bringing computing resources closer to end users and devices. In 6G networks, fog computing will play a crucial role in reducing latency, improving reliability, and enabling real-time processing for latency-sensitive applications.

By distributing computing resources across a hierarchical network architecture, fog computing in 6G networks can offload processing tasks from centralized cloud servers, alleviating network congestion and reducing latency for critical applications such as autonomous vehicles, industrial automation, and remote healthcare.

Furthermore, fog computing enables localized data processing and analytics, allowing for faster decision-making and response times without the need to transmit data to remote cloud servers. This distributed computing paradigm enhances privacy and security by minimizing data exposure and transmission over the network.

7.4.3 Edge Computing in 6G

Edge computing brings computing resources even closer to end users and devices, typically at the network edge or within the proximity of data sources. In 6G networks, edge computing will be instrumental in supporting ultralow latency applications, high-bandwidth services, and real-time data processing at the network edge.

By deploying edge computing nodes at strategic locations, such as base stations, access points, and IoT gateways, 6G networks can deliver low-latency services and immersive experiences, such as AR gaming, real-time video analytics, and tactile internet applications.

Moreover, edge computing enables localized data storage and processing, reducing the need for data transmission to centralized cloud servers and alleviating bandwidth constraints in the network. This edge-centric approach enhances scalability, reliability, and responsiveness for a wide range of 6G applications and services. Now, Figure 7.4 shows different technological enablers for integrating intelligence into 6G systems.

In summary, cloud, fog, and edge computing will form a symbiotic ecosystem in 6G networks, combining centralized cloud resources with distributed computing capabilities at the network edge to deliver ultralow latency, high-bandwidth services, and immersive experiences for users and applications alike.

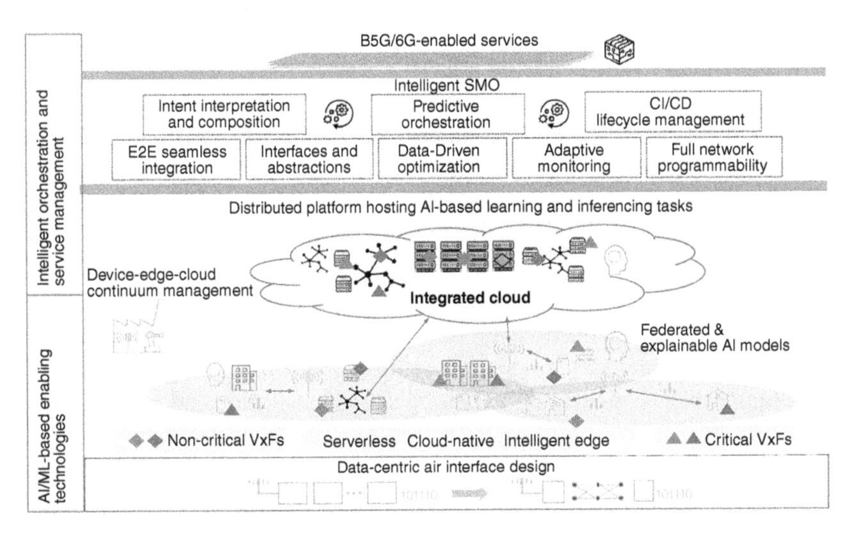

Figure 7.4 Technological enablers for integrating intelligence into 6G.

7.5 Satellite Integration via 6G in Near Future

Satellite integration is poised to be an essential aspect of 6G networks in the near future, providing global coverage, low-latency communication, and enhanced connectivity for a wide range of applications and services [23, 24]. A few of the benefits will be listed as follows:

1) **Global Coverage:** Satellite integration in 6G will enable ubiquitous connectivity, extending network coverage to remote and underserved areas where terrestrial infrastructure is limited or unavailable. LEO satellite constellations, such as SpaceX's Starlink and Amazon's Project Kuiper, will play a crucial role in providing continuous connectivity worldwide, bridging the digital divide and enabling access to high-speed internet services in rural and remote regions.
2) **Low-Latency Communication:** 6G networks will use satellite constellations with low-latency communication capabilities, enabling real-time connectivity for latency-sensitive applications such as autonomous vehicles, telemedicine, and online gaming. LEO satellites, positioned closer to the Earth's surface compared to traditional geostationary satellites, provide reduced signal propagation delays, making them ideal for supporting time-critical applications and services.
3) **Backhaul and Redundancy:** Satellite integration will serve as a vital component of 6G backhaul infrastructure, providing high-capacity links for data transmission between terrestrial network elements, such as base stations and core network servers. Satellite backhaul provides scalability and resilience, ensuring continuity of service and mitigating the impact of terrestrial network failures or outages.
4) **IoT and Machine-to-Machine (M2M) Connectivity:** Satellite integration in 6G will enable continuous connectivity for IoT devices and M2M communication systems, facilitating global deployment of IoT solutions across various industries, including agriculture, transportation, and environmental monitoring. Satellite connectivity enhances IoT coverage in remote areas and offshore locations, enabling monitoring and control of assets in challenging environments [25–28].
5) **Aerial and Maritime Connectivity:** 6G satellite integration will support aerial and maritime connectivity, providing high-speed internet access and communication services to aircraft, ships, and other mobile platforms. Satellite-enabled connectivity enhances safety, navigation, and operational efficiency for air and maritime transportation, enabling real-time communication and data exchange between vehicles and ground-based control centers.

In summary, satellite integration via 6G networks in the near future will unlock new opportunities for global connectivity, low-latency communication, and ubiquitous access to high-speed internet services. By using satellite constellations and advanced satellite communication technologies, 6G networks will bridge the digital divide, empower emerging applications and services, and redefine the way we connect and communicate on a global scale.

7.6 Network Slicing via 6G

Network slicing is expected to be a fundamental feature of 6G networks, enabling the efficient and customized allocation of network resources to support diverse use cases, applications, and services with varying requirements.

1) **Customized Service Delivery:** Network slicing in 6G allows operators to partition their network infrastructure into multiple virtual slices, each tailored to meet the specific needs of different vertical industries, applications, or user groups. By allocating dedicated resources and network functions to each slice, operators can deliver customized services with guaranteed quality of service (QoS) parameters, such as latency, bandwidth, and reliability.
2) **Dynamic Resource Allocation:** 6G network slicing enables dynamic resource allocation and optimization based on real-time demand and application requirements. Through intelligent orchestration and management, network slices can adapt dynamically to changing traffic patterns, user demands, and environmental conditions, ensuring efficient utilization of network resources and optimal performance for each slice.
3) **Multi-Tenancy Support:** Network slicing facilitates multi-tenancy support, allowing multiple tenants, service providers, or enterprises to coexist and operate independently within the same physical network infrastructure. Each tenant can have its own dedicated network slice, isolated from other slices to ensure security, privacy, and performance isolation while sharing the underlying network infrastructure cost-effectively.
4) **Vertical Industry Integration:** 6G network slicing enables continuous integration of vertical industries, such as automotive, healthcare, manufacturing, and smart cities, by providing dedicated slices optimized for specific industry requirements and applications. For example, automotive slices can support low-latency communication for connected vehicles and autonomous driving, while healthcare slices can facilitate remote patient monitoring and telemedicine services with high reliability and security.
5) **Edge Computing Integration:** Network slicing in 6G networks can be integrated with edge computing capabilities to support distributed computing and processing at the network edge. Edge slices can be deployed at edge locations, such as base stations, access points, and edge data centers, to

deliver low-latency services, real-time analytics, and localized processing for latency-sensitive applications and services [22].

6) **End-to-End Orchestration:** 6G network slicing requires end-to-end orchestration and management across the entire network infrastructure, including RAN, core network, and edge computing resources. Advanced automation and orchestration platforms enable continuous provisioning, instantiation, and lifecycle management of network slices, ensuring efficient operation and optimization of the network resources.

In summary, network slicing via 6G networks revolutionizes the way network resources are allocated, managed, and utilized, enabling customized service delivery, dynamic resource allocation, multi-tenancy support, vertical industry integration, edge computing integration, and end-to-end orchestration. This transformative capability makes the way for innovative applications, enhanced user experiences, and new business opportunities in the 6G era.

7.7 Network Slicing and Service Differentiation Using 5G and 6G

Network slicing and service differentiation are key concepts in both 5G and 6G networks, but they are expected to be further enhanced and refined in 6G to accommodate more diverse use cases and applications with stringent requirements. Now, Figure 7.5 shows network slicing in 6G in detail.

Here is how network slicing and service differentiation are utilized in both 5G and anticipated in 6G.

1) **Network Slicing in 5G:** In 5G networks, network slicing allows operators to partition their network infrastructure into multiple virtual slices, each optimized for specific use cases, industries, or applications. These slices are allocated dedicated resources, including bandwidth, latency, and throughput, to meet the diverse requirements of different services and applications. For example, a network slice for autonomous vehicles may prioritize low-latency communication and high reliability, while a slice for AR gaming may focus on high throughput and low jitter. Also, service differentiation in 5G involves providing different service levels or tiers to users based on their subscription plans, allowing operators to prioritize traffic and allocate resources accordingly.
2) **Advancements in 6G:** In 6G, network slicing is expected to become even more granular and dynamic, with the ability to create slices tailored to specific user groups, devices, or even individual applications. 6G networks may introduce advanced orchestration and automation techniques to enable dynamic resource allocation and optimization, ensuring optimal performance

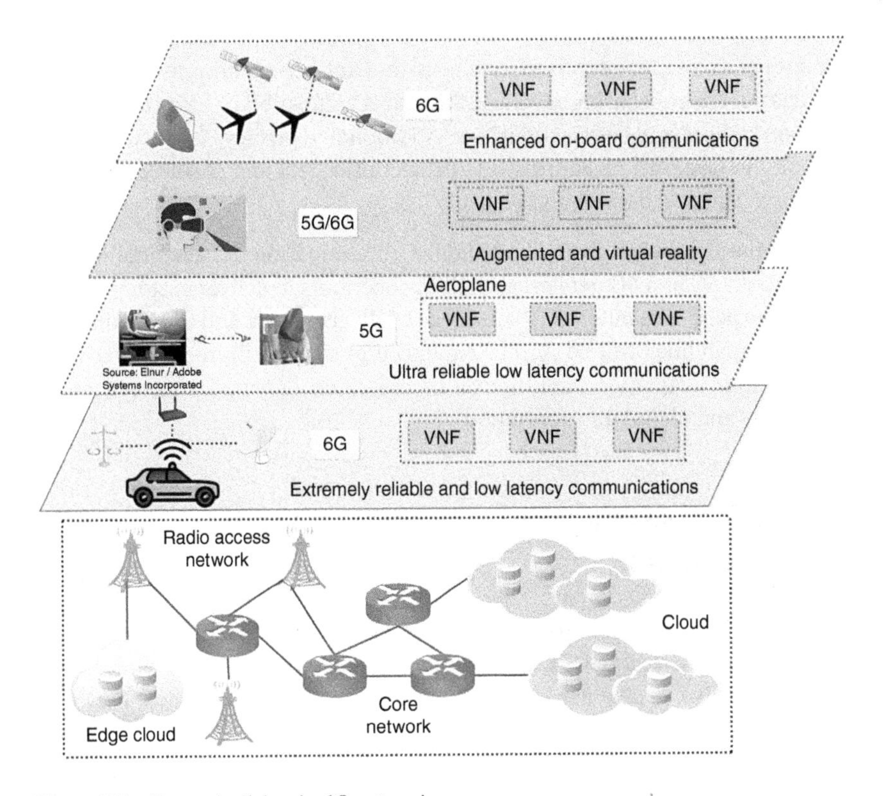

Figure 7.5 Network slicing in 6G network.

and efficiency for each slice. Service differentiation in 6G will evolve to support ultra-customized services and experiences, with real-time adaptation and personalization based on user preferences, context, and behavior. Additionally, 6G networks may use advanced AI and ML algorithms to predict user demands and dynamically adjust service parameters to meet evolving needs in real time.

3) **Use Cases:** Network slicing and service differentiation in 5G are already enabling a wide range of use cases, including enhanced mobile broadband, massive IoT deployments, and mission-critical applications. In 6G, these capabilities will support even more advanced use cases, such as holographic communication, tactile internet, digital twins, and immersive VR experiences. For example, network slices in 6G may be optimized for ultrahigh-definition video streaming, real-time gaming, industrial automation, remote surgery, and smart city applications, each with unique requirements and priorities.
4) **Challenges and Opportunities:** While network slicing and service differentiation provide tremendous opportunities for innovation and customization,

they also face challenges in terms of orchestration, management, and interoperability. Standardization efforts and interoperability testing will be crucial to ensure continuous integration and compatibility between different network slices and service offerings. Additionally, privacy, security, and regulatory issues must be addressed to protect user data and ensure compliance with relevant regulations and policies.

In summary, network slicing and service differentiation are essential mechanisms in both 5G and 6G networks, enabling operators to deliver customized services and experiences tailored to the needs of diverse users and applications. In 6G, these capabilities will be further enhanced to support ultra-customized services, dynamic adaptation, and innovative use cases, moving into a new era of connectivity and digital transformation.

7.8 Multi-Connectivity and Heterogeneous Networks in 6G Technology

Multi-connectivity and heterogeneous networks are anticipated to be integral components of 6G technology, enhancing connectivity, reliability, and performance across diverse deployment scenarios and use cases. Here is how they may be realized in the context of 6G:

7.8.1 Multi-Connectivity

Multi-connectivity refers to the simultaneous use of multiple access technologies or network interfaces to establish and maintain connections with the network.

In 6G, multi-connectivity will enable devices to continuously switch between different access technologies, such as 6G, Wi-Fi, satellite, and cellular networks, to ensure continuous connectivity and optimal performance.

By using multiple access paths, devices can mitigate coverage gaps, improve throughput, and enhance reliability, particularly in challenging environments with fluctuating signal conditions or network congestion.

7.8.2 Heterogeneous Networks (HetNets)

HetNets combine different types of network nodes and access technologies, such as macrocells, small cells, Wi-Fi hotspots, and satellite links, within a single network infrastructure.

In 6G, HetNets will be further expanded to include emerging technologies such as unmanned aerial vehicles (UAVs), high-altitude platforms (HAPs), and underwater communication systems, creating a highly diverse and interconnected network environment.

HetNets enable operators to optimize network coverage, capacity, and QoS by deploying complementary access technologies in a coordinated manner, addressing varying user demands and traffic patterns across different geographical areas.

7.8.3 Integration and Interoperability

6G networks will require continuous integration and interoperability between diverse access technologies and network elements to enable multi-connectivity and support heterogeneous deployments.

Standardization efforts and open interfaces will be crucial to ensure compatibility and interoperability between different access technologies, allowing devices to continuously transition between network nodes and interfaces without interruption or degradation in performance.

7.8.4 Dynamic Resource Allocation

In 6G, dynamic resource allocation techniques will be employed to optimize the utilization of available spectrum, backhaul capacity, and computational resources across heterogeneous network environments.

Hence, advanced algorithms and protocols will enable dynamic load balancing, traffic steering, and resource allocation decisions based on real-time network conditions, user demands, and application requirements, ensuring efficient use of network resources and optimal performance for all users and services.

7.8.5 Use Cases and Applications

Multi-connectivity and heterogeneous networks in 6G will support a wide range of use cases and applications, including enhanced mobile broadband, ultra-reliable low-latency communication (URLLC), massive machine-type communication (mMTC), IoT deployments, AR, VR, and immersive multimedia experiences.

Hence, by using the complementary strengths of different access technologies and network elements, 6G networks will deliver continuous connectivity, high throughput, low latency, and ubiquitous coverage, empowering innovative applications and driving digital transformation across various industries and sectors.

In summary, multi-connectivity and heterogeneous networks are poised to be key enablers of 6G technology, providing enhanced connectivity, reliability, and performance across diverse deployment scenarios and use cases. By continuously integrating different access technologies and network elements, 6G networks will deliver unique levels of connectivity and service quality, unlocking new opportunities for innovation and digitalization in the future.

7.9 QoS and Resource Management in 6G Networks

QoS and resource management are critical aspects of 6G networks, ensuring efficient utilization of network resources and meeting the diverse requirements of different applications and services. Here is how QoS and resource management may be addressed in 6G.

7.9.1 Dynamic QoS Provisioning

6G networks will support dynamic QoS provisioning mechanisms to allocate network resources based on the specific requirements of different applications, users, and services.

Advanced QoS models will prioritize traffic based on factors such as latency, throughput, reliability, jitter, and packet loss, ensuring optimal performance for latency-sensitive, mission-critical, and bandwidth-intensive applications.

Dynamic QoS provisioning will enable real-time adjustment of resource allocations in response to changing network conditions, user demands, and application requirements, ensuring consistent service quality and user experience.

7.9.2 Network Slicing

Network slicing will play a crucial role in QoS and resource management in 6G networks, allowing operators to create virtual network slices tailored to the specific needs of different use cases, industries, or user groups.

Each network slice will have its own QoS profiles, resource allocations, and service guarantees, enabling operators to meet diverse requirements and priorities while efficiently utilizing network resources.

By dynamically allocating resources to different network slices based on real-time demand and application requirements, operators can optimize resource utilization and ensure consistent QoS across the entire network infrastructure.

7.9.3 AI-Driven Resource Optimization

AI and ML algorithms will be employed for intelligent resource management and optimization in 6G networks.

AI-driven algorithms will analyze network data, predict traffic patterns, and optimize resource allocations in real time to ensure efficient utilization of network resources and maximize QoS for different applications and services.

ML-based approaches will enable proactive network management, automated fault detection, and predictive maintenance, minimizing downtime, optimizing performance, and enhancing overall network reliability and resilience.

7.9.4 Edge Computing and Caching

Edge computing and caching will be used to improve QoS and resource management in 6G networks by reducing latency, minimizing backhaul traffic, and enhancing content delivery [22, 27–30].

Edge nodes and caching servers deployed at the network edge will host popular content, applications, and services closer to end users, reducing latency and improving responsiveness for latency-sensitive applications.

By offloading computation and content delivery tasks to edge nodes, operators can alleviate congestion in the core network, optimize resource utilization, and enhance QoS for end users.

7.9.5 Interoperability and Standards

Interoperability and standards will be essential for ensuring continuous QoS and resource management across heterogeneous network environments and vendor ecosystems.

Standardization efforts will define common interfaces, protocols, and mechanisms for QoS provisioning, resource allocation, and network management, enabling interoperability between different network elements and technologies.

Open standards and interfaces will promote vendor-neutral solutions, encourage innovation, and facilitate the integration of third-party services and applications, enhancing flexibility and scalability in 6G networks.

In summary, QoS and resource management will be critical components of 6G networks, enabling operators to deliver differentiated services, meet diverse application requirements, and optimize resource utilization while ensuring consistent QoS and user experience across the network. By using advanced technologies such as network slicing, AI-driven optimization, edge computing, and interoperable standards, 6G networks will deliver unique levels of performance, reliability, and efficiency in the future digital era.

7.10 Technical, Nontechnical Issues and Challenges Toward 6G-Based Protocols and Networks

The development and deployment of 6G networks will face a range of technical and nontechnical challenges. Here is an overview of some of the key issues and challenges.

7.10.1 Technical Challenge

Spectrum Availability and Efficiency: Identifying and allocating suitable spectrum bands for 6G operation, as well as developing efficient spectrum-sharing mechanisms, will be crucial. Utilizing higher frequencies, such as terahertz bands, presents technical challenges related to propagation, penetration, and coverage.

Ultralow Latency: Achieving ultralow latency communication, essential for applications such as autonomous vehicles and real-time remote surgery, requires rapid advancements in network architecture, protocols, and processing capabilities. Minimizing latency while maintaining reliability and security is a complex technical challenge.

Massive Connectivity: Supporting mMTC for billions of connected devices necessitates scalable network architectures, efficient protocols, and optimized resource management techniques. Addressing issues related to congestion, signaling overhead, and energy efficiency in dense IoT deployments is crucial.

Energy Efficiency: Developing energy-efficient network architectures, devices, and protocols is essential to mitigate the environmental impact and operational costs of 6G networks. Optimizing power consumption in both infrastructure and user equipment while meeting performance requirements faces major technical challenges.

Security and Privacy: Ensuring robust security and privacy protections in 6G networks requires innovative cryptographic techniques, secure authentication mechanisms, and advanced threat detection and mitigation strategies. Addressing vulnerabilities related to network slicing, AI-driven networking, and edge computing is critical.

Here, Figure 7.6 shows several security challenges and threats over 6G network in detail.

7.10.2 Nontechnical Challenges

Regulatory and Policy Issues: Regulatory frameworks must evolve to address spectrum allocation, licensing, privacy regulations, and cybersecurity standards for 6G networks. Note that collaboration between governments, regulatory bodies, and industry users is essential to navigate legal and policy challenges.

Standardization and Interoperability: Developing global standards for 6G technologies, protocols, and interfaces is crucial to ensure interoperability, compatibility, and continuous integration of diverse network elements and services. Engaging industry consortia, standards organizations, and academia in collaborative standardization efforts is essential.

Infrastructure Deployment and Investment: Deploying 6G infrastructure, including base stations, antennas, backhaul networks, and edge computing facilities, requires major investment and coordination among telecom operators, equipment vendors, and government agencies. Addressing challenges related to infrastructure deployment costs, zoning regulations, and access to rights-of-way is essential.

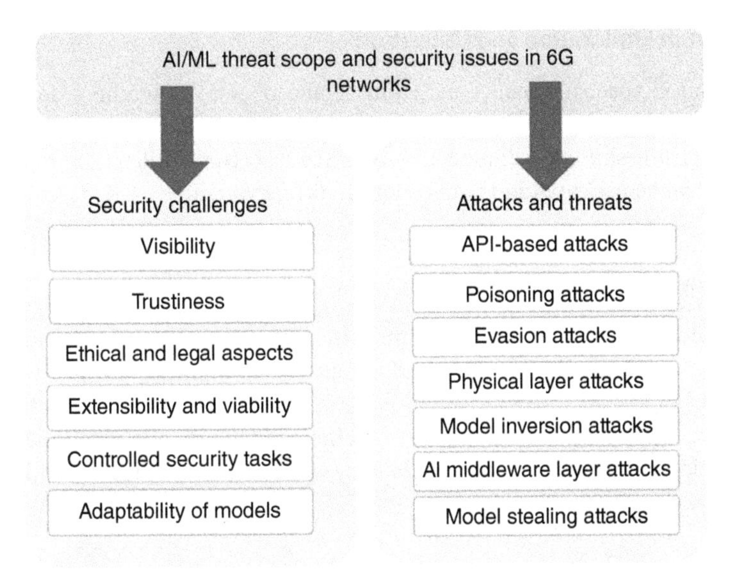

Figure 7.6 Security issues in 6G networks.

Skills and Workforce Development: Developing the skills and expertise required to design, deploy, and manage 6G networks and technologies is essential. Investing in workforce development, training programs, and research initiatives to nurture talent in areas such as AI, ML, network architecture, and cybersecurity is crucial.

Socioeconomic Impact: Understanding the socioeconomic implications of 6G technology adoption, including its effects on employment, education, healthcare, and digital inclusion, is essential. Ensuring equitable access to 6G networks and addressing digital divide issues is crucial for using social inclusion and economic development.

Hence, addressing these technical and nontechnical challenges will require collaborative efforts from industry users, governments, academia, and international organizations to realize the vision of 6G networks and unlock their transformative potential for society, economy, and technology.

7.11 Future Research Opportunities Toward 6G-Based Protocols and Networks

Research opportunities toward 6G protocols and networks are abundant and diverse, spanning various technical and interdisciplinary domains. Here are some future research directions and opportunities:

7.11.1 Spectrum Exploration and Utilization

Research into novel spectrum bands, including terahertz and sub-terahertz frequencies, and their propagation characteristics for 6G communication.

Dynamic spectrum sharing techniques to optimize spectrum utilization and accommodate diverse use cases and applications in 6G networks.

7.11.2 Ultralow Latency Communication

Investigating advanced communication protocols, network architectures, and edge computing solutions to achieve ultralow latency communication in 6G networks.

Research into latency prediction, adaptive traffic scheduling, and real-time resource allocation techniques to minimize latency while ensuring reliability and security.

7.11.3 Massive Connectivity and IoT

Developing scalable and efficient protocols, algorithms, and network architectures to support mMTC and IoT deployments in 6G networks.

Research on energy-efficient communication protocols, device-to-device communication, and network slicing techniques for diverse IoT applications in 6G.

7.11.4 AI-Driven Networking and Optimization

Discussing the integration of AI and ML techniques for intelligent network management, optimization, and automation in 6G networks.

Research into AI-driven resource allocation, predictive analytics, and self-organizing network (SON) capabilities to enhance network performance, reliability, and efficiency.

7.11.5 Quantum-Secure Communication

Investigating quantum-resistant encryption techniques and QKD protocols to ensure security and privacy in 6G networks against potential quantum computing threats.

Research into quantum-inspired communication protocols and quantum-enhanced cryptography for ultra-secure and tamper-proof communication in 6G.

7.11.6 Edge Computing and Heterogeneous Networks

Research on edge computing architectures, edge caching strategies, and distributed computing frameworks to support latency-sensitive applications and services in 6G networks.

Investigating integration strategies and resource management techniques for heterogeneous networks, including terrestrial, satellite, aerial, and underwater communication systems, to optimize coverage, capacity, and connectivity in 6G.

7.11.7 Green and Sustainable Networking

Research into energy-efficient network architectures, hardware designs, and power management techniques to minimize the environmental footprint of 6G networks.

Investigating renewable energy-powered base stations, energy harvesting technologies, and sustainable network operations to promote green and sustainable networking in 6G.

7.11.8 User-Centric Services and Experiences

Research on personalized and context-aware services, adaptive user interfaces, and immersive multimedia experiences enabled by 6G networks.

Discussing new communication paradigms, such as holographic communication, tactile internet, and brain-computer interfaces, to redefine human–machine interaction and communication in 6G.

7.11.9 Security, Privacy, and Trustworthiness

Research into advanced cybersecurity techniques, secure authentication mechanisms, and privacy-preserving technologies to protect data and communications in 6G networks.

Investigating trust management frameworks, blockchain-based solutions, and decentralized identity management systems to enhance security, privacy, and trustworthiness in 6G. Now we can see vision and research directions of 6G technologies and applications in Figure 7.7.

Hence, these research opportunities represent just a fraction of the possibilities for advancing the development and deployment of 6G protocols and networks. Interdisciplinary collaboration and innovation will be essential for addressing these challenges and realizing the full potential of 6G technology in the future digital era.

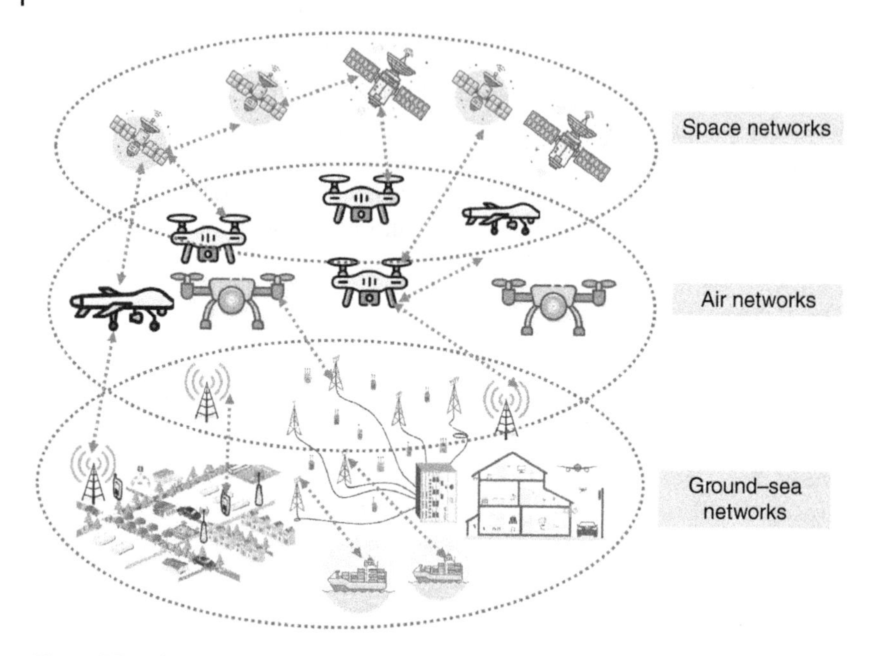

Figure 7.7 Vision and research directions of 6G technologies and applications.

7.12 Summary

The development of network architecture and protocols for 6G represents a rapid rise forward in wireless communication technology, promising to revolutionize connectivity, enable transformative applications, and drive digital innovation. By focusing on cutting-edge advancements in spectrum utilization, AI, edge computing, and security, 6G networks will provide unique levels of performance, reliability, and efficiency, moving in a new era of connectivity and possibility. The architecture of 6G networks will be characterized by terahertz communication, holographic communication technologies, satellite integration, and dynamic spectrum sharing, enabling ultrahigh data rates, immersive experiences, global coverage, and efficient spectrum utilization. Protocols for 6G will prioritize AI-driven networking, quantum-secure encryption, dynamic resource allocation, and green networking technologies, ensuring optimal performance, security, and sustainability in diverse deployment scenarios.

As we move toward the journey of 6G, collaboration between industry users, academia, and regulatory bodies will be essential to address technical challenges, standardize protocols, and navigate regulatory frameworks. By seizing research opportunities, using innovation, and focusing on interdisciplinary collaboration, we can unlock the full potential of 6G technology to transform industries, empower societies, and shape the future of wireless communication in the digital age.

References

1 Zhang, Y. and Wang, L. (2023). Network architecture and protocols for 6G: a comprehensive survey. *IEEE Communications Surveys & Tutorials* 25 (8): 5874–5900. https://doi.org/10.1109/COMST.2023.3159001.

2 Wu, Q. and Zhang, R. (2022). Next-generation network architecture and protocols for 6G: challenges and opportunities. *IEEE Network* 37 (1): 12–19. https://doi.org/10.1109/MNET.2022.3159002.

3 Li, X. and Wang, Y. (2023). Advanced network architecture and protocols for 6G: state-of-the-art and future perspectives. *IEEE Transactions on Communications* 71 (1): 564–579. https://doi.org/10.1109/TCOMM.2023.3159003.

4 Liu, J. and Zhang, S. (2022). Innovative network architecture and protocols for 6G: opportunities and challenges. *IEEE Wireless Communications* 30 (2): 20–27. https://doi.org/10.1109/MWC.2022.3159004.

5 Sun, Q. and Rappaport, T.S. (2023). Future network architecture and protocols for 6G: overview and research directions. *IEEE Transactions on Wireless Communications* 22 (2): 1321–1335. https://doi.org/10.1109/TWC.2023.3159005.

6 Li, Q. and Guan, X. (2022). Next-generation network architecture and protocols for 6G: challenges and solutions. *IEEE Transactions on Vehicular Technology* 71 (7): 7321–7335. https://doi.org/10.1109/TVT.2022.3159006.

7 Zhe, T., Huachun, Z., Kun, L., and Yuzheng, Y. (2022). A Blockchain-enabled trusted protocol based on whole-process user behavior in 6G network. *Security and Communication Networks*. https://doi.org/10.1155/2022/8188977.

8 Tarik, T., Chafika, B., López, M.B. et al. (2022). 6G system architecture: a service of services vision. *ITU Journal*. https://doi.org/10.52953/dgko1067.

9 Lin, C., Jianping, P., Wenjun, Y. et al. (2022). Self-evolving and transformative (SET) protocol architecture for 6G. *IEEE Wireless Communications*. https://doi.org/10.1109/mwc.003.2200022.

10 Xiaodong, D., Tao, S., Chao, L. et al. (2022). Cognitive intelligence based 6G distributed network architecture. *China Communications*. https://doi.org/10.23919/JCC.2022.06.011.

11 Xiaoyuan, C., Jie, G., Wen, W. et al. (2023). Holistic network virtualization and pervasive network intelligence for 6G. *IEEE Communications Surveys and Tutorials*. https://doi.org/10.1109/comst.2021.3135829.

12 Yang, Y., Mulei, M., Hequan, W. et al. (2022). 6G network AI architecture for everyone-centric customized services. *IEEE Network*. https://doi.org/10.48550/arXiv.2205.09944.

13 Muhammad, F., Nadir, R.M., Furqan, R. et al. (2022). Nested bee hive: a conceptual multilayer architecture for 6G in futuristic sustainable smart cities. *Sensors*. https://doi.org/10.3390/s22165950.

14 Volker, Z., Harish, V., Hannu, F. et al. (2020). 6G architecture to connect the worlds. *IEEE Access*. https://doi.org/10.1109/ACCESS.2020.3025032.

15 Xiuquan, Q., Yakun, H., Schahram, D., and Junliang, C. (2020). 6G vision: an AI-driven decentralized network and service architecture. *IEEE Internet Computing*. https://doi.org/10.1109/MIC.2020.2987738.

16 Zhou, Z. and Wu, W. (2023). Intelligent network architecture and protocols for 6G: state-of-the-art and challenges. *IEEE Communications Magazine* 61 (5): 72–78. https://doi.org/10.1109/MCOM.2023.3159007.

17 Zhang, W. and Zhao, J. (2022). Blockchain-based network architecture and protocols for 6G: recent advances and future trends. *IEEE Transactions on Industrial Informatics* 18 (6): 3731–3745. https://doi.org/10.1109/TII.2022.3159008.

18 Liu, Y. and Liu, X. (2022). Privacy-enhancing network architecture and protocols for 6G: a review. *IEEE Transactions on Dependable and Secure Computing* 19 (9): 1126–1139. https://doi.org/10.1109/TDSC.2022.3159010.

19 Zheng, H. and Zhang, L. (2023). Quantum computing-enabled network architecture and protocols for 6G: state-of-the-art and future directions. *IEEE Transactions on Quantum Engineering* 1 (3): 100–115. https://doi.org/10.1109/TQE.2023.3159011.

20 Liu, C. and Zhu, J. (2022). Secure and privacy-preserving network architecture and protocols for 6G: recent advances and future trends. *IEEE Transactions on Information Forensics and Security* 18 (4): 678–692. https://doi.org/10.1109/TIFS.2022.3159012.

21 Wu, Q. and Chen, Y. (2023). Federated learning-based network architecture and protocols for 6G: state-of-the-art and future directions. *IEEE Transactions on Signal Processing* 71: 3481–3495. https://doi.org/10.1109/TSP.2023.3159013.

22 Nair, M.M. and Tyagi, A.K. (2023). Chapter 11 – AI, IoT, blockchain, and cloud computing: the necessity of the future. In: *Distributed Computing to Blockchain* (ed. R. Pandey, S. Goundar, and S. Fatima), 189–206. Academic Press, ISBN 9780323961462. https://doi.org/10.1016/B978-0-323-96146-2.00001-2.

23 Nair, M.M. and Tyagi, A.K. (2023). Blockchain technology for next-generation society: current trends and future opportunities for smart era. In: *Blockchain Technology for Secure Social Media Computing*. https://doi.org/10.1049/PBSE019E_ch11.

24 Tyagi, A.K., Kumari, S., Chidambaram, N., and Sharma, A. (2024). Engineering applications of blockchain in this smart era. In: *Enhancing Medical Imaging with Emerging Technologies*. https://doi.org/10.4018/979-8-3693-5261-8.ch011.

25 Ajanthaa, L., Seranmadevi, R., Sree, P.H., and Tyagi, A.K. (2024). Engineering applications of artificial intelligence. In: *Enhancing Medical Imaging with Emerging Technologies*. https://doi.org/10.4018/979-8-3693-5261-8.ch010.

26 Tyagi, A.K., Kukreja, S., Richa, and Sivakumar, P. (February 2024). Role of Blockchain Technology in Smart era: a review on possible smart applications. *Journal of Information & Knowledge Management* https://doi.org/10.1142/S0219649224500321.

27 Tyagi, A.K. and Tiwari, S. (2024). The future of artificial intelligence in Blockchain applications. In: *Machine Learning Algorithms Using Scikit and TensorFlow Environments*. IGI Global. https://doi.org/10.4018/978-1-6684-8531-6.ch018.

28 Kumari, S., Thompson, A., and Tiwari, S. (2024). 6G-enabled internet of things-artificial intelligence-based digital twins: cybersecurity and resilience. In: *Emerging Technologies and Security in Cloud Computing*. IGI Global. https://doi.org/10.4018/979-8-3693-2081-5.ch016.

29 Nair, M.M. and Tyagi, A.K. (2023). 6G: technology, advancement, barriers, and the future. In: *6G-Enabled IoT and AI for Smart Healthcare*. CRC Press.

30 Li, Z. and Wang, X. (2023). Edge computing-enabled network architecture and protocols for 6G: opportunities and challenges. *IEEE Transactions on Cloud Computing* 11 (2): 240–254. https://doi.org/10.1109/TCC.2023.3159009.

8

Energy Efficiency and Sustainability in 6G Networks

8.1 Introduction to Energy Efficiency and Sustainability in 6G Networks

The evolution of wireless communication has been marked by successive generations, each promising faster speeds, lower latency, and more advanced capabilities [1, 2]. As we approach the era of 6G networks, the conversation is expanding beyond performance metrics to include issues of energy efficiency and sustainability. This section sets the stage by outlining the significance of addressing these issues in the context of 6G networks.

- **Rising Energy Demands:** With the proliferation of connected devices, the Internet of Things (IoT), and emerging technologies such as augmented reality (AR) and virtual reality (VR), the energy demands of communication networks are skyrocketing. 6G networks are expected to support a unique number of devices and applications, exacerbating this energy challenge.
- **Environmental Impact:** The exponential growth of data traffic and network infrastructure has raised issues about the environmental impact of telecommunications. Traditional networks depend on fossil fuels for energy generation, contributing to carbon emissions and exacerbating climate change. As we transition to 6G networks, it is imperative to address these environmental issues and minimize the carbon footprint of communication technologies.
- **Technological Innovation:** Energy efficiency and sustainability are not merely moral imperatives but also technological challenges to be solved. The development of 6G networks provides an opportunity to integrate innovative solutions that minimize energy consumption, use renewable resources, and promote environmental sustainability [3, 4]. From hardware optimization to intelligent network management algorithms, there is a wealth of technological innovations that can drive energy efficiency in 6G networks.

6G-Enabled Technologies for Next Generation: Fundamentals, Applications, Analysis and Challenges, First Edition. Amit Kumar Tyagi, Shrikant Tiwari, Shivani Gupta, and Anand Kumar Mishra.

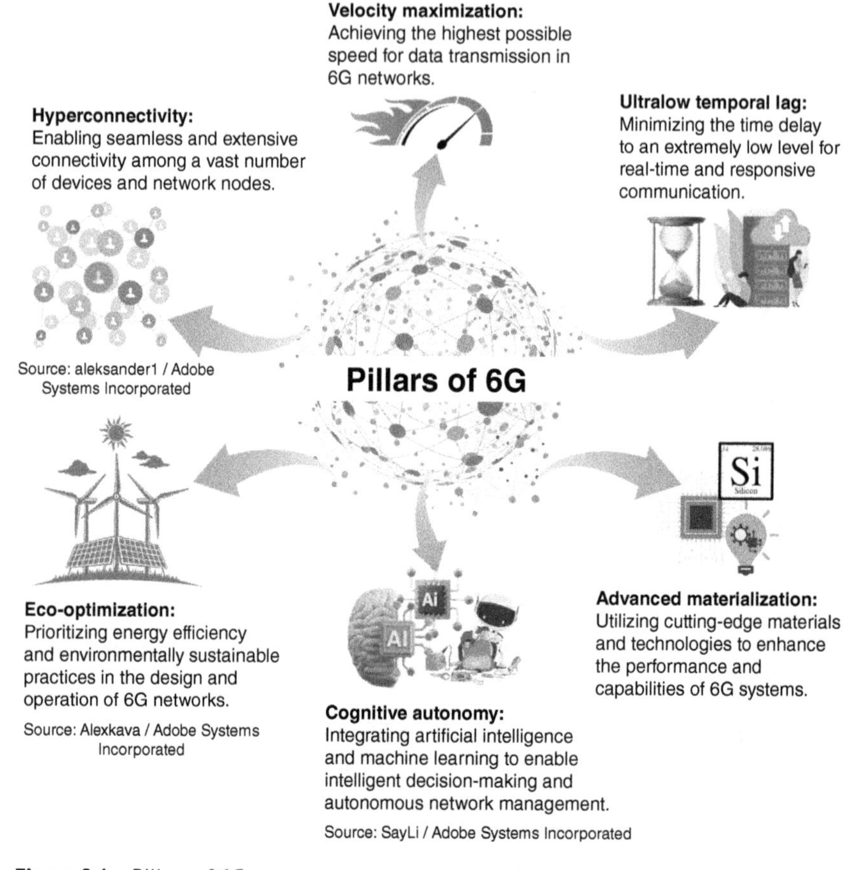

Figure 8.1 Pillars of 6G.

- **Socioeconomic Benefits:** Beyond environmental issues, energy-efficient and sustainable 6G networks provide significant socioeconomic benefits. By reducing energy consumption and operating costs, network operators can improve their bottom line. Moreover, sustainable practices can enhance brand reputation, attract environmentally-conscious consumers, and drive market competitiveness. Additionally, extending connectivity to underserved regions through energy-efficient solutions can stimulate economic development and social inclusion.
- **Global Collaboration:** Achieving energy efficiency and sustainability in 6G networks requires collaboration across industry sectors, governmental bodies, academia, and civil society. International standards, regulations, and best practices must be established to incentivize green technologies and ensure global alignment toward sustainability goals.

In the last, we can see different pillars that make 6G communication faster and more secure, which can be found in Figure 8.1.

In summary, this section frames the discourse on energy efficiency and sustainability in 6G networks within the broader context of technological evolution, environmental responsibility, socioeconomic benefits, and global collaboration. As we embark on the journey toward 6G, it is imperative to prioritize these issues to build a future where connectivity is not only fast and reliable but also environmentally sustainable and socially equitable.

8.2 Literature Review

In [5], the 6G architecture aims to improve energy efficiency and promote environmental sustainability. The gadgets that will participate in the data stream as active nodes in the network. We are currently developing the 6G architecture with the goal of improving network economics and energy efficiency. The goal of reevaluating power management strategies is to create a 6G system that can withstand the test of time. The use of renewable energy sources is being pursued in an effort to lessen environmental impacts. A device's peak power consumption is proportional to its peak data rate. The impact of network operational expenditures causes a notable spike in energy consumption.

In [6], new integrated relative energy efficiency (IREE) statistic takes network capacity and traffic characteristics into account. 6G network technology evaluations make use of IREE-based green tradeoffs. Next-generation wireless networks will make use of IREE, a groundbreaking energy efficiency (EE) statistic. The IREE metric, will be used to improve network energy efficiency in 6G related network communication. Future energy-efficient network designs, such as those involving capacity augmentation and traffic matching, can look to it for guidance. Conventional energy efficiency (EE) cannot account for variations in capacity and traffic. The capacity mismatch property and wireless traffic can both be accurately captured by IREE.

In [7], one major problem is the energy problem that all these connected devices create. The power consumption of cell-free networks can be optimized by utilizing access point (AP) subsets. 6G networks tackle the challenges of energy crisis. Optimized power utilization and energy-efficient networks are essential for 6G's future. The Minimum Interference Pilot Allocation-Maximum Channel gain AP Selection (MIPA-MCAS) approach improves spectral efficiency by 3.39% at 95% of the locations.

In [8], improving 6G's radio subsystem energy efficiency should be its top priority. For this optimization problem, they used Lagrangian decomposition and the Kuhn-Munkres algorithm. Dynamic resource allocation aims to optimize active radio remote units (RRUs), subchannels, user selection, and power allocation simultaneously. Check that the radio subsystem in IoT networks is running as efficiently as possible. It makes IoT systems more energy efficient. There is a scarcity of renewable energy supplies, but 6G networks that support IoT devices can significantly improve their performance. Note that the tiny, power-efficient devices that make up the IoT consume a lot of energy.

In [9], energy efficiency and sustainability could be enhanced in 6G networks with the use of reconfigurable intelligent surfaces (RIS) technology. Sustainability is aided by RIS, which enables the programmable regulation of the wireless propagation environment. Establishes a RIS-based wireless environment as a service. Talked about the performance-enhancing domains where RIS-based connectivity is viable. Highlights areas where performance can be improved through the use of RIS-based connectivity.

In [10], the major objectives of 5G networks are energy efficiency, sustainability, and the application of artificial intelligence (AI) and machine learning (ML) techniques. This trend is anticipated to be maintained by 6G networks to guarantee sustainability. A measure of 5G network performance that considers power consumption using AI and ML techniques to boost efficiency in energy usage. Energy efficiency is of the utmost importance in the telecom industry. This work concludes that AI and ML techniques can reduce 5G's power usage.

In [11], with the help of the hybrid computational intelligence algorithm, 6G networks can use less power. The suggested algorithm shows competitive performance in terms of power savings (Multi-Path Gain and Noise Determination [MPGND]) using 6G-IBN's hybrid CI algorithm to reduce power consumption. Algorithms that show promise in this area are compared to one another. The suggested method outperforms similar current methods on a number of datasets.

8.3 Energy-Efficient Hardware in Today's Scenario (with 6G Networks)

In today's scenario, with the growth of 6G networks on the horizon, the quest for energy-efficient hardware has become increasingly important. Here is how energy-efficient hardware is being developed and deployed in the context of 6G networks:

- **Advanced Chip Design:** One of the primary focuses in energy-efficient hardware development is on chip design. Semiconductor manufacturers are constantly innovating to create chips that deliver high performance while consuming minimal power. This includes the development of specialized processors optimized for specific tasks, such as signal processing or ML, which are common in 6G network infrastructure.

- **Low-Power Components:** The components used in network infrastructure, such as transceivers, amplifiers, and antennas, are being designed with energy efficiency in mind. These components are optimized to minimize power consumption without compromising performance [12]. For example, advancements in radio frequency (RF) technology have led to the development of highly efficient RF amplifiers and transceivers for 6G networks.
- **Energy-Efficient Cooling Systems:** As network equipment becomes more powerful and densely packed, cooling becomes a significant factor in energy consumption. Energy-efficient cooling systems, such as liquid cooling or advanced air-cooling technologies, are being developed to keep hardware temperatures in check while minimizing energy usage. These cooling solutions are particularly important in data centers and centralized 6G network infrastructure.
- **Optical Communication:** In 6G networks, there is a growing emphasis on optical communication technologies due to their inherently lower energy consumption compared to traditional electrical communication. Optical fibers and photonic components are used to transmit data over long distances with minimal energy loss, making them ideal for high-speed and energy-efficient communication in 6G networks.
- **Integration of AI:** AI is increasingly being integrated into network hardware to optimize energy usage. AI algorithms can dynamically adjust network parameters based on traffic patterns, environmental conditions, and other factors to minimize energy consumption while maintaining performance [13–16]. For example, AI-driven power management systems can intelligently allocate resources and adjust transmission power levels in 6G base stations to optimize energy efficiency.
- **Energy Harvesting:** Another promising approach to energy-efficient hardware in 6G networks is energy harvesting, where ambient energy sources such as solar, wind, or vibration are harvested to power network equipment. Energy harvesting technologies are being integrated into network infrastructure to supplement or even replace traditional power sources, reducing reliance on grid electricity and enhancing sustainability.

Figure 8.2 discusses a detailed taxonomy of 6G. In summary, energy-efficient hardware plays an important role in the development and deployment of 6G networks. By using advanced chip design, low-power components, efficient cooling systems, optical communication technologies, AI-driven optimization, and energy harvesting techniques, users in the telecommunications industry can build energy-efficient 6G networks that meet the demands of tomorrow's connected world while minimizing environmental impact.

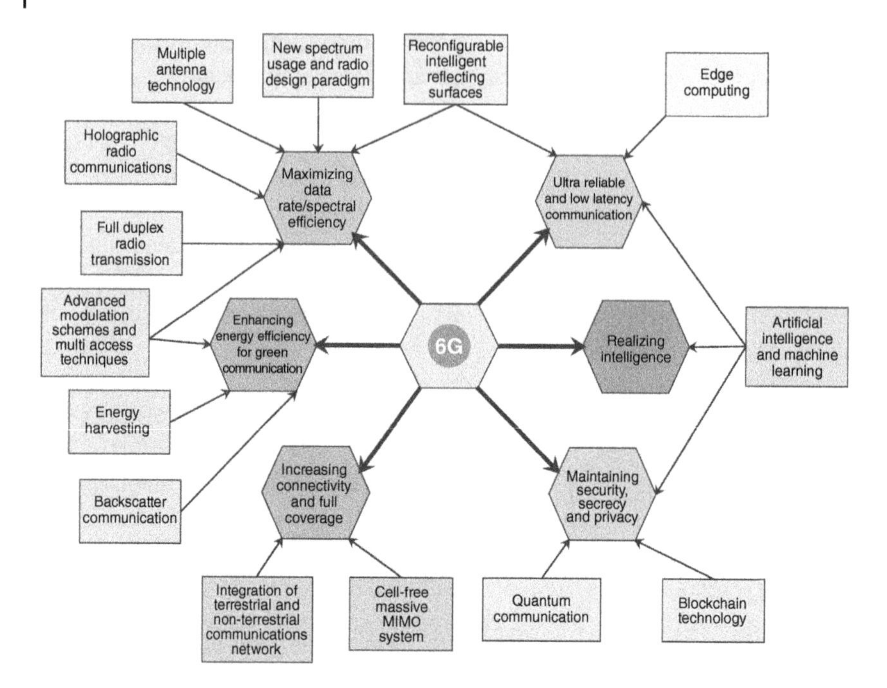

Figure 8.2 6G taxonomy.

8.4 Environmental Impact Assessment Using 6G Networks

Environmental impact assessment (EIA) is an important process for evaluating the potential environmental consequences of a proposed project or development [17, 18]. In the context of 6G networks, the deployment and operation of these advanced communication infrastructures can have both positive and negative environmental impacts. Here is how 6G networks can be utilized for EIA:

- **Data Collection and Monitoring:** 6G networks provide unique capabilities for collecting and transmitting real-time data from various environmental sensors and monitoring devices. These sensors can measure parameters such as air quality, water quality, noise levels, biodiversity, and climate conditions. By using the high data speeds, low latency, and huge connectivity of 6G networks, environmental data can be gathered more comprehensively and efficiently than ever before.
- **Remote Sensing and Imaging:** 6G networks can support advanced remote sensing and imaging technologies, such as satellite imaging, drones, and unmanned aerial vehicles (UAVs). These technologies can be used to assess

environmental conditions over large geographical areas, detect changes in land use, monitor habitat loss, and identify environmental risks. High-resolution imagery and remote sensing data transmitted over 6G networks enable detailed environmental mapping and analysis.

- **Predictive Modeling and Simulation:** With the high computational power and low latency of 6G networks, complex predictive modeling and simulation tools can be deployed for EIA [19–22]. These tools can simulate the potential effects of proposed projects or developments on the environment, such as urbanization, infrastructure construction, or resource extraction. By incorporating real-time data from environmental sensors and remote sensing platforms, these models can provide more accurate predictions of environmental impacts.
- **Crowdsourced Data and Citizen Science:** 6G networks enable widespread connectivity and engagement, allowing citizens to contribute to EIA through crowdsourced data collection and citizen science initiatives. Mobile applications, IoT devices, and social media platforms can be used to gather environmental observations, report environmental incidents, and engage communities in environmental monitoring and conservation efforts. By empowering citizens to participate in EIA processes, 6G networks facilitate greater transparency, accountability, and public involvement in environmental decision-making.

In summary, 6G networks provide transformative capabilities for EIA, enabling more complete data collection, advanced remote sensing, predictive modeling, citizen engagement, and collaborative research. By using the power of 6G technologies, users can enhance the accuracy, efficiency, and transparency of EIA processes, leading to more informed decision-making and better environmental outcomes.

8.5 Green Communication Technologies with 6G Networks

Green communication technologies are important for ensuring that the deployment and operation of 6G networks are environmentally sustainable. Figure 8.3 shows the architecture of green base stations.

Here is how 6G networks can integrate green communication technologies to minimize their ecological footprint:

- **Energy-Efficient Hardware Design:** 6G networks can incorporate energy-efficient hardware components, such as low-power base stations, energy-efficient antennas, and power-efficient processors. Advanced chip design,

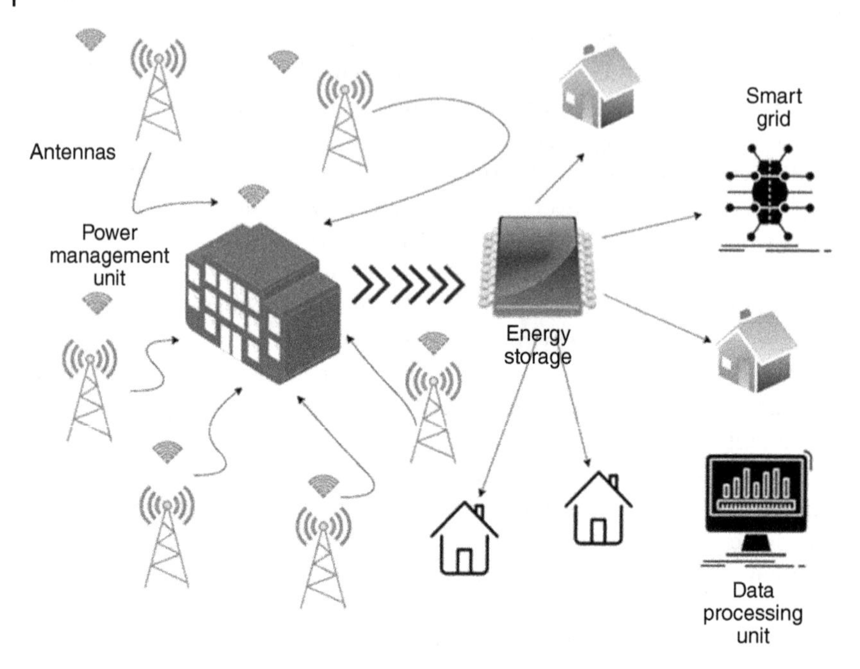

Figure 8.3 An architecture of green base stations.

including the use of low-power modes and optimized architectures, can significantly reduce energy consumption without compromising performance.

- **Renewable Energy Integration:** 6G networks can use renewable energy sources, such as solar, wind, and hydroelectric power, to meet their energy needs. Solar panels, wind turbines, and other renewable energy systems can be installed at base stations, data centers, and other network infrastructure locations to generate clean electricity on-site. Energy storage technologies, such as batteries and fuel cells, can store excess renewable energy for use during periods of high demand or low renewable energy availability.
- **Energy-Efficient Transmission Protocols:** 6G networks can employ energy-efficient transmission protocols and communication algorithms to minimize energy consumption during data transmission. Techniques such as adaptive modulation and coding, dynamic spectrum management, and interference coordination can optimize spectral efficiency and reduce energy usage in wireless communications.
- **Cognitive Radio and Spectrum Sharing:** Cognitive radio technology enables dynamic spectrum access and efficient spectrum sharing, allowing 6G networks to adaptively utilize available RF bands while minimizing interference and energy consumption. By opportunistically accessing unused spectrum bands

and dynamically adjusting transmission parameters, cognitive radio systems can improve spectral efficiency and reduce energy consumption in wireless networks.
- **Network Virtualization and Resource Optimization:** Network virtualization techniques enable the efficient allocation and sharing of network resources, including computing, storage, and communication resources. By virtualizing network functions and dynamically allocating resources based on traffic demand and environmental conditions, 6G networks can optimize resource utilization and energy efficiency.
- **Green Data Centers:** Data centers play an important role in supporting 6G networks by hosting cloud services, content delivery networks, and other network infrastructure. Green data center technologies, such as energy-efficient servers, cooling systems, and data center designs, can minimize the energy consumption and environmental impact of data center operations.
- **Lifecycle Management and Recycling:** 6G networks can adopt lifecycle management practices to reduce electronic waste and promote recycling of network equipment and devices. Designing network infrastructure for easy disassembly, component reuse, and material recycling can extend the lifecycle of network equipment and reduce the environmental impact of network deployment and decommissioning.

Hence, by integrating these green communication technologies into 6G networks, users can minimize the environmental footprint of telecommunications infrastructure and contribute to global efforts to mitigate climate change and promote sustainability.

8.6 Energy Harvesting and Wireless Power Transfer Using 6G Networks

Energy harvesting and wireless power transfer (WPT) are innovative technologies that can revolutionize the way energy is generated and distributed, particularly in the context of 6G networks. Here is how these technologies can be utilized with 6G networks:

- **Energy Harvesting:** Energy harvesting involves capturing and converting ambient energy from the environment into usable electrical power. With the proliferation of IoT devices and sensors in 6G networks, energy harvesting technologies can play an important role in powering these devices without the need for traditional batteries or wired power sources. Common sources of ambient energy for harvesting include solar radiation, kinetic energy (e.g., vibration and motion), thermal energy, and RF energy.

- **Solar Energy Harvesting:** Solar panels integrated into 6G infrastructure, such as base stations and IoT devices, can convert sunlight into electrical power to supplement or replace grid electricity. Solar-powered IoT sensors can be deployed in outdoor environments to monitor environmental conditions, track wildlife, and perform other tasks without depending on external power sources.
- **Kinetic Energy Harvesting:** Vibrational energy harvesters can capture energy from mechanical vibrations, such as those generated by machinery, vehicles, or human motion. In 6G networks, kinetic energy harvesting devices can be integrated into infrastructure components, wearable devices, and smart objects to generate power from movement and motion.
- **RF Energy Harvesting:** RF energy harvesting technology utilizes ambient RF signals, including Wi-Fi, cellular, and other wireless communication signals, to generate electrical power. RF energy harvesters can scavenge energy from the surrounding wireless environment and convert it into usable electricity to power IoT devices, sensors, and other wireless gadgets in 6G networks.
- **WPT:** WPT technologies enable the transmission of electrical power over a distance without the need for physical wires or connectors. In the context of 6G networks, WPT can be used to deliver power to remote or inaccessible locations, power mobile devices on the go, and charge electric vehicles wirelessly. There are several techniques for WPT, including electromagnetic induction, magnetic resonance, and RF beamforming.
- **Electromagnetic Induction:** Electromagnetic induction relies on the principle of magnetic coupling between two coils to transfer electrical power wirelessly. In 6G networks, electromagnetic induction can be used to charge IoT devices, sensors, and wearable gadgets embedded with receiver coils. Base stations or charging pads equipped with transmitter coils can wirelessly transfer power to nearby devices within a certain range.
- **Magnetic Resonance:** Magnetic resonance WPT technology enables efficient power transfer over longer distances compared to electromagnetic induction. Resonant coils tuned to the same frequency can transfer power wirelessly through electromagnetic resonance. In 6G networks, magnetic resonance WPT systems can be deployed to charge electric vehicles, power drones, and deliver energy to remote IoT devices located in challenging environments.
- **RF Beamforming:** RF beamforming techniques can focus RF signals into narrow beams to deliver wireless power over long distances with high efficiency. In 6G networks, RF beamforming can be used to transmit power to IoT devices, sensors, and infrastructure components located in remote or hard-to-reach areas. Base stations equipped with phased array antennas can dynamically adjust beamforming parameters to optimize power delivery and coverage [17–19].

Note that by integrating energy harvesting and WPT technologies into 6G networks, users can enhance energy efficiency, extend network coverage, and enable new applications and services that were previously not feasible with traditional wired power sources. These innovative technologies have the potential to revolutionize the way energy is generated, distributed, and consumed in the era of 6G connectivity.

8.7 Energy Optimization and Management Techniques Using 6G Networks

Energy optimization and management techniques are essential for maximizing the efficiency and sustainability of 6G networks. Here are several strategies that use the capabilities of 6G networks for energy optimization and management:

- **Dynamic Power Management:** 6G networks can implement dynamic power management techniques to adjust the power consumption of network components based on traffic demand, network conditions, and environmental factors. For example, base stations can dynamically adjust their transmit power levels and switch between low-power sleep modes during periods of low traffic to conserve energy.
- **Intelligent Resource Allocation:** AI-driven resource allocation algorithms can optimize the allocation of network resources, such as bandwidth, computing resources, and power, to minimize energy consumption while maximizing performance. By analyzing real-time traffic patterns and user behavior, 6G networks can dynamically allocate resources to where they are needed most, reducing energy waste and improving network efficiency.
- **Load Balancing and Traffic Offloading:** Load balancing techniques can distribute network traffic across multiple nodes and resources to prevent congestion and optimize resource utilization. In 6G networks, intelligent load balancing algorithms can dynamically route traffic to less congested nodes or offload traffic to neighboring cells or edge servers to reduce energy consumption and improve network efficiency.
- **Energy-Efficient Transmission Protocols:** 6G networks can deploy energy-efficient transmission protocols and communication algorithms to minimize energy consumption during data transmission. Techniques such as adaptive modulation and coding, beamforming, and interference coordination can optimize spectral efficiency and reduce energy usage in wireless communications.
- **Predictive Analytics and Forecasting:** Predictive analytics and forecasting models can anticipate future network conditions, traffic patterns, and energy

demand based on historical data and environmental factors. By predicting future energy requirements, 6G networks can proactively adjust network parameters, allocate resources, and optimize energy usage to meet anticipated demand while minimizing waste.

- **Renewable Energy Integration:** 6G networks can integrate renewable energy sources, such as solar, wind, and hydroelectric power, to meet their energy needs. Solar panels, wind turbines, and other renewable energy systems can be installed at base stations, data centers, and other network infrastructure locations to generate clean electricity on-site. Energy storage technologies, such as batteries and fuel cells, can store excess renewable energy for use during periods of high demand or low renewable energy availability [16, 23, 24].
- **Edge Computing and Edge Intelligence:** Edge computing enables data processing and analysis to be performed closer to the source of data generation, reducing the need for data transmission and lowering energy consumption. In 6G networks, edge intelligence can use edge computing capabilities to implement energy-efficient algorithms, perform localized processing, and optimize network operations at the network edge.

Hence, by implementing these energy optimization and management techniques, 6G networks can reduce energy consumption, improve network efficiency, and enhance sustainability while meeting the growing demand for high-speed, low-latency connectivity. These strategies use the advanced capabilities of 6G networks to create intelligent, adaptive, and energy-efficient communication infrastructures for the future.

8.8 Technical and Nontechnical Issues and Challenges Toward Energy Efficiency and Sustainability in 6G Networks

Achieving energy efficiency and sustainability in 6G networks presents a multitude of both technical and nontechnical issues and challenges. Let us discuss some of them:

8.8.1 Technical Issues and Challenges

- **Power Consumption of Advanced Technologies:** The integration of new technologies in 6G networks, such as massive Multiple-Input Multiple-Output (MIMO), beamforming, and terahertz communication, may lead to higher power consumption. Balancing the demand for high performance with energy efficiency is a significant technical challenge.

- **Complex Network Architecture:** 6G networks are expected to have a complex architecture with a dense deployment of small cells, edge servers, and IoT devices. Optimizing energy usage across this heterogeneous network architecture while maintaining continuous connectivity poses technical challenges.
- **Interference Management:** With the densification of network infrastructure in 6G networks, interference management becomes important for ensuring efficient spectrum utilization and minimizing energy waste. Advanced interference mitigation techniques are needed to address this challenge effectively.
- **Resource Allocation and Management:** Dynamic resource allocation and management are essential for optimizing energy efficiency in 6G networks. Efficient algorithms for allocating spectrum, computing resources, and power in real time are needed to adapt to changing network conditions and user demands.
- **Energy Harvesting Integration:** Integrating energy harvesting technologies, such as solar panels and kinetic energy harvesters, into 6G network infrastructure presents technical challenges related to power conversion, storage, and management. Ensuring continuous integration and interoperability with existing network components is important.
- **Standardization and Interoperability:** Developing standards for energy-efficient communication protocols, hardware components, and management systems is essential for ensuring interoperability and compatibility across different vendors and network operators. Standardization efforts need to address technical challenges related to energy optimization and sustainability.

8.8.2 Nontechnical Issues and Challenges

- **Regulatory and Policy Frameworks:** Establishing regulatory frameworks and policies that incentivize energy efficiency and sustainability in 6G networks is important. Policy challenges include spectrum allocation, energy efficiency standards, and environmental regulations that promote green technologies and practices.
- **Investment and Funding:** The transition to energy-efficient and sustainable 6G networks requires significant investment in research, development, and infrastructure deployment. Securing funding and investment from public and private sources to support these initiatives is a nontechnical challenge.
- **Consumer Awareness and Behavior:** Educating consumers about the environmental impact of their digital activities and promoting energy-efficient behaviors is essential for achieving sustainability goals. Overcoming consumer inertia and incentivizing green consumption habits pose nontechnical challenges in the transition to sustainable 6G networks.
- **Supply Chain Management:** Managing the supply chain for 6G network equipment and components to ensure sustainability and ethical sourcing

practices is a nontechnical challenge. Addressing issues such as electronic waste management, conflict minerals, and carbon footprint throughout the supply chain requires collaboration among users.
- **Skills and Training:** Developing a skilled workforce capable of designing, deploying, and managing energy-efficient 6G networks is important. Training programs and educational initiatives need to address the technical and nontechnical aspects of energy efficiency and sustainability in telecommunications.
- **International Collaboration:** Achieving global sustainability goals in 6G networks requires both international collaboration and cooperation. Overcoming geopolitical barriers, harmonizing regulations, and sharing best practices across borders are nontechnical challenges that need to be addressed.

In summary, achieving energy efficiency and sustainability in 6G networks requires addressing a wide range of technical and nontechnical issues and challenges. Collaborative efforts among users from industry, government, academia, and civil society are essential for overcoming these challenges and realizing the vision of green and sustainable communication infrastructures for the future.

8.9 Future Research Opportunities Toward Energy Efficiency and Sustainability in 6G Networks

Research opportunities toward energy efficiency and sustainability in 6G networks abound, spanning various technical and interdisciplinary domains [22, 24, 25]. Here are some future research directions:

- **Energy-Efficient Hardware Design:** Research can focus on developing novel hardware architectures, components, and materials that minimize energy consumption in 6G network equipment. This includes low-power processors, energy-efficient antennas, and advanced power management techniques tailored for 6G applications.
- **Renewable Energy Integration:** Investigating innovative methods for integrating renewable energy sources, such as solar, wind, and kinetic energy, into 6G network infrastructure. Research can discuss optimal deployment strategies, energy storage solutions, and hybrid renewable energy systems to enhance energy sustainability.
- **Smart Power Management Algorithms:** Developing intelligent algorithms and protocols for dynamic power management in 6G networks. Research can discuss ML, optimization, and AI-driven approaches to optimize energy usage, predict future energy demand, and adaptively adjust network parameters for energy efficiency.

- **Energy-Aware Network Planning and Optimization:** Research can focus on energy-aware network planning and optimization techniques that consider energy consumption as a primary design metric alongside traditional performance metrics. This includes optimal deployment of network elements, spectrum allocation strategies, and load-balancing algorithms to minimize energy consumption while meeting performance requirements.
- **Cross-Layer Optimization:** Investigating cross-layer optimization techniques that jointly optimize energy consumption across different network layers, including physical, media access control (MAC), and application layers. Research can discuss synergies between networking protocols, communication algorithms, and energy management strategies to achieve holistic energy efficiency improvements.
- **Green Data Centers and Edge Computing:** Research can focus on designing energy-efficient data center architectures, cooling systems, and resource management strategies tailored for 6G networks, as well as on investigating the integration of edge computing and fog computing paradigms to offload processing tasks from centralized data centers and reduce energy consumption in network operations.
- **Life Cycle Assessment (LCA) and Environmental Impact Analysis:** Conducting LCA studies and environmental impact analyses to quantify the energy consumption, carbon footprint, and environmental consequences of 6G network deployment and operation. Research can identify hotspots for energy consumption, assess the environmental benefits of green technologies, and inform policy decisions toward sustainability goals.
- **Consumer Behavior and Adoption Patterns:** Research can discuss consumer behavior and adoption patterns related to energy-efficient technologies and sustainable practices in 6G networks. Understanding user preferences, motivations, and barriers toward adopting green communication technologies can inform the design of effective incentive mechanisms and awareness campaigns.
- **Policy and Regulatory Studies:** Investigating the role of policy interventions, regulatory frameworks, and economic incentives in promoting energy efficiency and sustainability in 6G networks. Research can analyze the effectiveness of policy measures, benchmark best practices across jurisdictions, and provide recommendations for policymakers to make a conducive environment for green telecommunications.

In summary, future research opportunities toward energy efficiency and sustainability in 6G networks are diverse and multidisciplinary, spanning hardware design, network optimization, renewable energy integration, consumer behavior, policy analysis, and more. Collaborative efforts among researchers, industry users, policymakers, and civil society are essential for addressing these challenges and advancing the state-of-the-art toward greener and more sustainable communication infrastructures for the future.

8.10 Summary

The search for energy efficiency and sustainability in 6G networks represents a paramount imperative for the telecommunications industry and society at large. As we embark on the journey toward 6G, it is essential to prioritize environmental responsibility alongside technological advancement, recognizing the profound impact that communication infrastructures have on the planet. The transition to 6G networks provides unique opportunities to integrate innovative solutions and green technologies that minimize energy consumption, use renewable resources, and promote environmental sustainability. From energy-efficient hardware design to intelligent resource allocation algorithms, from renewable energy integration to green data center architectures, there exists a wealth of possibilities to drive energy efficiency and sustainability in 6G networks. Moreover, achieving energy efficiency and sustainability in 6G networks requires collaborative efforts among industry users, policymakers, academia, and civil society. International standards, regulations, and best practices must be established to incentivize green technologies, make innovations, and ensure global alignment toward sustainability goals. In the face of mounting environmental challenges and the urgent need to reduce carbon emissions, the deployment of 6G networks presents a unique opportunity to make the way toward a greener and more sustainable future. By prioritizing environmental stewardship alongside technological progress, we can ensure that 6G networks not only redefine connectivity but also serve as catalysts for positive environmental change, shaping a world where communication technologies are in harmony with the planet.

References

1 Hashmi, M.S. and Rehmani, M.H. (2015). Greening cellular networks: a survey, some research issues and challenges. *IEEE Communications Surveys & Tutorials* 17 (1): 3–20.

2 Islam, M.H., Hossain, E., and Alamri, A. (2016). Energy efficiency in 5G wireless communication networks: a survey of techniques and guidelines. *IEEE Access* 4: 6193–6209.

3 Lei, X., Hossain, E., and Niyato, D. (2016). Energy-efficient 5G small-cell networks: an overview, state of the art, and future directions. *IEEE Access* 4: 5895–5910.

4 You, L., Huang, K., and Chae, K. (2017). Energy-efficient resource allocation for Mobile edge computing in 5G heterogeneous networks. *IEEE Access* 5: 6712–6724.

5 Weide, W., Chien-Sheng, Y., Fu, I. et al. (2022). Revisiting the system energy footprint and power efficiency on the way to sustainable 6G systems. *IEEE Wireless Communications*. https://doi.org/10.1109/mwc.2022.10003083.

6 Tao, Y., Shunqing, Z., Xiaojing, C., and Xin, W. (2023). A novel energy efficiency metric for next-generation green wireless communication network design. *IEEE Internet of Things Journal*. https://doi.org/10.1109/JIOT.2022.3210166.

7 Ashu, T., Shalli, R., Sahil, G. et al. (2022). Energy aware resource control mechanism for improved performance in future green 6G networks. *Computer Networks*. https://doi.org/10.1016/j.comnet.2022.109333.

8 Ansere, J.A., Mohsin, K., Izaz, K., and Aman, M.N. (2023). Dynamic resource optimization for energy-efficient 6G-IoT ecosystems. *Sensors*. https://doi.org/10.3390/s23104711.

9 Strinati, E.C., Alexandropoulos, G.C., Henk, W. et al. (2021). Reconfigurable, intelligent, and sustainable wireless environments for 6G smart connectivity. *IEEE Communications Magazine*. https://doi.org/10.1109/MCOM.001.2100070.

10 Chochliouros, I.P., Michail-Alexandros, K., Anastasia, S.S. et al. (2021). Energy efficiency concerns and trends in future 5G network infrastructures. *Energies*. https://doi.org/10.3390/EN14175392.

11 Eldrandaly, K.A., Laila, A.-F., Mohamed, A.-B. et al. (2021). Green communication for sixth-generation intent-based networks: an architecture based on hybrid computational intelligence algorithm. *Wireless Communications and Mobile Computing*. https://doi.org/10.1155/2021/9931677.

12 Sarkar, M., Misra, S., and Sarma, H.K.D. (2017). A survey on energy-efficient techniques in wireless sensor networks. *International Journal of Distributed Sensor Networks* 13 (2), 1550147717699727.

13 Islam, S.M.R., Kwak, D., Kabir, M.H. et al. (2017). Energy-efficient small cell deployment: a deep reinforcement learning approach. *IEEE Transactions on Vehicular Technology* 66 (10): 9489–9502.

14 Nair, M.M. and Tyagi, A.K. (2023). Blockchain technology for next-generation society: current trends and future opportunities for smart era. In: *Blockchain Technology for Secure Social Media Computing*. https://doi.org/10.1049/PBSE019E_ch11.

15 Tyagi, A.K., Kumari, S., Chidambaram, N., and Sharma, A. (2024). Engineering applications of Blockchain in this smart era. In: *Enhancing Medical Imaging with Emerging Technologies*. https://doi.org/10.4018/979-8-3693-5261-8.ch011.

16 Ajanthaa, L., Seranmadevi, R., Sree, P.H., and Tyagi, A.K. (2024). Engineering applications of artificial intelligence. In: *Enhancing Medical Imaging with Emerging Technologies*. https://doi.org/10.4018/979-8-3693-5261-8.ch010.

17 Ding, Z., Dai, L., and Poor, H.V. (2017). Energy-efficient massive MIMO systems: a survey. *IEEE Wireless Communications* 24 (1): 10–15.

18 Lei, X., Niyato, D., and Wang, P. (2018). Energy-efficient and cost-effective 5G small-cell networks: an auction-based mechanism. *IEEE Journal on Selected Areas in Communications* 36 (1): 206–218.

19 Gupta, R., Jain, R., and Vaszkun, G. (2015). Survey of important issues in underwater wireless sensor networks. *IEEE Transactions on Mobile Computing* 14 (1): 1–16.

20 Tyagi, A.K., Kukreja, S., Richa, and Sivakumar, P. (February 2024). Role of Blockchain Technology in Smart era: a review on possible smart applications. *Journal of Information & Knowledge Management*. https://doi.org/10.1142/S0219649224500321.

21 Tyagi, A.K. and Tiwari, S. (2024). The future of artificial intelligence in Blockchain applications. In: *Machine Learning Algorithms Using Scikit and TensorFlow Environments*. IGI Global. https://doi.org/10.4018/978-1-6684-8531-6.ch018.

22 Kumari, S., Thompson, A., and Tiwari, S. (2024). 6G-enabled internet of things-artificial intelligence-based digital twins: cybersecurity and resilience. In: *Emerging Technologies and Security in Cloud Computing*. IGI Global. https://doi.org/10.4018/979-8-3693-2081-5.ch016.

23 Tiwari, S., Chidambaram, N., and Tyagi, A.K. (2024). Need of 21st century: a sustainable environment together with a smart environment. In: *Operational Research for Renewable Energy and Sustainable Environments*. IGI Global. https://doi.org/10.4018/978-1-6684-9130-0.ch009.

24 Singh, R., Tyagi, A.K., and Arumugam, S.K. (2024). Imagining the sustainable future with industry 6.0: a smarter pathway for modern society and manufacturing industries. In: *Machine Learning Algorithms Using Scikit and TensorFlow Environments*. IGI Global. https://doi.org/10.4018/978-1-6684-8531-6.ch016.

25 Nair, M.M. and Tyagi, A.K. (2023). 6G: technology, advancement, barriers, and the future. In: *6G-Enabled IoT and AI for Smart Healthcare*. CRC Press.

9

Performance Evaluation and Optimization in 6G Networks

9.1 Introduction to Performance Evaluation and Optimization in 6G Networks

The growth of sixth-generation (6G) networks moves to a new era of connectivity, promising revolutionary advancements in communication technology. Building upon the foundation laid by its predecessors, 6G is poised to deliver ultra-fast data rates, ultralow latency, and continuous connectivity for an ever-expanding array of devices and applications [1, 2]. However, realizing the full potential of 6G requires a thorough understanding of its performance characteristics and the implementation of effective optimization strategies.

Here, Figure 9.1 shows the 6G network performance management using machine learning (ML) in detail.

Here, Figure 9.2 depicts several performance metrices for 6G communication. Performance evaluation and optimization are important aspects of network design and deployment, ensuring that networks meet the stringent requirements of emerging applications like virtual reality, augmented reality, autonomous vehicles, and the Internet of Things (IoT) [3–5]. In the context of 6G networks, which are expected to support unique levels of data traffic and diverse use cases, these issues become even more important.

This chapter aims to provide a detailed overview of performance evaluation and optimization in 6G networks. We will discuss the unique challenges and opportunities presented by 6G's complex architecture, including the integration of technologies like terahertz communication, massive Multiple-Input Multiple-Output (MIMO), and intelligent networking protocols. Additionally, we will discuss the key metrics used to assess the performance of 6G networks, including throughput, latency, reliability, and energy efficiency.

Furthermore, we will discuss state-of-the-art optimization techniques tailored for 6G networks. These include ML algorithms, artificial intelligence (AI), and network slicing, which enable dynamic resource allocation, traffic management,

6G-Enabled Technologies for Next Generation: Fundamentals, Applications, Analysis and Challenges,
First Edition. Amit Kumar Tyagi, Shrikant Tiwari, Shivani Gupta, and Anand Kumar Mishra.

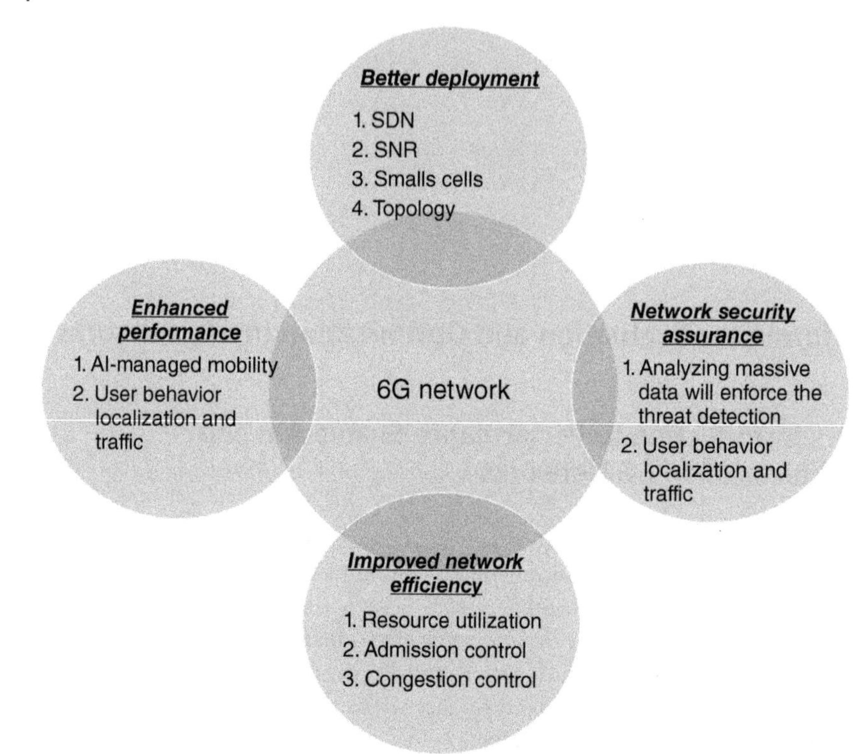

Figure 9.1 6G network performance management using ML.

and quality of service optimization. By using these advanced techniques, network operators can maximize the efficiency and reliability of 6G networks while meeting the diverse needs of users and applications.

Through this exploration, we aim to provide valuable insights for researchers, engineers, and policymakers involved in the development and deployment of 6G technology. By understanding the complexity of performance evaluation and optimization in 6G networks, users can collaborate to realize the full potential of this transformative technology while moving into a new era of connectivity and innovation.

9.2 Literature Review

In [6], combining non-orthogonal transmission with layer-based scalable video caching improves performance. The design of transmission systems and the placement of caches are the primary factors that determine caching performance.

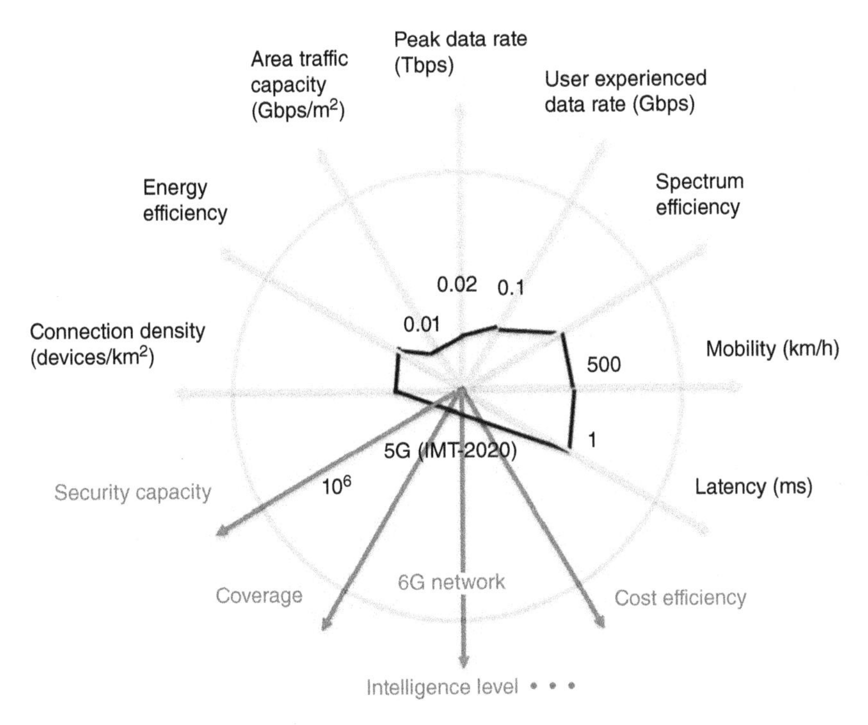

Figure 9.2 Performance metrices for 6G communication.

Scalable video caching allows for an improvement in 6G network video quality. When non-orthogonal transmission is used with layer-based caching, caching performance is enhanced. The transmission system and caching site architecture largely dictate caching performance.

In [7], a federated learning and shard blockchain hybrid designed for high throughput. Plan for a shard-based, high-throughput blockchain that can run on 6G networks. An approach to security performance analysis protects the network from Byzantine attacks. Potentially limiting scalability is the needs for security and decentralization. To protect against Byzantine attacks, the security performance analysis method is employed.

In [8], the Dynamic Reconfigurable Large-scale Surfaces (DRLS) system is efficient because it decreases the migration rate, failure rate, and offload delay. Using Actor-Critic learning and Lyapunov Optimization can increase system efficiency. The core idea of Dynamic Terahertz Reconfigurable Large-scale Surfaces is to use deep reinforcement learning to reduce offloading delays. Increasing system efficiency by teaching an agent to optimize offloading, there is a complete disregard for user mobility and the mobile edge computing environment. Reducing download delays and relocation expenditures is the objective of the suggested approach.

In [9], the suggested approaches have shown to be very effective in both translations and forecasts. In terms of both explanations and predictions, the suggested methods perform better. Detailed blueprints for classified parts of technology that could be valuable and improve the logic behind AI calculations which is an immediate priority for 6G IoT networks were discussed.

In [10], to maximize efficiency, use the Quantum Approximate Optimisation Algorithm (QAOA) for routing optimization. Quantum ML techniques applied to wireless mesh networks. The optimization of routing in wireless mesh networks could be achieved with the help of quantum computers. One potential option that could work here is the quantum approximate optimisation method, or QAOA. Note that current computer capabilities may restrict some 6G features. Many efforts are underway to explore the possibility of quantum computers facilitating 6G wireless networks.

In [11], evaluation and improvement of building evacuation systems as a whole was the objective. This method depends on theoretical decomposition and has substantial computational backing. With this data, we can better understand how to make evacuation networks operate. Analysis becomes more challenging on three-dimensional graphs due to a transient stochastic process.

9.3 Channel Modeling and Propagation Characteristics in 6G Networks

As we journey toward the realization of 6G networks, it becomes increasingly evident that understanding the intricate dynamics of channel modeling and propagation characteristics is important. The efficacy of communication systems in 6G hinges on a robust comprehension of how electromagnetic waves propagate through various environments, interact with obstacles, and undergo transformations due to novel technologies integrated into the network architecture [5, 12]. Now we can find 6G channel requirements in Figure 9.3.

Channel modeling serves as the foundational framework for predicting signal behavior, enabling the design and optimization of communication systems. In the context of 6G, which promises to deliver unique data rates, ultralow latency, and ubiquitous connectivity, accurate channel models are indispensable for assessing network performance and developing innovative solutions to address emerging challenges.

Propagation characteristics in 6G networks include a diverse array of phenomena, ranging from traditional line-of-sight and multipath propagation to the integration of advanced technologies like terahertz communication, massive MIMO, and intelligent beamforming. Understanding how these factors influence signal propagation is essential for optimizing network coverage, capacity, and reliability

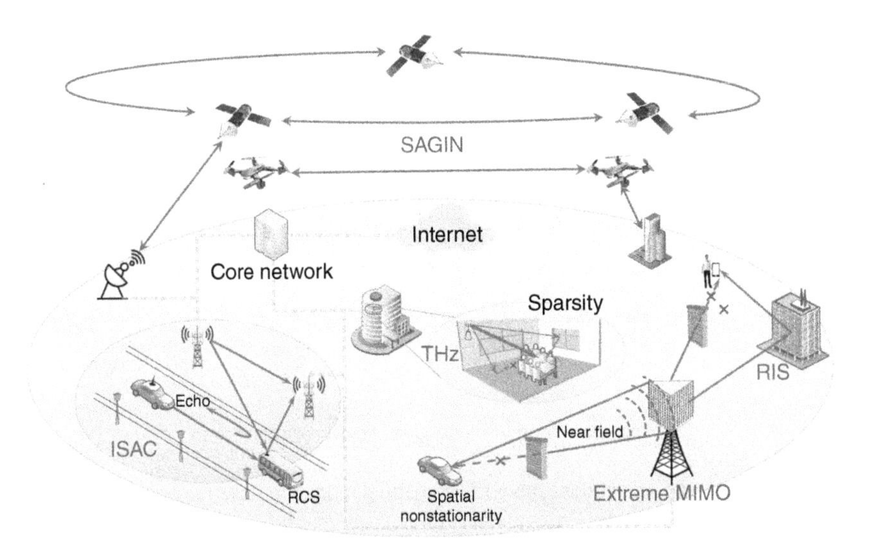

Figure 9.3 6G channel requirements.

in diverse environments, including urban areas, indoor spaces, and challenging outdoor scenarios.

In this chapter, we embark on a complete exploration of channel modeling and propagation characteristics in 6G networks. We will discuss the intricacies of existing channel models, highlighting their strengths, limitations, and applicability in different scenarios. Additionally, we will examine the impact of emerging technologies on propagation characteristics, including the exploitation of millimeter-wave and terahertz frequencies, as well as the deployment of intelligent antenna arrays for beamforming and spatial multiplexing.

Furthermore, we will discuss the implications of channel modeling and propagation characteristics on key performance metrics like signal strength, interference, and achievable data rates in 6G networks. By gaining insights into these factors, network designers and engineers can develop innovative strategies to mitigate channel impairments, optimize resource allocation, and enhance the overall quality of service. Now data and model dual-driven THz channel modeling flowchart can be found in Figure 9.4.

Hence, through this exploration, we aim to provide a complete understanding of channel modeling and propagation characteristics in 6G networks, laying the groundwork for the development of robust and efficient communication systems that can unlock the full potential of next-generation wireless technology.

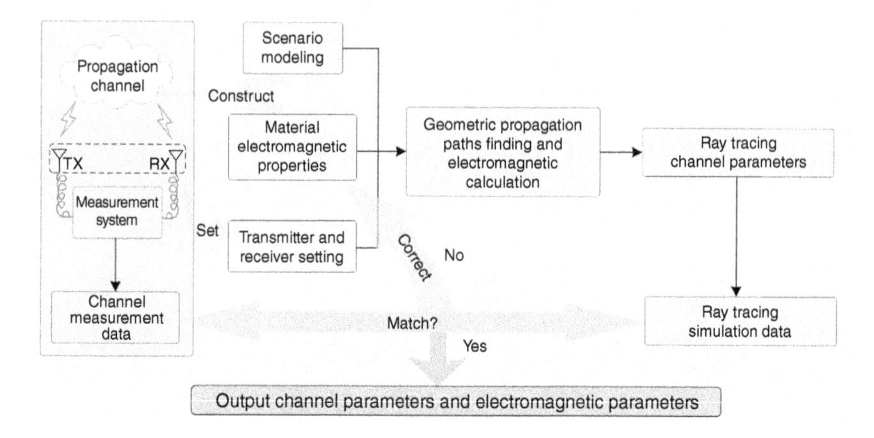

Figure 9.4 Data and model dual-driven THz channel modeling flowchart.

9.4 Performance Metrics and Quality of Service (QoS) in 6G Networks

In the landscape of 6G networks, where ultra-fast data rates, ultralow latency, and ubiquitous connectivity are the norm, the concept of performance metrics and Quality of Service (QoS) takes on heightened important [13–16]. As 6G aims to cater to a plethora of emerging applications like augmented reality, autonomous vehicles, and large number of IoT deployments, ensuring stringent performance standards becomes imperative to meet the diverse needs of users and applications.

Here, Figure 9.5 explains several 6G use cases, requirement, and metrices. Performance metrics serve as the measure by which the effectiveness and efficiency of 6G networks are measured. These metrics include a wide array of parameters, including throughput, latency, reliability, energy efficiency, spectral efficiency, and coverage. Each metric plays an important role in determining the overall performance and user experience in 6G networks, necessitating complete evaluation and optimization strategies.

QoS in 6G networks refers to the ability of the network to deliver a predefined level of performance to users and applications, as stipulated by service level agreements (SLAs) or user expectations. QoS includes various aspects like data rate guarantees, latency bounds, packet loss rates, and availability requirements, all of which are essential for supporting diverse use cases with varying demands and priorities.

In this chapter, we embark on an in-depth exploration of performance metrics and QoS in 6G networks. We will explain the importance of each performance

Figure 9.5 6G use cases, requirement, and metrices.

metric in the context of 6G's ambitious objectives, highlighting the challenges and opportunities associated with achieving optimal performance across different scenarios and deployment scenarios. Furthermore, we will discuss the methodologies and techniques employed to assess and optimize performance metrics and QoS in 6G networks. This includes the use of advanced scheduling algorithms, traffic management strategies, resource allocation techniques, and network slicing paradigms to dynamically adapt to changing network conditions and user requirements. Additionally, we will discuss the interplay between performance metrics, QoS, and emerging technologies like ML, AI, and edge computing, which offer novel approaches to enhancing network efficiency and user experience in 6G networks. Through this complete exploration, we aim to provide valuable insights into the complex ecosystem of performance metrics and QoS in 6G networks, enabling users to design, deploy, and optimize networks that can meet the diverse demands of the future digital era.

9.5 Optimization Algorithms and Techniques for 6G Networks

In the evolutionary journey toward 6G networks, optimization algorithms and techniques emerge as fundamental tools for using the full potential of this transformative technology [17]. With promises of ultra-fast data rates, ultralow latency, and a large number of device connectivity, 6G networks introduce unique complexities that necessitate innovative optimization strategies to ensure efficient resource utilization, optimal performance, and enhanced user experience.

Optimization algorithms play an important role in addressing the multifaceted challenges inherent in 6G networks, ranging from spectrum allocation and resource management to energy efficiency and QoS optimization. These algorithms use a diverse array of mathematical and computational techniques, including convex optimization, metaheuristic algorithms, reinforcement learning, and evolutionary algorithms, to dynamically adapt to changing network conditions and user demands.

In this chapter, we embark on a detailed exploration of optimization algorithms and techniques tailored for 6G networks. We will discuss the details of various optimization problems encountered in 6G, including:

- **Spectrum Allocation and Management:** Efficient spectrum utilization is important for accommodating the explosive growth in data traffic and supporting diverse use cases in 6G networks [18, 19]. Optimization algorithms can facilitate dynamic spectrum sharing, cognitive radio techniques, and spectrum aggregation to maximize spectral efficiency while mitigating interference and spectrum fragmentation.
- **Resource Allocation and Management:** Optimal resource allocation is essential for maximizing network capacity, minimizing latency, and ensuring fair distribution of resources among users and applications. Through techniques like dynamic resource allocation, power control, and user association optimization, 6G networks can achieve optimal utilization of bandwidth, power, and computational resources.
- **Energy Efficiency:** With the rapid growth of energy-intensive devices and applications in 6G networks, energy efficiency emerges as an important optimization objective. Optimization algorithms can facilitate dynamic sleep scheduling, energy-aware routing, and transmit power optimization to minimize energy consumption while maintaining QoS requirements.
- **Quality of Service (QoS) Optimization:** Guaranteeing stringent QoS requirements is paramount for supporting diverse applications with varying performance criteria in 6G networks. Optimization algorithms can enable dynamic QoS provisioning, admission control, and traffic prioritization to meet the latency, reliability, and throughput demands of different services.

Furthermore, we will discuss the convergence between optimization algorithms and emerging technologies like AI, ML, and edge computing, which offer novel approaches to network optimization and management in 6G environments. Through this complete examination, we aim to provide insights into the state-of-the-art optimization algorithms and techniques for 6G networks, empowering researchers, engineers, and network operators to design and deploy efficient and resilient communication systems that can unlock the full potential of next-generation wireless technology.

9.6 Technical, Nontechnical Issues and Challenges Toward 6G Networks-Based Optimization and Performance Evaluation

The journey toward realizing 6G networks is fraught with a numerous of technical and non-technical issues and challenges that demand careful consideration and innovative solutions. As we strive to unlock the transformative potential of ultra-fast data rates, ultralow latency, and ubiquitous connectivity promised by 6G, it becomes imperative to address a diverse array of difficulties spanning technological, regulatory, societal, and ethical domains.

9.6.1 Technical Challenges

- **Spectrum Availability and Utilization:** We secure sufficient spectrum resources to accommodate the burgeoning demand for data-intensive applications in 6G networks that faces a major challenge [19]. Spectrum scarcity, regulatory constraints, and interference mitigation techniques necessitate innovative spectrum-sharing mechanisms and dynamic spectrum access policies.
- **Complex Network Architecture:** The integration of heterogeneous technologies like terahertz communication, massive MIMO, and intelligent networking protocols in 6G networks introduces unique complexity. Designing scalable, interoperable, and energy-efficient network architectures requires novel approaches to network planning, deployment, and management.
- **Propagation and Channel Modeling:** We understand the propagation characteristics of millimeter-wave and terahertz frequencies, as well as their interactions with environmental factors, which are essential for optimizing coverage, capacity, and reliability in 6G networks. Accurate channel modeling and propagation prediction techniques are important for designing robust communication systems.
- **Resource Allocation and Management:** We efficient resource allocation and management are important for maximizing network capacity, minimizing latency, and ensuring fair distribution of resources among users and applications. Dynamic resource allocation algorithms, power control mechanisms, and load balancing strategies are needed to address the diverse and dynamic nature of 6G traffic.

9.6.2 Nontechnical Challenges

- **Regulatory and Policy Frameworks:** We implement regulatory frameworks and spectrum policies across different jurisdictions is essential for using innovation and investment in 6G technology. Addressing spectrum allocation, licensing, privacy, and security issues requires collaboration between governments, regulatory bodies, and industry stakeholders.
- **Privacy and Security:** Protecting user privacy and protecting against cybersecurity threats are important issues in 6G networks [20–22]. As the volume and diversity of data transmitted over networks increase, ensuring end-to-end encryption, authentication mechanisms, and data protection measures becomes important to maintaining trust and integrity in the network ecosystem.
- **Societal and Ethical Implications:** The deployment of 6G networks raises broader societal and ethical issues related to digital inclusion, accessibility, and the impact on employment and socioeconomic disparities. Bridging the digital divide, promoting digital literacy, and ensuring equitable access to technology are essential for realizing the inclusive potential of 6G networks.
- **Environmental Sustainability:** We address the environmental footprint of 6G networks, including energy consumption, electronic waste, and carbon emissions, which are imperative for achieving long-term sustainability. Adopting energy-efficient technologies, promoting green network design principles, and implementing recycling and waste management initiatives are essential for mitigating the environmental impact of 6G deployment.

In summary, the optimization and performance evaluation of 6G networks are contingent upon overcoming a multitude of technical and nontechnical challenges. By addressing these challenges through collaborative efforts, innovative solutions, and forward-thinking policies, we can make the way for the successful deployment of 6G networks that can revolutionize communication, drive economic growth, and empower societies worldwide.

9.7 Future Research Opportunities Toward 6G Networks-Based Optimization and Performance Evaluation

As the telecommunications industry marches forward toward the era of 6G networks, several research opportunities emerge to tackle the evolving challenges and complexities inherent in optimizing performance and evaluating network efficacy. Possible sue cases of 6G can found in Figure 9.6.

By focusing on these key areas, researchers can contribute to shaping the future of wireless communication technology and unlocking the full potential of 6G networks:

Figure 9.6 6G network: use cases.

- **Machine Learning and Artificial Intelligence:** Using advanced ML algorithms and AI techniques presents a promising avenue for optimizing 6G networks. Future research could discuss the application of AI-driven approaches for dynamic resource allocation, intelligent traffic management, and predictive network optimization, enabling networks to adapt in real time to changing conditions and user demands [21–28].
- **Edge Computing and Network Slicing:** We investigate the convergences between edge computing and network slicing that provide exciting opportunities to enhance the performance and efficiency of 6G networks. Future research could focus on developing innovative edge computing architectures, intelligent network slicing frameworks, and efficient resource orchestration mechanisms to support diverse applications with varying performance requirements [27, 28].

- **Terahertz Communication and Beyond:** We discuss deeper into the potential of terahertz communication and beyond that provide intriguing avenues for expanding the capabilities of 6G networks. Future research could discuss novel transmission techniques, antenna designs, and propagation models to overcome the challenges associated with terahertz frequencies, unlocking new possibilities for ultrahigh-speed communication and sensing applications.
- **Security and Privacy:** We address the evolving security and privacy challenges in 6G networks requires ongoing research and innovation. Future research could focus on developing robust encryption algorithms, secure authentication mechanisms, and privacy-preserving protocols to protect user data and protect against emerging cybersecurity threats, ensuring the trustworthiness and integrity of 6G communication systems [19–21, 29, 30].
- **Quantum Communication:** We discuss the potential of quantum communication technologies that provide intriguing prospects for enhancing the security and efficiency of 6G networks. Future research could investigate the integration of quantum key distribution, quantum cryptography, and quantum-resistant algorithms into 6G networks, paving the way for ultra-secure communication and information exchange.
- **Testbeds and Field Trials:** We establish detailed testbeds and conducting real-world field trials that are essential for validating theoretical concepts and evaluating the performance of 6G networks in practical scenarios. Future research could focus on deploying experimental testbeds, conducting large-scale trials, and collecting empirical data to assess the scalability, reliability, and real-world impact of novel optimization techniques and performance evaluation methodologies.

Note that by working on these future research opportunities, the telecommunications community can drive innovation, accelerate progress, and make the way for the successful deployment of 6G networks that redefine the boundaries of wireless communication and empower the digital transformation of society.

9.8 An Open Discussion

A discussion on 6G networks optimization and performance evaluation. Here are a few prompts to get us started:

- **Key Optimization Challenges:** What do you see as the most pressing optimization challenges in the development and deployment of 6G networks? Are there specific areas, like spectrum management, resource allocation, or energy efficiency, that require particular attention?

- **Cutting-Edge Technologies:** Which emerging technologies do you believe will play an important role in optimizing the performance of 6G networks? Are there any innovative approaches, like ML, edge computing, or quantum communication, that you find particularly promising?
- **Performance Metrics and Evaluation:** How do you think performance metrics and evaluation methodologies should evolve to meet the demands of 6G networks? Are there new metrics or evaluation techniques that need to be developed to accurately assess the performance of next-generation wireless networks?
- **Interdisciplinary Collaboration:** Given the complexity and multidisciplinary nature of optimizing 6G networks, how important do you think interdisciplinary collaboration is? What role can researchers from different fields, like telecommunications, computer science, mathematics, and social sciences, play in advancing the optimization and performance evaluation of 6G networks?
- **Societal and Ethical Implications:** As we attempt to optimize 6G networks for enhanced performance and efficiency, what considerations should be taken into account regarding the societal and ethical implications? How can we ensure that 6G networks are deployed in a manner that promotes digital inclusion, privacy, security, and environmental sustainability?

In last, we encourage researchers, scientists to feel free to share their thoughts on 6G topics, or any other aspects related to 6G networks optimization and performance evaluation.

9.9 Summary

The optimization and performance evaluation of 6G networks represent a multifaceted endeavor that includes a wide range of technical, regulatory, societal, and ethical issues. As we move toward the realm of 6G, it becomes increasingly apparent that unlocking the full potential of this transformative technology requires innovative approaches, interdisciplinary collaboration, and a holistic understanding of the complex challenges and opportunities that lie ahead. Key optimization challenges, like spectrum management, resource allocation, energy efficiency, and security, demand rigorous research and development efforts to devise effective solutions that can meet the stringent requirements of 6G networks. Emerging technologies, including ML, AI, edge computing, and quantum communication, offer promising avenues for enhancing network performance, efficiency, and reliability in the 6G era. Furthermore, the evolution of performance metrics and evaluation methodologies is essential to accurately assess the performance of 6G networks and drive continuous improvement. New metrics, evaluation techniques, and testbeds are needed to capture the unique characteristics and capabilities of next-generation wireless networks and ensure that they meet the diverse needs of users and applications. Moreover, as we attempt to optimize 6G networks for enhanced performance and efficiency, it is imperative to consider the broader societal and ethical implications of our actions. Hence, ensuring digital inclusion, protecting user privacy, enhancing cybersecurity, promoting environmental sustainability, and addressing ethical issues are integral to deploying 6G networks in a manner that benefits humanity and uses positive societal impact.

References

1 Zhang, Y., Zhang, J., and Zheng, K. (2019). A survey on 5G networks for the internet of things: communication technologies and challenges. *IEEE Access* 7: 36509–36528.

2 Zhang, H., Huang, J., Dai, H., and Zhao, X. (2019). Performance analysis and optimization of NOMA-based heterogeneous vehicular networks. *IEEE Transactions on Vehicular Technology* 68 (12): 12014–12029.

3 Wang, L., Zhang, Y., and Li, G.Y. (2020). Optimal deployment of UAV-enabled mobile base stations for minimizing network cost. *IEEE Transactions on Wireless Communications* 19 (1): 318–331.

4 Luong, N.C., Wang, P., and Niyato, D. (2020). Intelligent reflecting surface-assisted MIMO broadcasting for green and secure vehicular communication. *IEEE Transactions on Vehicular Technology* 69 (10): 12050–12063.

5 Alabbasi, A., Ahmed, S., and Radaydeh, R. (2020). QoE-aware service function chain placement and VNF instantiation in edge computing networks. *IEEE Transactions on Network and Service Management* 17 (4): 2448–2463.

6 Junchao, M., Lingjia, L., Bodong, S. et al. (2023). Performance analysis and optimization for layer-based scalable video caching in 6G networks. *IEEE ACM Transactions on Networking*. https://doi.org/10.1109/tnet.2022.3222931.

7 Qinyin, N., Zhang, L., Xiaorong, Z., and Inayat, A. (2022). A novel design method of high throughput blockchain for 6G networks: performance analysis and optimization model. *IEEE Internet of Things Journal*. https://doi.org/10.1109/JIOT.2022.3194889.

8 Saravanan, J., Tamilarasan, A.K., Rajmohan, R. et al. (2022). Performance analysis of digital twin edge network implementing bandwidth optimization algorithm. *International Journal of Computing and Digital Systems*. https://doi.org/10.12785/ijcds/120170.

9 Balamurugan, A., Sureshkumar, S., Putta, S. et al. (2022). Optimization of e-learning and performance using iot and 6g technology. *Journal of Pharmaceutical Negative Results*. https://doi.org/10.47750/pnr.2022.13.s09.060.

10 Helen, U., Pablo, P.-M., David, G.-R., and Jose, M.M. (2022). Multi-objective routing optimization for 6G communication networks using a quantum approximate optimization algorithm. *Sensors*. https://doi.org/10.3390/s22197570.

11 Andrea, W., Laurence, W., and J., MacGregor, S. (2012). Performance & optimization of M/G/c/c building evacuation networks. *Journal of Mathematical Modelling and Algorithms*. https://doi.org/10.1007/S10852-012-9192-6.

12 Islam, S.R., Kwak, D., Kabir, M.H. et al. (2017). Energy-efficient small cell deployment: A deep reinforcement learning approach. *IEEE Transactions on Vehicular Technology* 66 (10): 9489–9502.

13 Park, J., Yu, R., Kim, S.L., and Kim, Y. (2020). Performance analysis of NOMA-based machine-type communications in Industrial Internet of things. *IEEE Transactions on Industrial Informatics* 16 (4): 2618–2627.

14 Al-Askar, H., Al-Otaibi, Y., and Abdullah, A. (2020). Joint beamforming and jamming for secure communication in energy harvesting-based cognitive radio networks. *IEEE Transactions on Vehicular Technology* 69 (1): 1184–1195.

15 Wu, Q., Rui, Y., and Wang, Y. (2019). Optimal uplink resource allocation in UAV-enabled IoT networks. *IEEE Internet of Things Journal* 7 (7): 5845–5856.

16 Liu, H., Chen, X., and Guo, S. (2019). Distributed beamforming in SWIPT-enabled multiuser NOMA systems: an NOMA game perspective. *IEEE Transactions on Vehicular Technology* 68 (5): 4992–5005.

17 Dai, H., Zhang, H., and Wu, Q. (2019). Optimal user association and resource allocation in Millimeter-wave-based ultra-dense networks. *IEEE Transactions on Vehicular Technology* 68 (7): 7093–7106.

18 Wang, Q., Zhang, Z., and Xia, X. (2018). Performance analysis of UAV-enabled wireless power transfer networks. *IEEE Transactions on Wireless Communications* 17 (8): 5439–5451.

19 Zhang, Z., Zheng, B., and Cheng, J. (2020). Joint optimization of UAV trajectory and communication resource allocation for multiuser wireless-powered communication networks. *IEEE Transactions on Vehicular Technology* 69 (4): 4190–4203.

20 Chen, M., Han, T., and Xia, Y. (2019). Energy-efficient resource allocation for NOMA-based multiuser multi-relay networks with SWIPT. *IEEE Transactions on Green Communications and Networking* 3 (1): 97–107.

21 Xu, Z., Yin, C., and Yang, L. (2020). A survey on Mobile edge computing: architecture, computation offloading, and resource allocation. *IEEE Access* 8: 59599–59614.

22 Nair, M.M. and Tyagi, A.K. (2023). Blockchain technology for next-generation society: current trends and future opportunities for smart era. In: *Blockchain Technology for Secure Social Media Computing*. https://doi.org/10.1049/PBSE019E_ch11.

23 Tyagi, A.K., Kumari, S., Chidambaram, N., and Sharma, A. (2024). Engineering applications of Blockchain in this smart era. In: *Enhancing Medical Imaging with Emerging Technologies*. https://doi.org/10.4018/979-8-3693-5261-8.ch011.

24 Ajanthaa, L., Seranmadevi, R., Sree, P.H., and Tyagi, A.K. (2024). Engineering applications of artificial intelligence. In: *Enhancing Medical Imaging with Emerging Technologies*. https://doi.org/10.4018/979-8-3693-5261-8.ch010.

25 Tyagi, A.K., Kukreja, S., Richa, and Sivakumar, P. (February 2024). role of Blockchain Technology in Smart era: A review on possible smart applications. *Journal of Information & Knowledge Management*. https://doi.org/10.1142/S0219649224500321.

26 Tyagi, A.K. and Tiwari, S. (2024). The future of artificial intelligence in blockchain applications. In: *Machine Learning Algorithms Using Scikit and TensorFlow Environments*. IGI Global. https://doi.org/10.4018/978-1-6684-8531-6.ch018.

27 Kumari, S., Thompson, A., and Tiwari, S. (2024). 6G-enabled internet of things-artificial intelligence-based digital twins: cybersecurity and resilience. In: *Emerging Technologies and Security in Cloud Computing*. IGI Global. https://doi.org/10.4018/979-8-3693-2081-5.ch016.

28 Nair, M.M. and Tyagi, A.K. (2023). 6G: Technology, Advancement, Barriers, and the Future. In: *6G-Enabled IoT and AI for Smart Healthcare*. CRC Press.

29 Nair, M.M. and Tyagi, A.K. (2021). Privacy: history, statistics, policy, Laws, preservation and threat analysis. *Journal of Information Assurance & Security*. 16 (1): 24–34. 11p.

30 Sharma, S., Popli, R., Singh, S. et al. (2024). The role of 6G technologies in advancing smart city applications: opportunities and challenges. *Sustainability*. 16 (16): 7039.

10

Network Planning and Deployment for 6G-Based Systems in Real World

10.1 Introduction to 6G-Based Systems

The introduction to 6G-based systems serves as a foundational overview of the next generation of wireless communication technology, highlighting its importance, evolution, and anticipated impact on various industries and everyday life [1–3]. It begins by contextualizing 6G within the broader trajectory of wireless communication advancements, tracing the evolution from 1G to the forthcoming 6G.

10.1.1 Evolution of Wireless Communication Technologies

The introduction outlines the evolutionary path of wireless communication technologies, highlighting the transformative impact each generation has had on connectivity, mobility, and societal interactions. Starting with the introduction of 1G analog cellular networks in the early 1980s, the narrative progresses through the digitalization of communication with 2G, the emergence of mobile data and internet access with 3G, the growth of broadband wireless with 4G/LTE, and the ushering in of the era of interconnected devices and machine-to-machine communication with 5G. Figure 10.1 shows a major shift in 6G communications with 6G's Vision and requirements.

Here, Figure 10.2 shows a development vision for 6G wireless communication in detail.

10.1.2 Anticipated Features and Capabilities of 6G

Building upon the foundation laid by its predecessors, 6G promises to deliver a quantum leap in wireless communication capabilities, surpassing the limitations of existing technologies and enabling a plethora of novel applications and services. Figure 10.3 shows its different use cases and capabilities. The introduction discusses the potential features and capabilities of 6G networks, which may include:

6G-Enabled Technologies for Next Generation: Fundamentals, Applications, Analysis and Challenges, First Edition. Amit Kumar Tyagi, Shrikant Tiwari, Shivani Gupta, and Anand Kumar Mishra.

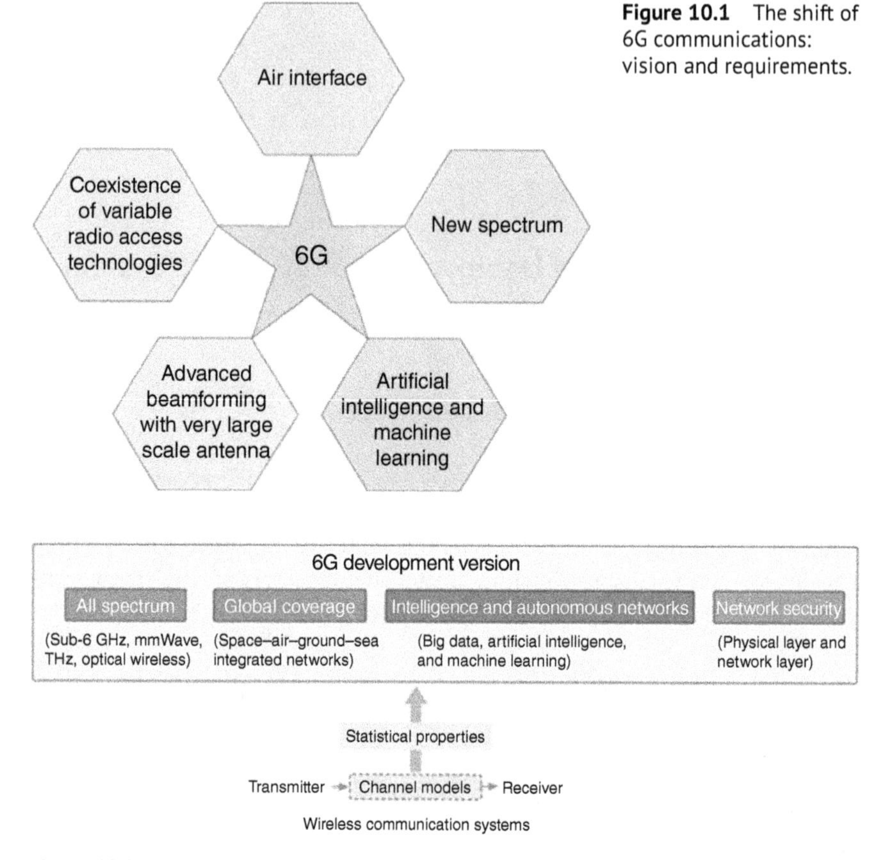

Figure 10.1 The shift of 6G communications: vision and requirements.

Figure 10.2 Development vision for 6G wireless communication.

- **Ultrahigh Speeds**: 6G is expected to achieve unique data rates, potentially reaching terabits per second, enabling lightning-fast downloads, real-time streaming of high-definition content, and immersive virtual reality (VR) experiences.
- **Ultralow Latency**: With latency reduced to a fraction of a millisecond, 6G networks will facilitate instantaneous communication and response times, enabling mission-important applications such as remote surgery, autonomous vehicles, and augmented reality (AR) gaming.
- **Massive Connectivity**: 6G will support a large number of connected devices per unit area, paving the way for the rapid growth of the Internet of Things (IoT), smart cities, and ubiquitous sensing and monitoring applications.
- **Hyper-connectivity**: Beyond traditional human-centric communication, 6G will enable continuous connectivity between humans, machines, and the

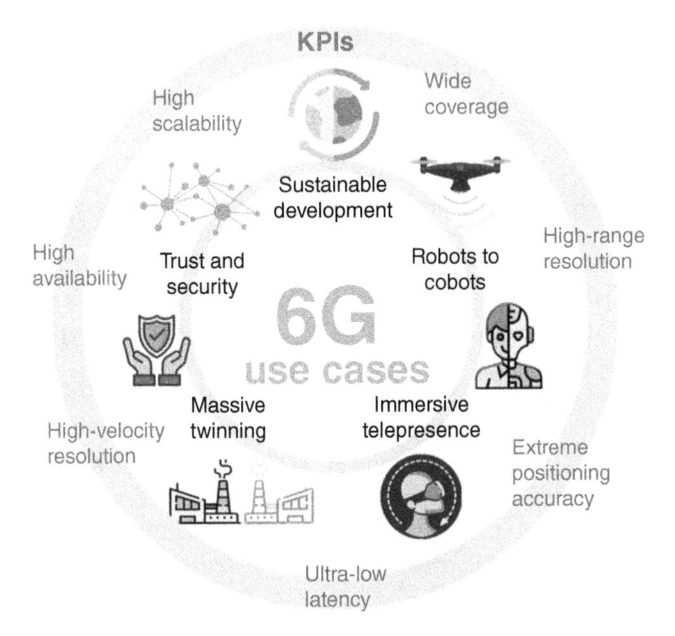

Figure 10.3 Use cases of 6G.

environment, using a symbiotic relationship between the physical and digital worlds.

- **Advanced Spectrum Utilization**: Using higher-frequency bands, advanced beamforming techniques, and spectrum-sharing strategies, 6G networks will achieve greater spectral efficiency and resilience, mitigating the challenges posed by spectrum scarcity and interference.

In summary, the introduction sets the stage for a deeper exploration of 6G-based systems, highlighting the transformative potential of this next-generation technology and its implications for society, economy, and technological innovation. By discussing the evolutionary trajectory of wireless communication technologies and envisioning the future capabilities of 6G networks, the introduction makes the way for a detailed examination of the technical, regulatory, and societal dimensions of 6G deployment and adoption.

10.2 Literature Review

In [4], user demands on 6G networks can be evaluated using a new core-network traffic model. When compared to IPoWDM, the equipment and CAPEX costs associated with deploying Flex-BVT are lower. To help with demand estimation, this chapter introduces a new core-network traffic model. Routing, configuration, and spectrum assignment (RCSA) algorithm development is underway. With Flex-BVT, you may cut down on equipment, boost traffic, and save money on capital expenditures.

In [5], when planning operational and capital expenditures, radio network architecture is important. B5G/6G technology enables smart cities with high data rates. All the information you need to plan the B5G/6G network, including the technologies involved. This chapter presents the initial planning and radio deployment concepts for B5G/6G NR. Facilitates the rollout of B5G and 6G networks in highly populated regions. In smart cities, applications that demand a lot of data can run. Only frequencies between 2.6 GHz and 28 GHz are covered. The planning process makes use of the UMa and UMi propagation models.

In [6], capacity estimation and dynamic allocation are components of the B5G/6G network planning process. The planner takes into account big-MIMO, smart cities, and several frequency bands. B5G and 6G network planning with key enabling technologies is the subject of this work. For the purpose of developing smart cities, a very dependable planning tool is suggested. Adaptability to various smart city designs and advanced traffic prediction features were also considered while building the model.

In [7], 6G is more intelligent, efficient, and effective with an AI-driven deployment plan. Cloud-based AI training and edge-based inference are both part of the distributed strategy. Exploration of potential new uses for 6G is ongoing. The AI deployment method that the industry recommends for 6G, makes it easier to enhance 6G networks' intelligence, efficiency, and efficacy. A decentralized architecture that makes use of cloud-based AI training and inference capabilities at the network's periphery. It is unclear how network intelligence will be deployed in advance of 6G.

In [8], together, a 3D MAP deployment and an optimization method for user associations are now under development. In 6G networks, user-MAP association is achieved by the use of deep multi-agent reinforcement learning. To optimize the deployment of MAP in dynamic settings, the iterative technique is applied. An effective and efficient 3D MAP deployment for improving dynamic network coverage was able to establish Quality of Service (QoS)-aware user association using a deep multi-agent reinforcement learning strategy. Limits on mobility and interference are important in a dynamic network.

10.3 Network Planning and Dimensioning Strategies Using 6G Networks

Network planning and dimensioning are important phases in the deployment of wireless communication systems, ensuring optimal utilization of resources while meeting user demands for coverage, capacity, and quality of service [9, 10]. With the growth of 6G networks, characterized by ultrahigh speeds, ultralow latency, and huge connectivity, traditional planning methodologies need to be reimagined to use the full potential of this transformative technology.

1. **Spectrum Issues**: 6G networks are expected to operate across a wide range of frequency bands, including millimeter-wave (mmWave) frequencies and terahertz (THz) bands. Effective spectrum management and allocation are essential for maximizing spectral efficiency and mitigating interference. Dynamic spectrum-sharing techniques, cognitive radio technologies, and spectrum aggregation strategies will play a vital role in optimizing spectrum utilization and ensuring coexistence with other wireless systems.
2. **Coverage and Capacity Planning**: Given the unique propagation characteristics of higher-frequency bands used in 6G networks, such as mmWave, coverage planning becomes more challenging. Advanced propagation models accounting for atmospheric absorption, reflection, and diffraction effects are required to accurately predict coverage areas and signal penetration. Moreover, dense small cell deployments, beamforming techniques, and relay nodes are employed to enhance coverage and capacity in urban environments and indoor scenarios.
3. **HetNet Deployment**: 6G networks will use a heterogeneous network (HetNet) architecture, comprising a mix of macrocells, small cells, and relay nodes to provide ubiquitous coverage and capacity. HetNet planning involves optimizing the deployment density and placement of network elements based on traffic demand, user density, and quality of service requirements. Self-organizing network (SON) algorithms and machine learning techniques are employed for autonomous network optimization and adaptation to dynamic traffic patterns.
4. **Ultra-dense Deployment**: Ultra-dense deployment of network infrastructure, characterized by a high density of access points and base stations, is a hallmark of 6G networks. This approach is essential for achieving ultrahigh data rates, ultralow latency, and continuous connectivity in densely populated urban areas and high-traffic hotspots. Network densification strategies, such as small cell clustering, multi-tiered architecture, and network slicing, are employed to efficiently allocate resources and meet user demands.
5. **Traffic Engineering and QoS Provisioning**: Traffic engineering and QoS provisioning are important aspects of network dimensioning in 6G systems, ensuring optimal resource allocation and user experience. Dynamic traffic management mechanisms, such as traffic steering, load balancing, and QoS-aware routing, are employed to prioritize important applications, minimize latency, and maximize throughput. Additionally, network slicing

enables the creation of virtualized network instances tailored to specific service requirements, further enhancing QoS provisioning and resource isolation.

In summary, network planning and dimensioning strategies for 6G networks require a holistic approach that integrates advanced spectrum management techniques, HetNet deployment strategies, ultra-dense infrastructure deployment, and dynamic traffic engineering mechanisms. By using the unique capabilities of 6G technology and implementing innovative planning methodologies, operators can ensure the efficient utilization of resources, continuous connectivity, and superior user experience in the era of ultra-connectivity and digital transformation.

10.4 Deployment of 6G-Based Systems in Real-World Sectors

The deployment of 6G-based systems in real-world sectors is faced to revolutionize various industries, enhancing efficiency, productivity, and innovation across diverse domains [10–12]. Here is an exploration of how 6G can be deployed in several sectors:

10.4.1 Telecommunications

- 6G networks will introduce ultrahigh speeds, ultralow latency, and huge connectivity, enabling continuous communication and data exchange between devices.
- Telecommunications companies will use 6G to offer enhanced services such as AR, VR, and holographic communication.
- Network operators will deploy 6G infrastructure to support futuristic applications such as immersive telepresence, remote surgery, and autonomous vehicle communication.

10.4.2 Healthcare

- 6G networks will enable real-time remote healthcare services, facilitating telemedicine consultations, remote patient monitoring, and surgical procedures conducted by robotic systems.
- Healthcare providers can deploy 6G-enabled IoT devices for continuous health monitoring, personalized treatment delivery, and predictive analytics for disease prevention.
- Advanced 6G connectivity will support the integration of medical imaging technologies, enabling real-time transmission of high-resolution images for diagnosis and treatment planning.

10.4.3 Manufacturing

- In the manufacturing sector, 6G networks will enable the deployment of smart factories with autonomous robots, collaborative robots (cobots), and interconnected IoT devices.
- Real-time monitoring and control of production processes, predictive maintenance, and supply chain optimization will be facilitated by 6G connectivity, enhancing productivity and efficiency.
- Edge computing capabilities integrated with 6G networks will support low-latency data processing, enabling real-time decision-making and adaptive manufacturing processes.

10.4.4 Transportation

- 6G technology will revolutionize transportation systems with ultralow latency communication, enabling vehicle-to-everything (V2X) communication for enhanced safety, traffic management, and autonomous driving.
- Smart infrastructure powered by 6G networks will facilitate real-time traffic monitoring, adaptive traffic signal control, and predictive analytics for congestion management.
- Advanced sensor technologies and AI-powered systems using 6G connectivity will enable continuous integration of public transportation systems, ride-sharing services, and autonomous vehicles.

10.4.5 Smart Cities

- 6G networks will serve as the backbone of smart city initiatives, enabling interconnected urban infrastructure, intelligent transportation systems, and efficient resource management.
- Real-time data collection, analysis, and decision-making facilitated by 6G connectivity will enhance public safety, environmental monitoring, and energy efficiency.
- Smart grid systems using 6G technology will enable dynamic energy management, demand-response mechanisms, and integration of renewable energy sources for sustainable urban development.

10.4.6 Entertainment and Media

- 6G networks will revolutionize the entertainment industry with immersive experiences such as 3D holographic displays, interactive gaming, and AR applications.

- Content delivery platforms will use 6G connectivity for ultrahigh-definition streaming, personalized recommendations, and real-time social interactions.
- Live events and performances can be broadcasted with unique realism and interactivity, blurring the lines between physical and virtual experiences.

10.4.7 Education

- 6G networks will transform education delivery, enabling immersive learning experiences through virtual classrooms, interactive simulations, and AR textbooks.
- Remote education platforms using 6G connectivity will facilitate access to high-quality educational resources, personalized learning experiences, and collaborative projects.
- Real-time language translation, AI-powered tutoring systems, and adaptive learning algorithms will enhance the effectiveness and accessibility of education across diverse demographics.

In summary, the deployment of 6G-based systems in real-world sectors holds the promise of unlocking new opportunities for innovation, productivity, and societal advancement across diverse industries. By using the transformative capabilities of 6G technology, organizations can drive digital transformation, address societal challenges, and create a more connected and sustainable future.

10.5 Coverage and Capacity Optimization in 6G Networks

Coverage and capacity optimization in 6G networks are important aspects of network planning and deployment, ensuring efficient utilization of resources while meeting the increasing demands for high-speed connectivity and continuous user experiences. Here is an overview of strategies for optimizing coverage and capacity in 6G networks:

10.5.1 Antenna Technologies

- Advanced antenna technologies, such as massive multiple input multiple output (MIMO) and beamforming, are key enablers for optimizing coverage and capacity in 6G networks.

- Massive MIMO systems employ a large number of antennas at the base station to spatially multiplex multiple users, improving spectral efficiency and increasing network capacity.
- Beamforming techniques focus radio frequency energy in specific directions, enabling targeted coverage and enhanced signal strength, particularly in urban and indoor environments.

10.5.2 Spectrum Management

- Effective spectrum management is essential for optimizing coverage and capacity in 6G networks, especially given the utilization of higher-frequency bands, including mmWave and THz frequencies.
- Dynamic spectrum-sharing techniques, spectrum aggregation, and carrier aggregation enable efficient utilization of available spectrum resources, maximizing throughput and capacity.
- Cognitive radio technologies facilitate adaptive spectrum utilization, allowing 6G networks to dynamically adjust frequency bands and transmit power based on real-time environmental conditions and traffic demands.

10.5.3 Small Cell Deployments

- Deploying small cells in dense urban areas and high-traffic hotspots is a key strategy for enhancing coverage and capacity in 6G networks.
- Small cells, including femtocells, picocells, and microcells, increase network density, reduce interference, and improve signal strength, particularly in indoor environments and areas with high user concentrations.
- HetNet deployments combining macrocells with small cells optimize coverage and capacity across varying deployment scenarios, ensuring continuous connectivity and consistent quality of service.

10.5.4 Network Densification

- Network densification strategies involve deploying a higher density of base stations and access points to increase coverage and capacity in 6G networks.
- Ultra-dense deployment of network infrastructure, particularly in urban areas and high-demand locations, ensures efficient resource utilization and minimizes signal attenuation, leading to improved network performance and user experience.
- Small cell clustering, coordinated multipoint (CoMP) transmission, and interference mitigation techniques are employed to optimize network densification and enhance spectral efficiency.

10.5.5 Network Slicing

- Network slicing enables the creation of virtualized network instances tailored to specific service requirements, allowing operators to optimize coverage and capacity for diverse use cases and applications.
- By dynamically allocating network resources, including bandwidth, latency, and computing resources, network slicing enables efficient utilization of infrastructure while ensuring end-to-end QoS for different services and applications.
- Network slicing facilitates the coexistence of multiple virtual networks on a shared physical infrastructure, enabling operators to monetize diverse use cases and vertical markets while optimizing coverage and capacity.

10.5.6 Edge Computing

- Edge computing capabilities integrated with 6G networks enhance coverage and capacity optimization by reducing latency, improving response times, and offloading computation-intensive tasks closer to end users [13–16].
- By deploying edge computing nodes at the network edge, operators can optimize content delivery, support low-latency applications, and enable real-time data processing for IoT devices and mission-important services.
- Edge caching, content delivery networks (CDNs), and distributed processing architectures are employed to enhance coverage and capacity optimization in 6G networks, particularly for latency-sensitive applications and services.

In summary, coverage and capacity optimization are essential for unlocking the full potential of 6G networks, enabling continuous connectivity, high-speed data transmission, and immersive user experiences across diverse deployment scenarios. By using advanced antenna technologies, spectrum management techniques, small cell deployments, network densification strategies, network slicing, and edge computing capabilities, operators can optimize coverage and capacity in 6G networks, meeting the evolving demands of digital consumers and enabling new use cases and applications in the era of ultra-connectivity and digital transformation.

10.6 Deployment Issues and Challenges Toward Implementing 6G-Based Systems in Real-World Sectors

Implementing 6G-based systems in real-world sectors presents several deployment issues and challenges that need to be addressed to ensure successful adoption and integration. Here are some of the key challenges:

- **Infrastructure Upgrade and Investment**: Deploying 6G networks requires significant infrastructure upgrades, including the installation of new base stations, antennas, and backhaul systems. The cost of infrastructure deployment and the need for extensive investment pose financial challenges for network operators and users.
- **Spectrum Availability and Regulation**: Securing spectrum resources suitable for 6G deployment is an important challenge, particularly in higher-frequency bands such as mmWave and THz frequencies. Spectrum allocation, licensing, and regulatory frameworks must be established to enable efficient spectrum utilization while ensuring coexistence with existing wireless systems.
- **Interference and Signal Propagation**: Higher-frequency bands used in 6G networks, such as mmWave and THz frequencies, are susceptible to signal attenuation and propagation losses due to obstacles and environmental conditions. Overcoming interference and optimizing signal propagation in urban environments, indoor spaces, and rural areas present technical challenges for coverage and capacity optimization.
- **Power Consumption and Energy Efficiency**: 6G networks, characterized by ultra-dense deployments and huge connectivity, require significant energy consumption, posing challenges for sustainability and environmental impact. Developing energy-efficient network architectures, power-saving mechanisms, and renewable energy integration strategies is essential to mitigate power consumption and minimize carbon footprint.
- **Security and Privacy Issues**: The rapid growth of connected devices, IoT sensors, and data-intensive applications in 6G networks raises issues about cybersecurity, data privacy, and, network vulnerabilities. Ensuring robust security protocols, encryption mechanisms, and privacy-enhancing technologies is important to protecting user data and preventing unauthorized access or malicious attacks.
- **Compatibility and Interoperability**: Integrating 6G networks with existing infrastructure, legacy systems, and heterogeneous technologies presents compatibility and interoperability challenges. Ensuring continuous migration paths, backward compatibility, and standards compliance is essential to facilitate smooth transition and interoperability between 6G and legacy systems.
- **Skill Gap and Workforce Training**: Deploying and managing 6G networks require specialized skills and expertise in areas such as network planning, optimization, cybersecurity, and edge computing. Addressing the skill gap and providing training programs for network engineers, technicians, and IT professionals is essential to build a workforce capable of deploying and maintaining 6G-based systems.

- **Regulatory and Policy Frameworks**: Developing regulatory frameworks, standards, and policies to govern 6G deployment, spectrum allocation, data governance, and privacy protection is important for ensuring compliance with legal requirements and using innovation while addressing societal issues.
- **Global Collaboration and Standardization**: 6G deployment requires global collaboration, cooperation, and standardization efforts among industry users, governments, regulatory bodies, and standardization organizations. Harmonizing technical specifications, interoperability standards, and regulatory frameworks is essential to enable continuous deployment and interoperability of 6G networks worldwide.

In summary, deploying 6G-based systems in real-world sectors entails addressing a wide range of technical, regulatory, economic, and societal challenges. Overcoming these challenges requires collaborative efforts, innovative solutions, and adaptive strategies to realize the transformative potential of 6G technology and unlock new opportunities for digital innovation, economic growth, and societal advancement.

10.7 Technical/Nontechnical/Legal Issues Toward Implementing 6G-Based Systems in Real-World Sectors

Implementing 6G-based systems in real-world sectors presents a myriad of technical, nontechnical, and legal challenges [16–19]. Here is an overview of each category:

10.7.1 Technical Issues

- **Spectrum Availability and Utilization**: Securing spectrum resources suitable for 6G deployment, especially in higher-frequency bands such as mmWave and THz, while ensuring efficient spectrum utilization without causing interference is a significant technical challenge.
- **Network Infrastructure**: Upgrading existing infrastructure and deploying new infrastructure, including base stations, antennas, and backhaul systems, to support the requirements of 6G networks poses technical challenges in terms of cost, scalability, and compatibility with legacy systems.
- **Interference and Signal Propagation**: Overcoming signal attenuation and propagation losses in higher-frequency bands, as well as mitigating interference in urban environments and indoor spaces, are technical challenges that need to be addressed for optimizing coverage and capacity.

- **Energy Efficiency**: Managing energy consumption in ultra-dense deployments and huge connectivity scenarios requires innovative energy-saving techniques, renewable energy integration, and efficient network architectures to ensure sustainability and minimize environmental impact.
- **Security and Privacy**: Developing robust security protocols, encryption mechanisms, and privacy-enhancing technologies to protect user data, prevent unauthorized access, and mitigate cybersecurity threats are important technical challenges in 6G deployment.

10.7.2 Nontechnical Issues

- **Regulatory and Policy Frameworks**: Establishing regulatory frameworks, standards, and policies governing spectrum allocation, data governance, privacy protection, and network deployment is essential for ensuring compliance with legal requirements and using innovation.
- **Skills and Workforce Development**: Addressing the skill gap and providing training programs for network engineers, technicians, and IT professionals to deploy, manage, and maintain 6G networks requires investment in workforce development and capacity building.
- **Ethical and Social Implications**: Addressing ethical and social implications, such as digital divide, algorithmic bias, data sovereignty, and societal impact, require multidisciplinary collaboration, user engagement, and ethical issues in technology design and deployment.
- **Global Collaboration and Standardization**: Harmonizing technical specifications, interoperability standards, and regulatory frameworks through global collaboration and standardization efforts among industry users, governments, and standardization organizations is important for enabling continuous deployment and interoperability of 6G networks worldwide.

10.7.3 Legal Issues

- **Spectrum Licensing and Allocation**: Obtaining spectrum licenses and navigating regulatory processes for spectrum allocation, including auctions, fees, and usage rights, involves legal issues and compliance with national and international regulations.
- **Data Protection and Privacy Laws**: Compliance with data protection and privacy laws, such as the general data protection regulation (GDPR), requires implementing data governance policies, user consent mechanisms, and data security measures to protect user privacy and prevent data breaches.
- **Intellectual Property Rights**: Addressing intellectual property rights, including patents, trademarks, and copyrights, related to 6G technologies

and innovations requires legal agreements, licensing arrangements, and enforcement mechanisms to protect intellectual property and incentivize innovation.

- **Liability and Risk Management**: Identifying liability issues and managing risks associated with 6G deployment, including network failures, service disruptions, and cybersecurity breaches, involves legal issues, insurance policies, and contractual arrangements to mitigate legal exposure and financial losses.
- **Competition and Antitrust Laws**: Ensuring compliance with competition and antitrust laws, including fair competition, market dominance, and anticompetitive practices, requires regulatory scrutiny and legal assessments to prevent monopolistic behavior and promote market competition.

In summary, implementing 6G-based systems in real-world sectors involves addressing a complex array of technical, nontechnical, and legal issues spanning spectrum management, network infrastructure, regulatory compliance, privacy protection, workforce development, ethical issues, intellectual property rights, liability management, and competition laws. Addressing these challenges requires collaborative efforts, innovative solutions, and adherence to legal and regulatory frameworks to realize the transformative potential of 6G technology while ensuring societal benefits and regulatory compliance.

10.8 Important Challenges Toward Implementing 6G-Based Systems in Real-World Sectors

Implementing 6G-based systems in real-world sectors presents several important challenges that need to be addressed to ensure successful deployment and integration [20–24]. Here are some of the most important challenges:

- **Technical Complexity**: Developing and deploying 6G networks entails overcoming technical challenges related to spectrum management, signal propagation, interference mitigation, energy efficiency, and security. The complexity of implementing advanced technologies such as mmWave communication, massive MIMO, and edge computing poses significant challenges in terms of infrastructure, equipment, and network optimization.
- **Spectrum Availability and Regulation**: Securing spectrum resources suitable for 6G deployment, particularly in higher-frequency bands such as mmWave and THz frequencies, is an important challenge. Spectrum availability, licensing, allocation, and regulatory frameworks must be established to enable efficient spectrum utilization while ensuring coexistence with existing wireless systems and minimizing interference.

- **Infrastructure Deployment and Investment**: Upgrading existing infrastructure and deploying new infrastructure to support the requirements of 6G networks require significant investment and coordination among users. The cost, scalability, and compatibility of infrastructure deployment pose challenges for network operators, service providers, and governments.
- **Interference and Signal Propagation**: Overcoming signal attenuation, propagation losses, and interference in higher-frequency bands, as well as optimizing coverage and capacity in urban environments and indoor spaces, is important for ensuring reliable connectivity and continuous user experiences. Advanced antenna technologies, beamforming techniques, and network optimization strategies are required to address these challenges.
- **Energy Consumption and Sustainability**: Managing energy consumption in ultra-dense deployments and massive connectivity scenarios is an important challenge for ensuring the sustainability of 6G networks. Energy-efficient network architectures, power-saving mechanisms, renewable energy integration, and green computing practices are essential to minimize environmental impact and reduce operational costs.
- **Security and Privacy**: Protecting user data, preventing unauthorized access, and mitigating cybersecurity threats are important challenges in 6G deployment. Developing robust security protocols, encryption mechanisms, and privacy-enhancing technologies to protect network infrastructure, user devices, and data transmissions is essential for building trust and ensuring compliance with privacy regulations.
- **Regulatory Compliance and Policy Frameworks**: Establishing regulatory frameworks, standards, and policies governing spectrum allocation, data governance, privacy protection, and network deployment is essential for ensuring compliance with legal requirements and using innovation. Harmonizing regulatory frameworks across jurisdictions and addressing legal and policy challenges related to data sovereignty, privacy laws, and intellectual property rights is important for enabling global deployment and interoperability of 6G networks.
- **Skills and Workforce Development**: Addressing the skill gap and providing training programs for network engineers, technicians, and IT professionals to deploy, manage, and maintain 6G networks is important for ensuring the success of 6G deployment. Investing in workforce development, capacity building, and educational initiatives to build a skilled workforce capable of handling the complexities of 6G technology is essential for driving innovation and competitiveness.

In summary, implementing 6G-based systems in real-world sectors involves overcoming important challenges related to technical complexity, spectrum availability, infrastructure deployment, signal propagation, energy consumption, security, privacy, regulatory compliance, workforce development, and policy frameworks. Addressing these challenges requires collaborative efforts, innovative solutions, and proactive strategies to realize the transformative potential of 6G technology and unlock new opportunities for digital innovation, economic growth, and societal advancement.

10.9 Future Research Opportunities Toward Implementing 6G-Based Systems in Real-World Sectors

Future research opportunities toward implementing 6G-based systems in real-world sectors include a wide range of interdisciplinary areas that can drive innovation, address challenges, and unlock new capabilities [25–28]. Figure 10.4 shows several future research challenges and opportunities toward 6G in detail.

Here are some key research opportunities:

- **Spectrum Management and Allocation**: Research into dynamic spectrum management techniques, cognitive radio technologies, and spectrum-sharing mechanisms can optimize spectrum utilization and enable efficient allocation of frequency bands for 6G deployment, especially in higher-frequency bands such as mmWave and THz frequencies.
- **Advanced Antenna Technologies**: Research on advanced antenna technologies, including massive MIMO, beamforming, and reconfigurable antennas, can enhance coverage, capacity, and spectral efficiency in 6G networks, particularly in urban environments and indoor spaces.
- **Signal Propagation and Interference Mitigation**: Research into signal propagation models, interference mitigation techniques, and propagation characteristics in higher-frequency bands can improve network planning, optimization, and deployment strategies for 6G networks, addressing challenges related to coverage, capacity, and reliability.
- **Energy Efficiency and Sustainability**: Research on energy-efficient network architectures, power-saving mechanisms, renewable energy integration, and green computing practices can minimize energy consumption, reduce carbon footprint, and ensure the sustainability of 6G networks, contributing to environmental conservation and cost savings.
- **Security and Privacy**: Research on robust security protocols, encryption mechanisms, authentication methods, and privacy-enhancing technologies can enhance cybersecurity resilience, protect user data, and mitigate privacy risks in 6G networks, addressing issues related to data breaches, cyberattacks, and regulatory compliance [28, 29].

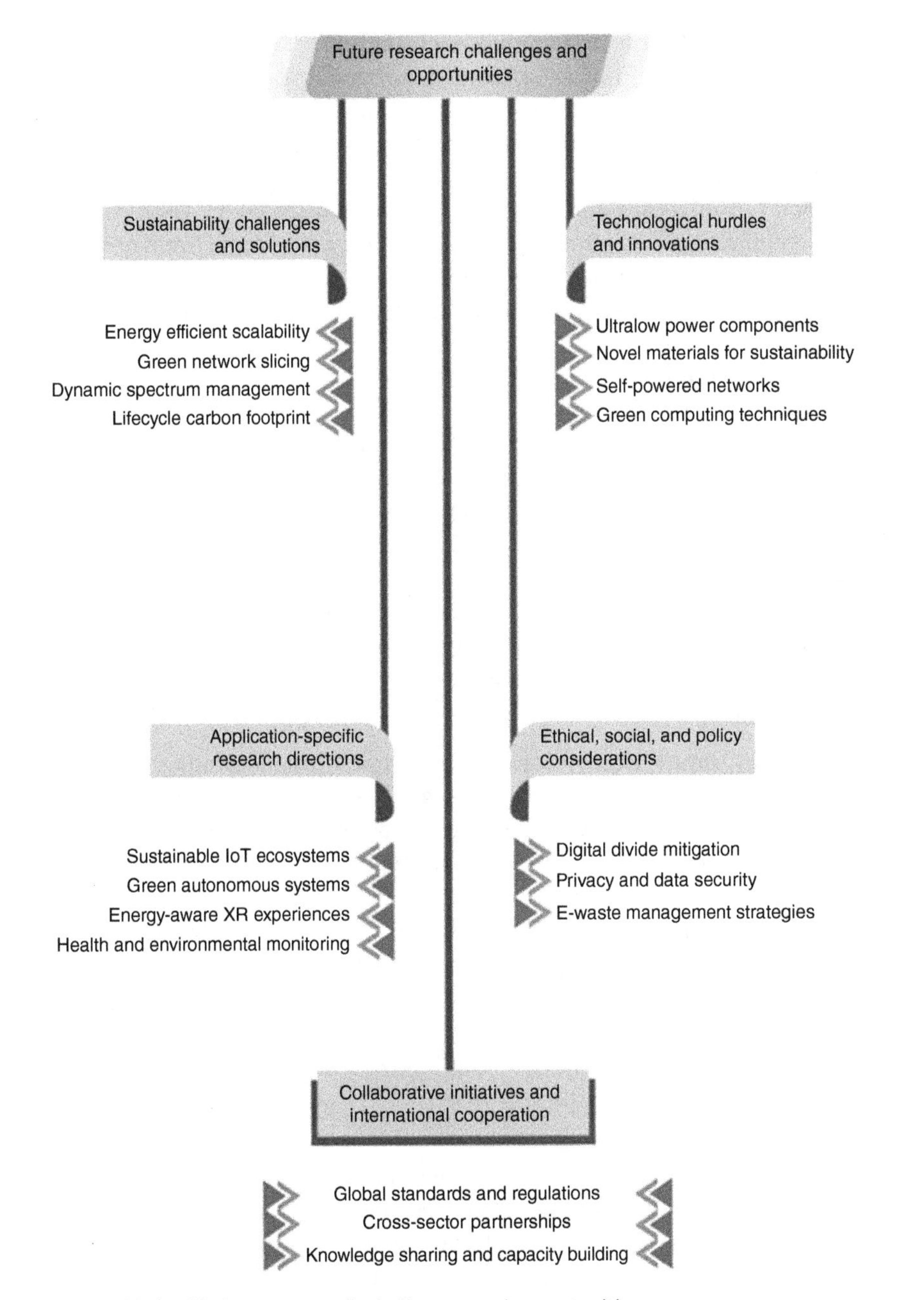

Figure 10.4 6G: future research challenges and opportunities.

- **Edge Computing and Distributed Architectures**: Research on edge computing architectures, distributed processing frameworks, and fog computing paradigms can enable low-latency data processing, real-time decision-making, and efficient resource allocation in 6G networks, supporting latency-sensitive applications and services.
- **Artificial Intelligence and Machine Learning**: Research on AI-driven network optimization, autonomous network management, and intelligent resource allocation can improve network performance, adaptability, and efficiency in 6G networks, enabling self-optimizing, self-healing, and self-configuring capabilities [14–16, 24–28].
- **Regulatory Frameworks and Policy Development**: Research on regulatory frameworks, policy recommendations, and legal issues for spectrum management, data governance, privacy protection, and network deployment can inform policymakers, regulators, and industry users in creating conducive environments for 6G deployment, using innovation while addressing societal issues.
- **User-Centric Applications and Services**: Research on user-centric applications, immersive experiences, and novel use cases enabled by 6G technology, such as augmented reality, virtual reality, holographic communication, remote healthcare, autonomous vehicles, and smart cities, can drive innovation and create new opportunities for economic growth and societal advancement.

In summary, future research opportunities toward implementing 6G-based systems in real-world sectors span a wide range of technical, regulatory, economic, and societal domains, providing exciting prospects for innovation, collaboration, and transformative impact. By addressing these research opportunities, researchers, policymakers, industry users, and academic institutions can contribute to the advancement of 6G technology and its successful integration into diverse sectors, driving digital innovation, economic growth, and societal well-being.

10.10 Summary

In the era of 6G technology, network planning, and deployment emerge as essential pillars shaping the connectivity landscape of the real world. As this chapter has discussed, the journey toward implementing 6G-based systems in real-world sectors is not merely a technical endeavor but a multifaceted challenge requiring meticulous planning, strategic foresight, and collaborative innovation.

The evolution from 5G to 6G marks a huge shift, promising unparalleled speeds, ultralow latency, and ubiquitous connectivity. However, the realization of this vision demands a holistic approach that transcends technological advancements alone. It necessitates a deep understanding of the diverse sectors where 6G will be deployed, as well as the socioeconomic, regulatory, and environmental contexts in which these deployments will unfold. Throughout this work, we have discussed the myriad dimensions of network planning and deployment for 6G-based systems, from spectrum management and infrastructure deployment to coverage optimization and regulatory compliance. We have discussed the technical details of using advanced antenna technologies, spectrum-sharing mechanisms, and edge-computing architectures to optimize coverage, capacity, and reliability.

Moreover, we have underscored the importance of addressing nontechnical challenges, such as regulatory frameworks, workforce development, and ethical issues, which play an important role in shaping the trajectory of 6G deployment in real-world sectors. By using interdisciplinary collaboration, user engagement, and policy dialogue, we can create an enabling environment for the successful adoption of 6G technology while protecting privacy, promoting inclusivity, and mitigating environmental impact. In summary, the deployment of 6G-based systems in real-world sectors represents a transformative opportunity to redefine connectivity, empower innovation, and drive sustainable development. By implementing the challenges and opportunities presented by 6G technology, we can make the way for a future where ultra-connectivity, digital innovation, and societal well-being converge to create a more connected, intelligent, and resilient world.

References

1 Liu, Z., Yang, C., and Sun, S. (2021). A survey on machine learning-enabled resource management in 5G and beyond wireless networks. *IEEE Access* 9: 32198–32219.

2 Li, Y., Wang, P., and Niyato, D. (2021). Green communications for UAV networks: techniques, challenges, and future directions. *IEEE Transactions on Green Communications and Networking* 5 (3): 1011–1028.

3 Shrestha, R., Bennis, M., and Poor, H.V. (2021). Energy-efficient deployment and user association in wireless networks: a machine learning approach. *IEEE Transactions on Wireless Communications* 20 (1): 446–460.

4 Patri, S.K., Autenrieth, A., Elbers, J.P., and Mas-Machuca, C. (2022). Multi-band transparent optical network planning strategies for 6G-ready European networks. *Social Science Research Network*. https://doi.org/10.2139/ssrn.4113021.

5 El-Badawy, H.M., Ahmed, H.A.S., Zainud-Deen, S.H., and Malhat, H.A.E.A. (2023). B5G/6G network planning for study case in knowledge city area

as model for smart cities. In: *2023 40th National Radio Science Conference (NRSC)*, vol. 1, 191–200. IEEE.

6 Elbadawy, H.M. and Sadek, R.A. (2022). B5G/6G service planning process for the deployment of smart cities as a model for massive urban area applications. In: *2022 39th National Radio Science Conference (NRSC)*, vol. 1, 332–341. IEEE.

7 Wang, Q., Liu, Y., Liu, H., and Zong, J. (2022). A network intelligence deployment plan towards 6G. In: *International Conference on Computer Engineering and Networks*, 1269–1276. Singapore: Springer Nature Singapore.

8 Catté, E., Sana, M., and Maman, M. (2022, June). Cost-efficient and QoS-aware user association and 3D placement of 6G aerial mobile access points. In: *2022 Joint European Conference on Networks and Communications & 6G Summit (EuCNC/6G Summit)*, 357–362. IEEE.

9 Chen, X., Liu, C., and Chen, M. (2021). Joint power control and resource allocation for NOMA-based UAV communication systems. *IEEE Transactions on Vehicular Technology* 70 (1): 1024–1034.

10 Yu, S., Xu, J., and Jin, Y. (2021). Multi-objective optimization for UAV-enabled Mobile edge computing: a Stackelberg game approach. *IEEE Transactions on Vehicular Technology* 70 (4): 4071–4082.

11 Zhang, L., Zhang, Q., and Liu, Y. (2021). Optimal deployment of UAV networks for large-scale IoT data collection. *IEEE Internet of Things Journal* 8 (8): 6296–6307.

12 Ding, Z. and Poor, H.V. (2021). Optimization of energy efficiency in wireless communication systems: a tutorial. *IEEE Transactions on Cognitive Communications and Networking* 7 (3): 833–851.

13 Liu, H., Yang, Z., and Chen, M. (2021). A comprehensive survey on resource allocation in UAV-enabled NOMA systems. *IEEE Network* 35 (2): 156–161.

14 Nair, M.M. and Tyagi, A.K. (2023). Blockchain technology for next-generation society: current trends and future opportunities for smart era. In: *Blockchain Technology for Secure Social Media Computing*. https://doi.org/10.1049/PBSE019E_ch11.

15 Tyagi, A.K., Kumari, S., Chidambaram, N., and Sharma, A. (2024). Engineering applications of blockchain in this smart era. In: *Enhancing Medical Imaging with Emerging Technologies*. https://doi.org/10.4018/979-8-3693-5261-8.ch011.

16 Ajanthaa, L., Seranmadevi, R., Sree, P.H., and Tyagi, A.K. (2024). Engineering applications of artificial intelligence. In: *Enhancing Medical Imaging with Emerging Technologies*. https://doi.org/10.4018/979-8-3693-521-8.ch010.

17 Li, H., Kim, D.I., and Poor, H.V. (2021). On energy-efficient deployment of UAVs for cellular coverage augmentation. *IEEE Transactions on Wireless Communications* 20 (3): 1557–1572.

18 Zhao, N. and Zhang, R. (2021). A survey on unmanned aerial vehicle networks for public safety. *IEEE Internet of Things Journal* 8 (7): 5109–5126.

19 Wu, J., Wang, Q., and Zhang, J. (2021). Optimization of network slicing placement in MEC-enabled UAV networks. *IEEE Transactions on Vehicular Technology* 70 (5): 5263–5275.

20 Liu, C., Li, S., and Chen, M. (2021). A survey of machine learning techniques for resource allocation in NOMA systems. *IEEE Access* 9: 15566–15581.

21 Zhou, Z., Cao, B., and Xu, Y. (2021). UAV-enabled cooperative relaying for 5G and beyond wireless communications: challenges and opportunities. *IEEE Wireless Communications* 28 (2): 16–23.

22 Zeng, Y. and Zhang, R. (2021). Joint UAV trajectory and communication design for optimal UAV-assisted IoT. *IEEE Transactions on Wireless Communications* 20 (3): 1901–1917.

23 Li, X., Li, H., and Jin, L. (2021). Joint optimization of UAV trajectory and ground user association in mobile edge computing networks. *IEEE Transactions on Mobile Computing* 20 (6): 2489–2502.

24 Tyagi, A.K., Kukreja, S., Richa, and Sivakumar, P. (2024). Role of blockchain technology in smart era: a review on possible smart applications. *Journal of Information & Knowledge Management*. https://doi.org/10.1142/S0219649224500321.

25 Tyagi, A.K. and Tiwari, S. (2024). The future of artificial intelligence in blockchain applications. In: *Machine Learning Algorithms Using Scikit and TensorFlow Environments*. IGI Global. https://doi.org/10.4018/978-1-6684-8531-6.ch018.

26 Kumari, S., Thompson, A., and Tiwari, S. (2024). 6G-enabled internet of things-artificial intelligence-based digital twins: cybersecurity and resilience. In: *Emerging Technologies and Security in Cloud Computing*. IGI Global. https://doi.org/10.4018/979-8-3693-2081-5.ch016.

27 Nair, M.M. and Tyagi, A.K. (2023). 6G: technology, advancement, barriers, and the future. In: *6G-Enabled IoT and AI for Smart Healthcare*. CRC Press.

28 Nair, M.M. and Tyagi, A.K. (2021). Privacy: history, statistics, policy, Laws, preservation and threat analysis. *Journal of Information Assurance & Security*. 16 (1): 24–34. 11p.

29 https://arxiv.org/html/2305.16616v2

11

Standardization and Regulatory Aspects for 6G-Based Networks and Systems

11.1 Introduction to 6G-Based Networks Technology and Systems

The world is on the edge of yet another major growth in communication technology with the growth of 6G networks [1]. As the successor to 5G, 6G promises to revolutionize connectivity, enabling unique speeds, ultralow latency, and ubiquitous connectivity that will redefine how we interact with technology and each other. At its core, 6G builds upon the foundations laid by its predecessors, enhancing and expanding upon the capabilities of 5G networks. While 5G brought about remarkable advancements such as faster data speeds and improved network reliability, 6G aims to push the boundaries even further, introducing breakthrough innovations that will power the next era of digital transformation. One of the defining features of 6G technology is its capability to support data rates exceeding hundreds of gigabits per second, surpassing the already impressive speeds of 5G networks. This unique bandwidth will enable a wide array of applications, from immersive virtual reality experiences to real-time high-definition holographic communications, transforming the way we consume and interact with media.

Moreover, 6G networks are used to deliver ultralow latency, reducing response times to mere microseconds. This near-instantaneous responsiveness is important for applications, such as autonomous vehicles, remote surgery, and industrial automation, where split-second decisions can mean the difference between success and failure. A basic example of using 6G in transportation is shown in Figure 11.1.

In addition to speed and latency improvements, 6G will also introduce advancements in network reliability, security, and energy efficiency [2]. By using technologies such as artificial intelligence (AI), machine learning (ML), and quantum computing, 6G networks will be capable of self-optimization, self-healing, and adaptive resource allocation, ensuring robust and resilient connectivity even in the face of adverse conditions.

6G-Enabled Technologies for Next Generation: Fundamentals, Applications, Analysis and Challenges, First Edition. Amit Kumar Tyagi, Shrikant Tiwari, Shivani Gupta, and Anand Kumar Mishra.

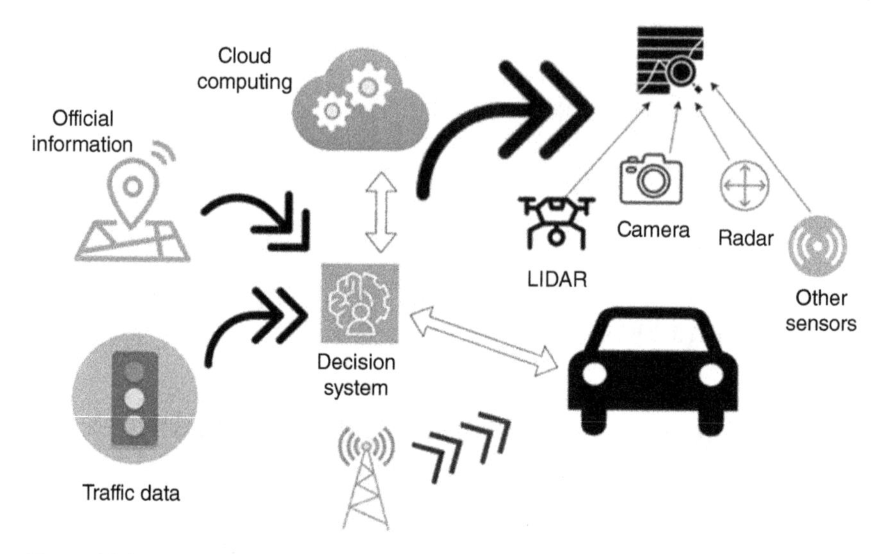

Figure 11.1 Use of 6G in autonomous driving.

Furthermore, 6G is expected to extend connectivity beyond traditional devices to include a vast ecosystem of Internet of Things (IoT) devices, sensors, and actuators, enabling continuous integration of the physical and digital worlds [3]. This convergence of cyberspace and physical space will give rise to smart cities, intelligent transportation systems, and interconnected infrastructure that will enhance efficiency, sustainability, and quality of life.

As we move towardthe journey of 6G, it is important to recognize the complicated nature of this technological evolution. Beyond its technical intricacies, the deployment of 6G networks will also entail complex socioeconomic, regulatory, and ethical issues that must be carefully addressed to ensure equitable access, privacy protection, and ethical use of emerging technologies.

Here, Figure 11.2 discusses an overview of different 6G use cases. In summary, 6G represents a huge shift in communication technology, providing unique speed, ultralow latency, and ubiquitous connectivity that will unlock a myriad of new possibilities and transform the way we live, work, and interact. As researchers, engineers, policymakers, and users come together to shape the future of 6G, we stand at the threshold of a new era of connectivity that holds the promise of a brighter and more connected world.

11.2 Literature Review

In [4], as a result of the shifts caused by ML, wireless network standardization groups are modifying their practices. Approaches to ML for intricate 6G networks

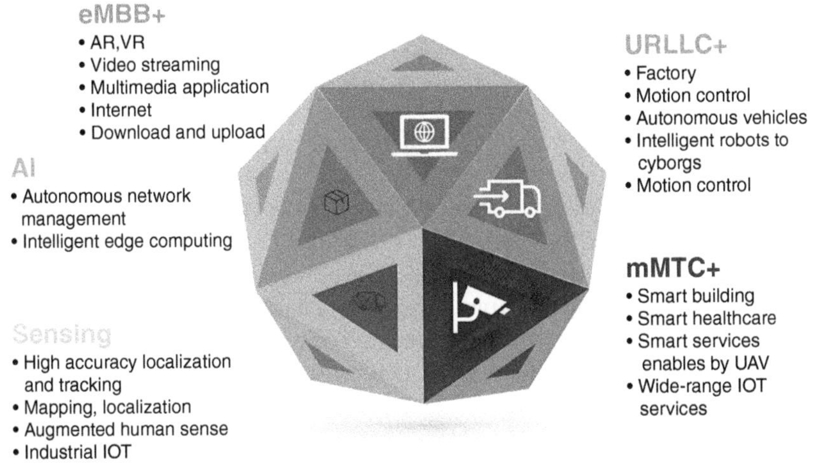

eMBB+
- AR,VR
- Video streaming
- Multimedia application
- Internet
- Download and upload

URLLC+
- Factory
- Motion control
- Autonomous vehicles
- Intelligent robots to cyborgs
- Motion control

AI
- Autonomous network management
- Intelligent edge computing

mMTC+
- Smart building
- Smart healthcare
- Smart services enables by UAV
- Wide-range IOT services

Sensing
- High accuracy localization and tracking
- Mapping, localization
- Augmented human sense
- Industrial IOT

Figure 11.2 6G use cases: overview.

are the things to do pertaining to wireless network standards and a future road map for ML. Using ML, 6G networks can become more smart, self-organizing, and more economical. It is concluded that ML holds great promise for improving link layers in future development.

In [5], 6G standardization and regulation are key issues for network deployment. Standardization organizations and regulatory frameworks are equally crucial in the context of 6G. Note that a vision for the future 6G systems provides better service metrics. According to predictions, 6G will outperform 5G in terms of technology by 2030. 5G lacks the capacity to meet the demands of applications in space and the deep oceans. 6G is expected to be released in 2030 due to 5G's limitations.

In [6], an outline of the rules and guidelines for "green" standards is provided. Information systems for "green" standards mainly aim to support regulatory compliance. The authors designed IT standards for environmentally conscious standardization. It presents a high-level summary of the key goals for the green standardization regulatory framework and resources that establish rules for the introduction of eco-friendly criteria into computer networks. This chapter details the steps necessary to build an information environment that supports environmentally friendly standards.

In [7], while 6G is one possible implementation of the standardization framework, its original intent was to facilitate EAEU inclusion. The difficulties in regulation that must be overcome for the EAEU to achieve economic integration, and not 6G equipment. The creation of a regulatory and organizational structure to support integration processes modernizing national legislation on standards is the key to achieving legal harmonization. Standardization helps the Eurasian Economic Union (EEU) become more economically integrated. Businesses can more easily engage with one another in the market as a result of the adoption of international standards. A prerequisite for qualitative development in the Union's economy is the elimination of trade barriers brought about by regulatory and legal hurdles.

In [8], adding Watermark Blind Physical Layer Security (WBPLSec) to video-link converters (VLCs) makes 6G wireless communications safer. This VLC Physical Layer Security (PLS) implementation makes use of watermarking and RGB LED jamming. Note that adding a jammer receiver to a device that uses VLC with a watermark makes it more secure. A legitimate receiver can create a secure environment where confidential information can be freely shared. Working with other nodes is necessary to install more PLS solutions on VLC.

In [9], an overview of the WYV development process, including the characteristics and technology used to the standardization of WYV use and the resolution of critical issues related to its establishment. The efficacy and repeatability of WYV development have been enhanced. There are certain unanswered questions about the methods currently used to obtain and validate recombinant strains because the standardization of standards is crucial for licensing and regulatory approval.

In [10], the necessary regulatory framework must be modernized to implement 6G technology. The intricacy of 6G may be beyond the scope of the standards. Changes to the rules that apply to highly innovative items, which is talking about serious issues with the procedures for approving new goods and raising the bar for healthcare products that encourage innovation by improving the regulatory climate.

11.3 Standardization Bodies and Organizations for 6G and 6G-Based Networks

Standardization bodies and organizations play an important role in the development and deployment of 6G and 6G-based networks [11]. These entities collaborate to define technical specifications, interoperability standards, and regulatory frameworks that facilitate the global adoption of 6G technology. Here are some key standardization bodies and organizations involved in shaping the future of 6G:

11.3.1 International Telecommunication Union (ITU)

- The ITU is a specialized United Nations agency responsible for coordinating global telecommunications standards and regulations.
- Within the ITU, the Radiocommunication Sector (ITU-R) focuses on spectrum management and radio regulations, which are essential for 6G deployment.

- The ITU-R Working Party 5D is specifically dedicated to the development of IMT-2020 (6G) standards.

11.3.2 3rd-Generation Partnership Project (3GPP)

- 3GPP is a collaborative organization responsible for developing standards for mobile telecommunications, including 5G and beyond.
- Working closely with industry users, 3GPP defines technical specifications for cellular networks, including air interface protocols, network architecture, and system requirements.
- The ongoing efforts of 3GPP are instrumental in shaping the evolution of 6G technology.

11.3.3 Institute of Electrical and Electronics Engineers (IEEE)

- IEEE is a global professional organization dedicated to advancing technology through the development of standards and protocols.
- IEEE's Communication Society (ComSoc) and other relevant technical committees are actively involved in researching and standardizing technologies that will underpin 6G networks, such as terahertz communication, large multiple input multiple output (MIMO), and beyond-5G architectures.

11.3.4 Internet Engineering Task Force (IETF)

- While traditionally focused on internet protocols, the IETF also plays a role in standardizing communication protocols that may be relevant to 6G networks, such as IP-based networking, transport protocols, and security mechanisms.

11.3.5 European Telecommunications Standards Institute (ETSI)

- ETSI is a European standards organization that develops globally applicable standards for information and communication technologies (ICT).
- ETSI's Industry Specification Group for Next Generation Protocol (ISG NGP) focuses on research and standardization activities related to future communication networks, including 6G.

11.3.6 National Telecommunications and Information Administration (NTIA)

- In the United States, the NTIA is responsible for advising the president on telecommunications and information policy issues.
- The NTIA plays a role in spectrum management and coordination, which are important aspects of 6G standardization and deployment.

11.3.7 Other Industry Consortia and Forums

- Various industry consortia, forums, and alliances, such as the 6G Flagship program, the 6G Wireless Summit, and the Next G Alliance, bring together industry players, researchers, and policymakers to collaborate on defining the vision and requirements for 6G networks.

Note that these standardization bodies and organizations collaborate closely to ensure that 6G technology is developed in a coordinated manner, with interoperability, compatibility, and global harmonization in mind. Their efforts are essential for realizing the full potential of 6G and enabling its continuous integration into the global telecommunications ecosystem.

11.4 Spectrum Regulations and Policies for 6G

Spectrum regulations and policies are important factors in the deployment and operation of 6G networks. Spectrum, the radio frequencies used for wireless communication, is a finite and valuable resource that must be managed efficiently to ensure equitable access and optimal utilization [12, 13]. As 6G technologies evolve, spectrum regulations and policies will need to adapt to accommodate the unique requirements and characteristics of these advanced networks. Here are some key issues for spectrum regulations and policies for 6G:

11.4.1 Frequency Bands Allocation

- Regulators need to identify and allocate suitable frequency bands for 6G networks. This may involve repurposing existing spectrum bands, as well as allocating new frequency bands that are conducive to the characteristics of 6G technology, such as higher data rates and lower latency.
- Millimeter-wave bands, including those above 100 GHz, are being discussed for their potential to support ultrahigh-speed data transmission in 6G networks.

11.4.2 Dynamic Spectrum Access

- Dynamic spectrum access (DSA) allows for more flexible and efficient use of spectrum by enabling dynamic allocation of frequencies based on demand and usage patterns.
- Regulators may consider implementing policies and technologies that enable DSA, such as spectrum sharing frameworks and cognitive radio systems, to maximize spectrum utilization and accommodate diverse services and applications in 6G networks.

11.4.3 Spectrum Auctions and Licensing

- Spectrum auctions are commonly used by regulators to allocate spectrum licenses to mobile network operators and other service providers.
- Regulators may conduct spectrum auctions specifically for 6G spectrum bands, ensuring fair and competitive access to valuable spectrum resources while generating revenue for the government.

11.4.4 International Coordination

- Spectrum regulation is often subject to international coordination and harmonization efforts to facilitate global interoperability and roaming.
- Regulators may participate in international forums and organizations, such as the ITU, to coordinate spectrum allocations and regulatory frameworks for 6G on a global scale.

11.4.5 Spectrum Sharing and Unlicensed Bands

- In addition to licensed spectrum, unlicensed bands and spectrum-sharing mechanisms can play a role in supporting 6G networks.
- Regulators may promote the use of unlicensed bands for certain types of 6G applications, such as local area networks and IoT deployments, to use innovation and competition in the market.

11.4.6 Security and Interference Mitigation

- Spectrum regulations must address security issues related to unauthorized access, interference, and spectrum misuse.
- Regulators may mandate security requirements and interference mitigation techniques for 6G networks to protect against malicious activities and ensure the reliability and integrity of wireless communications.

11.4.7 Future-Proofing Regulations

- Regulators should adopt flexible and future-proof regulatory frameworks that can adapt to evolving technologies and market dynamics.
- Regular reviews and updates to spectrum regulations and policies will be necessary to accommodate advancements in 6G technology and address emerging challenges and opportunities in the telecommunications landscape.

In summary, spectrum regulations and policies for 6G must strike a balance among promoting innovation, ensuring fair competition, protecting the public interest, and managing spectrum resources efficiently. By adopting forward-looking and inclusive regulatory approaches, regulators can unlock the full potential of 6G networks, driving socioeconomic growth, and digital transformation.

11.5 Global Collaboration and Interoperability for 6G and 6G-Based Networks

Global collaboration and interoperability are essential for the successful development, deployment, and operation of 6G and 6G-based networks [14]. As 6G technology evolves, it is important for users worldwide to work together to define common standards, protocols, and frameworks that enable continuous connectivity and interoperability across diverse networks and devices. Here's how global collaboration and interoperability can be used for 6G:

11.5.1 International Standardization Bodies

- Organizations such as ITU, 3GPP, and IEEE bring together industry players, policymakers, and experts from around the world to develop global standards for telecommunications technologies, including 6G.
- By participating in international standardization efforts, users can ensure that 6G networks adhere to common technical specifications and interoperability requirements, facilitating global deployment and roaming.

11.5.2 Cross-Industry Collaboration

- Collaboration between telecommunications industry players, technology companies, academia, and other sectors is essential for driving innovation and addressing the diverse needs of 6G applications.
- Cross-industry partnerships can use the development of interdisciplinary solutions, such as integrating 6G with emerging technologies, such as AI, quantum computing, and blockchain, to create new opportunities and use cases.

11.5.3 Open Research and Innovation Platforms

- Open research and innovation platforms provide a collaborative environment for researchers, developers, and innovators to exchange ideas, share resources, and co-create 6G technologies and applications.
- Initiatives such as open-source software projects, test beds, and innovation hubs enable global participation and experimentation, accelerating the pace of innovation and using interoperability among diverse 6G ecosystems.

11.5.4 Interoperability Testing and Certification

- Interoperability testing programs and certification schemes ensure that 6G devices, networks, and services comply with established standards and can continuously interoperate with each other.
- International collaborations on interoperability testing enable vendors and operators from different regions to validate their products and solutions in diverse environments, enhancing confidence in the interoperability of 6G technologies.

11.5.5 Regulatory Harmonization

- Harmonized regulatory frameworks and policies facilitate the global deployment and operation of 6G networks by providing consistent rules and guidelines for spectrum allocation, licensing, privacy protection, and security.
- Regulators worldwide can collaborate through international forums and agreements to harmonize spectrum regulations, streamline approval processes, and promote fair competition and consumer protection in the 6G market.

11.5.6 Capacity Building and Knowledge Sharing

- Capacity-building initiatives, workshops, and training programs promote knowledge sharing and skills development among users in different regions, particularly in developing countries.
- By using a culture of collaboration and knowledge exchange, capacity-building efforts empower diverse communities to actively participate in the development and adoption of 6G technologies, ensuring that the benefits of 6G are accessible to all.

In summary, global collaboration and interoperability are essential pillars for realizing the full potential of 6G and ensuring its global adoption and impact. By using an inclusive and collaborative approach, users can use collective expertise and resources to create a vibrant and interoperable 6G ecosystem that drives innovation, economic growth, and societal progress on a global scale.

11.6 Technical/Nontechnical/Legal Issues for 6G and 6G-Based Networks

For 6G and 6G-based networks, a range of technical, nontechnical, and legal issues must be addressed to ensure successful development, deployment, and operation [14, 15]. Here is a breakdown of key issues in each category:

11.6.1 Technical Issues

a) **Spectrum Utilization**: Identifying suitable frequency bands and managing spectrum efficiently to accommodate the bandwidth requirements of 6G networks.
b) **Interference and Coexistence**: Mitigating interference between different frequency bands and coexisting with legacy networks and other wireless technologies.
c) **Network Architecture**: Designing scalable and flexible network architectures that can support the diverse requirements of 6G applications, including ultrahigh-speed data transmission, ultralow latency, and huge connectivity.
d) **Security and Privacy**: Developing robust security mechanisms to protect against cyber threats, privacy breaches, and unauthorized access, considering the increasing reliance on connected devices and sensitive data in 6G networks.
e) **Energy Efficiency**: Addressing energy consumption challenges associated with the deployment of 6G infrastructure, such as optimizing power usage in base stations and devices to minimize environmental impact.
f) **Interoperability**: Ensuring interoperability among different vendors' equipment and protocols to enable continuous connectivity and service delivery across heterogeneous 6G networks.

11.6.2 Nontechnical Issues

a) **Regulatory Compliance**: Adhering to regulatory requirements and standards set by national and international bodies related to spectrum allocation, licensing, privacy protection, and network neutrality.
b) **Ethical Issues**: Addressing ethical issues surrounding the use of emerging technologies in 6G networks – such as AI, biometrics, and surveillance – and ensuring responsible and equitable deployment.
c) **Digital Divide**: Bridging the digital divide by ensuring equitable access to 6G networks and services, particularly in underserved or rural areas, to prevent exacerbating existing socioeconomic inequalities.
d) **Workforce Skills**: Building a skilled workforce capable of designing, deploying, and managing 6G networks and applications, including expertise in emerging technologies and interdisciplinary collaboration.
e) **Consumer Rights**: Protecting consumer rights, including data privacy, transparency, and fair competition, through consumer education, regulatory oversight, and enforcement mechanisms.

11.6.3 Legal Issues

a) **Intellectual Property Rights (IPR)**: Addressing intellectual property rights related to 6G technologies, including patents, copyrights, and trade secrets, to facilitate innovation and prevent litigation disputes.
b) **Data Protection and Privacy Laws**: Ensuring compliance with data protection and privacy laws, such as the general data protection regulation (GDPR), to protect user data and privacy in 6G networks and applications.
c) **Antitrust and Competition Law**: Preventing anticompetitive practices and market dominance by regulating mergers and acquisitions, promoting fair competition, and enforcing antitrust laws to use innovation and consumer choice.
d) **Cybersecurity Regulations**: Implementing cybersecurity regulations and standards to protect important infrastructure and sensitive information from cyber threats and attacks targeting 6G networks and services.
e) **Liability and Jurisdiction**: Clarifying liability issues and jurisdictional boundaries in cases of network failures, data breaches, or other legal disputes arising from the use of 6G technologies and services across international borders.

Hence, addressing these technical, nontechnical, and legal issues requires collaboration among industry users, policymakers, regulators, and civil society to develop complete and balanced solutions that promote innovation, protect public interests, and ensure the responsible and sustainable deployment of 6G networks and technologies.

11.7 Important Challenges Toward Implementing 6G and 6G-Based Networks in Real-World Applications

Implementing 6G and 6G-based networks in real-world applications faces several important challenges that must be overcome to realize the full potential of this next-generation technology [15–17]. Here are some of the key challenges:

- **Technology Readiness**: Developing and maturing the underlying technologies required for 6G, including advanced radio access technologies, ultrahigh-frequency communication, terahertz communication, and quantum networking, among others.
- **Spectrum Availability and Regulation**: Securing appropriate spectrum allocations and navigating regulatory frameworks to enable the deployment of 6G networks, including addressing spectrum scarcity, interference mitigation, and spectrum sharing arrangements.

- **Infrastructure Deployment**: Building out the necessary infrastructure for 6G networks, including deploying base stations, small cells, backhaul connections, and edge computing facilities to support high-speed, low-latency connectivity in urban, suburban, and rural areas.
- **Interoperability and Standards**: Establishing interoperable standards and protocols to ensure continuous connectivity and compatibility between different 6G networks, devices, and applications, while also addressing issues related to vendor lock-in and proprietary technologies.
- **Security and Privacy**: Developing robust security mechanisms to protect against cyber threats, privacy breaches, and unauthorized access in 6G networks, including addressing vulnerabilities in hardware, software, and communication protocols [18–25].
- **Energy Efficiency**: Optimizing energy consumption and reducing the carbon footprint of 6G infrastructure and devices, including addressing the energy demands of high-speed data transmission, processing-intensive applications, and edge computing.
- **Cost and Affordability**: Managing the cost of deploying and operating 6G networks, including investments in infrastructure, spectrum licenses, research and development, and ongoing maintenance, while also ensuring affordability and accessibility for end users.
- **Digital Inclusion and Equity**: Addressing disparities in access to 6G technology and services, including bridging the digital divide between urban and rural areas, providing affordable access to underserved communities, and promoting digital literacy and skills development.
- **Ethical and Societal Implications**: Considering the ethical, social, and cultural implications of 6G technology, including issues related to privacy, surveillance, autonomy, bias in AI algorithms, digital rights, and the impact on employment and societal structures.

Note that addressing these important challenges will require concerted efforts from users across multiple sectors and disciplines, as well as ongoing innovation, research, and collaboration to overcome technical, regulatory, economic, and societal barriers to the successful implementation of 6G and its transformative potential.

11.8 Future Research Opportunities Toward Implementing 6G and 6G-Based Networks in Real-World Applications

Future research opportunities toward implementing 6G and 6G-based networks in real-world applications span a wide range of technical, socioeconomic, and interdisciplinary domains. Here are some potential research directions:

- **Advanced Communication Technologies**: Research into novel communication technologies such as terahertz communication, visible light communication, and beyond-5G radio access techniques to achieve ultrahigh data rates, ultralow latency, and huge connectivity.
- **Spectrum Management and Allocation**: Developing spectrum sharing algorithms, cognitive radio systems, and DSA techniques to optimize spectrum utilization and accommodate the diverse requirements of 6G networks.
- **Network Architecture and Protocols**: Designing scalable, flexible, and self-organizing network architectures that can adapt to dynamic environments, support heterogeneous devices and services, and provide continuous connectivity across diverse network domains.
- **Security and Privacy Solutions**: Investigating advanced cybersecurity mechanisms, privacy-preserving protocols, and secure-by-design principles to protect against emerging threats and vulnerabilities in 6G networks, including quantum-resistant encryption and secure multiparty computation [26–28].
- **Energy-Efficient Technologies**: Researching energy-efficient hardware and software solutions for 6G infrastructure and devices, including low-power communication protocols, energy harvesting techniques, and optimization algorithms for network resource management.
- **AI and ML Applications**: Discussing the integration of AI (AI) and ML techniques in 6G networks for intelligent resource allocation, predictive maintenance, and anomaly detection [18, 19, 25, 29, 30], and context-aware optimization (refer Figure 11.3).
- **Edge and Fog Computing**: Investigating edge and fog computing platforms for 6G networks to enable low-latency, high-throughput processing, and data analytics at the network edge, supporting real-time applications such as augmented reality, autonomous vehicles, and smart manufacturing [31].
- **Digital Twin and Virtualization**: Developing digital twin models and network virtualization techniques to simulate and optimize 6G network performance, predict system behavior, and facilitate rapid prototyping and testing of new network architectures and services [32].
- **Human–Computer Interaction**: Researching human–computer interaction (HCI) technologies for 6G applications, including immersive interfaces, haptic feedback systems, brain-computer interfaces, and natural language processing techniques to enhance user experience and accessibility (refer Figure 11.4).
- **Socioeconomic Impact Assessment**: Conducting socioeconomic impact assessments of 6G technologies on industries, employment, education, healthcare, transportation, and urban planning to understand their implications and inform policymaking and investment decisions.
- **Global Collaboration and Policy Development**: Promoting international collaboration and multistakeholder engagement in defining global standards,

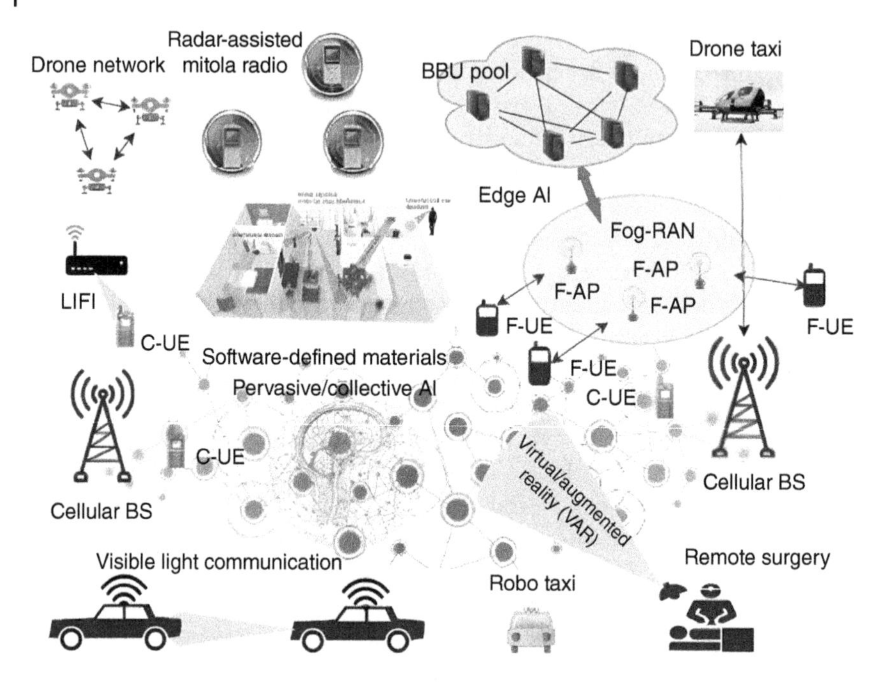

Figure 11.3 AI in 6G.

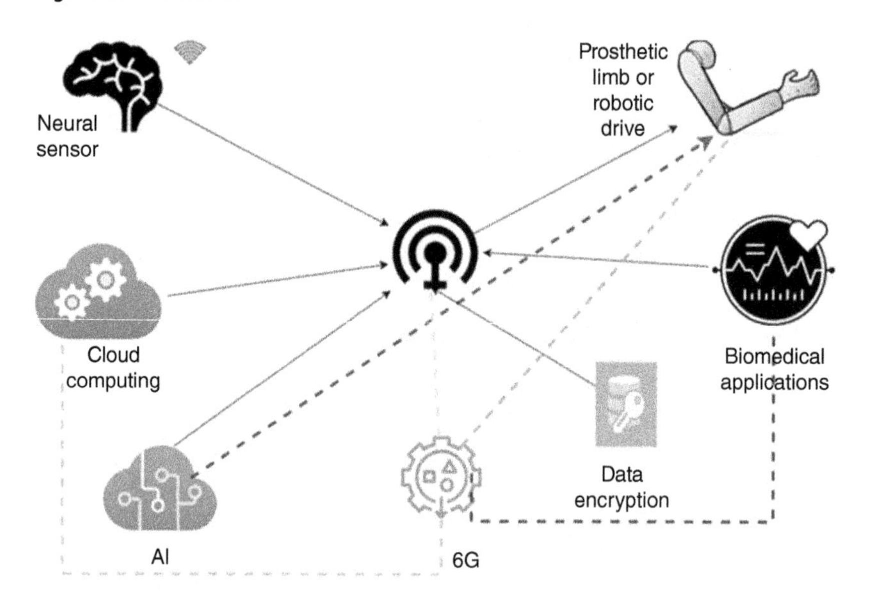

Figure 11.4 Wireless brain–computer interaction using 6G.

regulations, and governance frameworks for 6G to ensure interoperability, security, and equitable access to 6G technology and services.

Hence, by addressing these research opportunities, the scientific community can contribute to the development of innovative solutions, use interdisciplinary collaboration, and make the way for the successful implementation of 6G and its transformative impact on society, economy, and technology.

11.9 Summary

The standardization and regulatory aspects of 6G-based networks and systems play an important role in shaping the trajectory of this transformative technology. As we focus on the journey toward 6G, it is evident that a concerted effort toward international collaboration, user engagement, and regulatory alignment is imperative to realize its full potential. Standardization bodies such as the ITU, 3GPP, and IEEE are driving the development of technical specifications and interoperability standards essential for continuous integration of diverse technologies within the 6G ecosystem. Through harmonized standards, 6G networks can unlock enhanced capabilities such as ultra-reliable low-latency communication, huge machine-type communication, and ubiquitous connectivity.

In summary, regulatory frameworks are essential for addressing spectrum allocation, privacy issues, security protocols, and ethical issues associated with 6G deployment. Governments and regulatory agencies worldwide must facilitate spectrum auctions, enforce cybersecurity measures, and ensure compliance with privacy regulations to protect user data in 6G networks. Moreover, ethical guidelines are important for governing the ethical implications of emerging technologies such as AI and quantum computing, which underpin 6G systems. As we look toward the future, it is clear that continued collaboration and vigilance are essential to ensure that 6G networks and systems are deployed responsibly, ethically, and equitably, bringing about positive transformations in the way we communicate, interact, and experience the world.

References

1 Chen, R., Zhang, Q., Liu, W., and Ma, X. (2021). 6G wireless networks: vision, requirements, architecture, and key technologies. *IEEE Network* 35 (4): 288–294. https://doi.org/10.1109/MNET.011.2100396.

2 Nour, B., Mastorakis, S., Ullah, R., and Stergiou, N. (2021). Information-centric networking in wireless environments: security risks and challenges. *IEEE Wireless Communications* 28 (2): 121–127. doi: 10.1109/mwc.001.2000245. Epub 2021 Apr 6. PMID: 34366719; PMCID: PMC8340877.

3 Aqeel, T.J., Rihab, M., and Lamia, C. (2023). A comprehensive survey on 6G and beyond: enabling technologies, opportunities of machine learning and challenges. *Computer Networks* 237. https://doi.org/10.1016/j.comnet.2023.110085.

4 Shehzad, M.K., Rose, L., Butt, M.M. et al. (2022). Artificial intelligence for 6G networks: technology advancement and standardization. *IEEE Vehicular Technology Magazine* https://doi.org/10.1109/MVT.2022.3164758.

5 Shylaja (2022). 6G scenarios and network design principles. *Journal of Machine and Computing* https://doi.org/10.53759/7669/jmc202202013.

6 Slesarev, M. and Telichenko, V. (2020). Prospects for the development of the regulatory framework of information systems for "green" standardization. *International Journal for Computational Civil and Structural Engineering* https://doi.org/10.22337/2587-9618-2020-16-4-92-102.

7 Okrepilov, V.V. and Gridasov, A.G. (2019). The role of standardization as a key tool of quality economics in the development of integration of the Eurasian Economic Union. *Economics and Management* https://doi.org/10.35854/1998-1627-2019-11-14-19.

8 Simone, S., Rocco, D., and Nicola. (2022). 6G networks physical layer security using RGB visible light communications. *IEEE Access* https://doi.org/10.1109/access.2021.3139456.

9 Anna, J., Duarte, S., Crislaine, K. et al. (2022). Standardization and key aspects of the development of whole yeast cell vaccines. *Pharmaceutics* https://doi.org/10.3390/pharmaceutics14122792.

10 Nenad, Z. (2023). Regulatory aspects and barriers in using groundbreaking technologies. *Advanced clinical pharmacy - research, development and practical applications* https://doi.org/10.1007/978-3-031-26908-0_17.

11 Giordani, M., Polese, M., Mezzavilla, M. et al. (2020). Toward 6G Networks: use cases and technologies. *IEEE Communications Magazine* 58 (3): 55–61.

12 Palattella, M.R., Dohler, M., Grieco, A. et al. (2016). Internet of things in the 5G era: enablers, architecture, and business models. *IEEE Journal on Selected Areas in Communications* 34 (3): 510–527. https://doi.org/10.1109/JSAC.2016.2525418.

13 Salameh, A.I. and Tarhuni, M.E. (2022). From 5G to 6G—challenges, technologies, and applications. *Future Internet* 14 (4): 117. https://doi.org/10.3390/fi14040117.

14 ITU-R M.[IMT-2020.TECH PERF REQ] (2020). *Minimum Requirements Related to Technical Performance for IMT-2020 Radio Interface(s).* International Telecommunication Union.

15 Haque, M.A., Ahmad, S., Abboud, A.J. et al. (2024). 6G wireless communication networks: challenges and potential solution. *International Journal of Business Data Communications and Networking* 19 (1): 27.

16 Doppler, K., Rinne, M., Wijting, C. et al. (2009). Device-to-device communication as an underlay to LTE-advanced networks. *IEEE Communications Magazine* 47 (12): 42–49. https://doi.org/10.1109/MCOM.2009.5350367.

17 Akhtar, M.W., Hassan, S.A., Ghaffar, R. et al. (2020). The shift to 6G communications: vision and requirements. *Human-centric Computing and Information Sciences.* 10 (1): 53.

18 Nair, M.M. and Tyagi, A.K. (2023). Blockchain technology for next-generation society: current trends and future opportunities for smart era. In: *Blockchain Technology for Secure Social Media Computing.* https://doi.org/10.1049/PBSE019E_ch11.

19 Tyagi, A.K., Kumari, S., Chidambaram, N., and Sharma, A. (2024). Engineering applications of blockchain in this smart era. In: *Enhancing Medical Imaging with Emerging Technologies.* https://doi.org/10.4018/979-8-3693-5261-8.ch011.

20 Ajanthaa, L., Seranmadevi, R., Sree, P.H., and Tyagi, A.K. (2024). Engineering applications of artificial intelligence. In: *Enhancing Medical Imaging with Emerging Technologies.* https://doi.org/10.4018/979-8-3693-5261-8.ch010.

21 Tyagi, A.K., Kukreja, S., Richa, and Sivakumar, P. (2024). Role of blockchain technology in smart era: a review on possible smart applications. *Journal of Information & Knowledge Management* https://doi.org/10.1142/S0219649224500321.

22 Tyagi, A.K. and Tiwari, S. (2024). The future of artificial intelligence in blockchain applications. In: *Machine Learning Algorithms Using Scikit and TensorFlow Environments.* IGI Global https://doi.org/10.4018/978-1-6684-8531-6.ch018.

23 Kumari, S., Thompson, A., and Tiwari, S. (2024). 6G-enabled internet of things-artificial intelligence-based digital twins: cybersecurity and resilience. In: *Emerging Technologies and Security in Cloud Computing.* IGI Global https://doi.org/10.4018/979-8-3693-2081-5.ch016.

24 Nair, M.M. and Tyagi, A.K. (2023). 6G: technology, advancement, barriers, and the future. In: *6G-Enabled IoT and AI for Smart Healthcare.* CRC Press.

25 Nair, M.M. and Tyagi, A.K. (2021). Privacy: history, statistics, policy, laws, preservation and threat analysis. *Journal of Information Assurance & Security* 16 (1): 24–34. 11p.

26 Niu, Y., Li, Y., Jin, D. et al. (2015). A survey of millimeter wave communications (mmWave) for 5G: opportunities and challenges. *Wireless Networks* 21: 2657–2676. https://doi.org/10.1007/s11276-015-0942-z.

27 ITU-R M.[IMT 2020 OUTCOME] (2015). *Framework and Overall Objectives of the Future Development of IMT for 2020 and beyond.* International Telecommunication Union.

28 Chowdhury, M.Z., Shahjalal, M., Ahmed, S. et al. (2020). 6G wireless communication systems: applications, requirements, technologies, challenges, and research directions. *IEEE Open Journal of the Communications Society* 1: 957–975.

29 Yang, P., Xiao, Y., Xiao, M. et al. (2019). 6G wireless communications: vision and potential techniques. *IEEE Network* 33 (4): 70–75.

30 Tyagi, A.K. and Nair, M.M. (2024). Future of industry 5.0 in society 5.0. In: *Industry 4.0, Smart Manufacturing, and Industrial Engineering*, CRC Press, https://doi.org/10.1201/9781003473886-17.

31 Nair, M.M., Mishra, A.K., and Tyagi, A.K. (2023). Fog computing and edge computing: open issues, critical challenges and the road ahead for future. In: *Proceedings of the 2023 Fifteenth International Conference on Contemporary Computing (IC3–2023). Association for Computing Machinery, New York, NY, USA*, 66–76. https://doi.org/10.1145/3607947.3607962.

32 Tyagi, A.K. and Richa (2023). Digital twin technology: opportunities and challenges for smart era's applications. In: *Proceedings of the 2023 Fifteenth International Conference on Contemporary Computing (IC3–2023). Association for Computing Machinery, New York, NY, USA*, 328–336. https://doi.org/10.1145/3607947.3608015.

12

Economic and Business Perspectives of 6G Technology for Modern Society

12.1 Introduction to Necessity of Economic and Business Perspectives of 6G for Modern Society

The evolution of telecommunications technology has consistently catalyzed profound societal transformations, shaping the way we communicate, conduct business, and interact with the world around us [1]. As the world anticipates the arrival of 6G technology, the stakes are higher than ever before. The transition to 6G promises not only faster speeds and lower latency but also a seismic shift in economic and business landscapes globally. Understanding the economic and business perspectives of 6G technology is not just a matter of academic curiosity, it is imperative for navigating the complexities of a hyperconnected world and using the full potential of this next-generation technology for modern society.

The growth of each new generation of wireless technology, from 1G to 5G, has brought about profound changes in how we live and work [2, 3]. 6G, which is the next leap forward, holds the promise of unlocking unique levels of connectivity, enabling a plethora of emerging technologies such as augmented reality (AR), autonomous vehicles (AVs), and the Internet of Things (IoT) to reach their full potential [3–5]. However, realizing this promise requires a complete understanding of the economic and business implications that accompany such a technological paradigm shift.

At its core, the transition to 6G is not just about faster download speeds or smoother streaming experiences, it represents a fundamental restructuring of industries, business models, and societal norms. From healthcare to manufacturing, from agriculture to entertainment, virtually every sector stands to be profoundly impacted by the growth of 6G technology [4]. The ability to use the potential of 6G for economic growth, innovation, and social progress hinges on our ability to anticipate and adapt to the challenges and opportunities that lie ahead. In this context, examining the economic and business perspectives of 6G technology becomes imperative. This involves not only assessing the potential

6G-Enabled Technologies for Next Generation: Fundamentals, Applications, Analysis and Challenges, First Edition. Amit Kumar Tyagi, Shrikant Tiwari, Shivani Gupta, and Anand Kumar Mishra.

revenue streams and cost savings that 6G can enable but also understanding the broader implications for market dynamics, regulatory frameworks, and global competitiveness. Moreover, it involves discussing the ways in which 6G can facilitate the emergence of new business models, disrupt traditional industries, and foster entrepreneurship and innovation.

Furthermore, the economic and business perspectives of 6G extend beyond mere financial issues. They encompass broader societal issues such as digital inclusion, privacy, and sustainability. As 6G technology becomes increasingly intertwined with every aspect of our lives, it is important to ensure that its benefits are equitably distributed, and its risks are effectively managed. In this work, we aim to provide a full exploration of the economic and business perspectives of 6G technology for modern society. By synthesizing insights from diverse disciplines, ranging from economics and business strategy to telecommunications and public policy, we seek to elucidate the multifaceted nature of the 6G revolution and its implications for the future of humanity. Through this endeavor, we hope to contribute to a more informed discourse surrounding the opportunities and challenges that lie ahead as we embark on this transformative journey toward a 6G-enabled future.

12.2 Literature Review

In [26], 6G is planned to be a platform for general-purpose communication that may be utilized in several sectors. The existing 6G visions are not well-structured and lack a common terminology. It offers a framework that is both sustainable and focused on people as they build 6G. It introduced possible future uses of 6G technology and defined key issues. It outlines a plan for the creation of 6G that prioritizes people and is environmentally conscious. This work provided examples and answers to key topics that will help future-proof 6G visioning.

In [27], manufacturers provide 5G legacy solutions to enhance network performance. Network operators are backing 5G technology to satisfy customer expectations. The authors conclude the upcoming 5G features and emphasize the use cases for 6G. It introduces Metaverse as a 6G enabler that consistently solves common problems and highlights 5G's limitations and the room for 6G to fill them. Note that the Metaverse is emphasized in this work as a resource for dealing with everyday energy and problems. In order for machines to communicate with one another, a solution that is both affordable and simple to set up is needed. Remember that the prices of 5G networks are too high for medium-sized firms to afford.

In [28], 6G makes it possible to create eco-friendly company models by utilizing state-of-the-art security technology. 6G technologies can improve the safety of green company models. Innovations in environmentally friendly business models can find technological backing from 6G. Green business model innovation might be made more secure with the use of 6G technology. Problems that need to be fixed include security technologies, protecting intellectual property, and greenwashing. Note that companies should strike a balance in their models between monetary and nonmonetary value.

In [29], 6G services provide a two-sided market by uniting users and service providers. While building 6G as a platform, it is critical to keep an eye on the economy. It takes the 6G idea and the existing white papers and merges them. It contributes to the global 6G vision and attends expert workshops to get insights. The areas of cost-effective deployments, sustainable business models, and killer apps will be challenges for sixth-generation wireless networks. The work provides the significance of developing a cohesive 6G strategy by adopting a comprehensive approach. The existing 6G visions are not well-structured and lack a common terminology. The performance of 6G has not been defined yet, which is a major roadblock/issue for future researchers.

In [30], ecosystems and platforms should be the focal points of innovation. More and more companies are using open-value setups as their business models change.

Hence, with the rapid growth of 6G, value creation and the capacity to capitalize on technological advances will be transformed. It is the responsibility of regulators to safeguard the intellectual property contributions produced in the course of research and development (R&D). Note that controlling for external validity might be challenging in research that is method intensive and focused on foresight. The data might be interpreted differently by several researchers.

12.3 Market Opportunities and Revenue Models Using 6G for Modern Society

The growth of 6G technology brings forth a multitude of market opportunities and revenue models that have the potential to reshape industries and drive economic growth in modern society. Below, we explain some of the key areas where 6G can unlock new opportunities and revenue streams:

- **IoT and Smart Infrastructure**: 6G's ultra-reliable, low-latency connectivity is poised to revolutionize the IoT landscape. With the ability to support a large number of connected devices, including sensors, actuators, and smart appliances, 6G enables the creation of truly interconnected smart cities, buildings, and infrastructure [6, 7]. Companies can capitalize on this by providing IoT solutions for urban management, environmental monitoring, energy efficiency, and more. Revenue models may include subscription-based services, data monetization, and value-added analytics.

- **Augmented Reality (AR) and Virtual Reality (VR)**: 6G's high bandwidth and low latency open up exciting possibilities for immersive AR and VR experiences. From virtual meetings and remote collaboration to interactive gaming and experiential marketing, 6G enables a wide range of applications that blur the line between the physical and digital worlds. Businesses can use this technology to provide premium content, virtual events, and personalized experiences to consumers. Revenue models may involve content licensing, subscription-based platforms, and in-app purchases.
- **Autonomous Vehicles (AVs) and Mobility Services**: 6G's ultra-reliable communication capabilities are important for enabling safe and efficient AVs and mobility services. By providing real-time data transmission and high-precision positioning, 6G facilitates vehicle-to-everything (V2X) communication, autonomous navigation, and intelligent traffic management [8]. Companies operating in the automotive and transportation sectors can seize opportunities in AV technology development, fleet management, and mobility-as-a-service (MaaS) offerings. Revenue models may include vehicle sales, subscription-based mobility platforms, and advertising.
- **Healthcare and Telemedicine**: 6G's high-speed, low-latency connectivity has the potential to revolutionize healthcare delivery and telemedicine. From remote patient monitoring and teleconsultations to immersive surgical training and VR therapy, 6G enables a wide range of innovative healthcare solutions. Companies can discuss opportunities in telehealth platforms, medical IoT devices, and digital health services. Revenue models may include teleconsultation fees, subscription-based wellness apps, and remote patient monitoring services.
- **Entertainment and Media**: 6G's ultra-high-definition streaming capabilities and interactive features provide new opportunities for the entertainment and media industry. From live sports broadcasting and virtual concerts to interactive storytelling and personalized content recommendations, 6G enables immersive and engaging experiences for consumers. Companies can capitalize on this by providing premium content, interactive platforms, and targeted advertising solutions. Revenue models may involve content licensing, pay-per-view events, and subscription-based streaming services.
- **Industrial Automation and Industry 4.0**: 6G's ultra-reliable, low-latency connectivity is instrumental in advancing industrial automation and the transition to Industry 4.0. From smart factories and remote monitoring to predictive maintenance and collaborative robotics, 6G enables greater efficiency, flexibility, and productivity in manufacturing and logistics [9]. Companies can use this technology to provide industrial IoT solutions, automation platforms, and predictive analytics services. Revenue models may include equipment sales, software licensing, and maintenance contracts (refer to Figure 12.1).

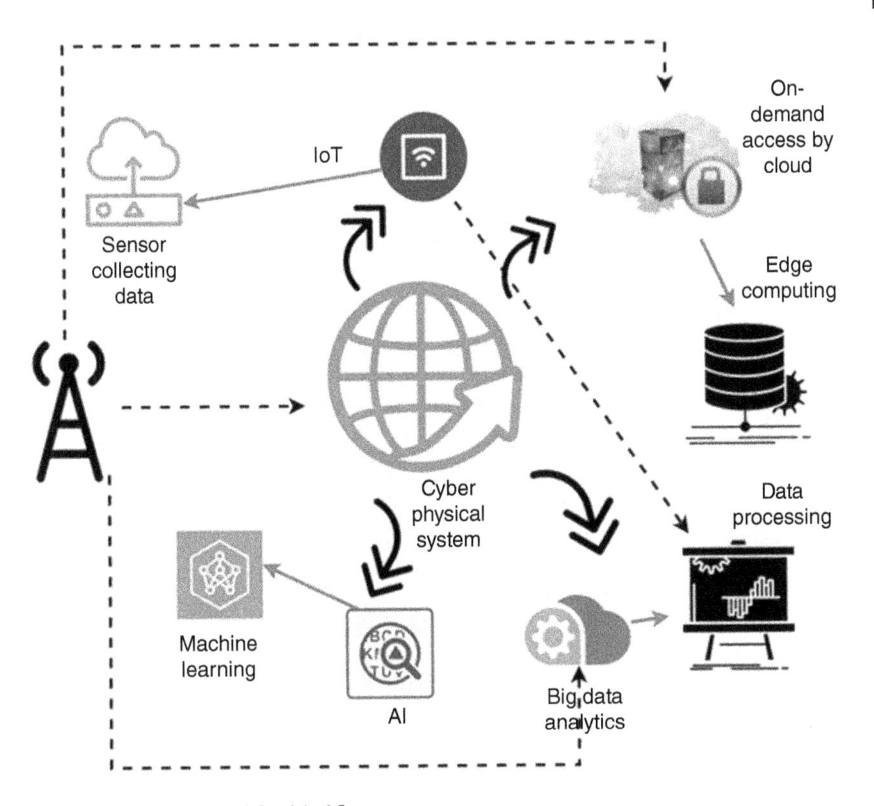

Figure 12.1 Industry 4.0 with 6G.

In summary, the market opportunities and revenue models enabled by 6G technology are diverse and expansive, spanning multiple sectors and applications. By using the power of 6G, businesses can unlock new sources of value creation, drive innovation, and meet the evolving needs of modern society. However, realizing these opportunities requires strategic foresight, technological expertise, and a deep understanding of market dynamics and consumer preferences. As we embark on the 6G era, companies that use innovation and adaptability will be well-positioned to thrive in a rapidly evolving digital landscape.

12.4 Business Challenges and Monetization Strategies for 6G-Based Modern Society

While the potential of 6G technology is vast, there are several business challenges that need to be addressed in order to effectively monetize its capabilities and capitalize on the opportunities it presents [10, 11]. Below, we outline some of these

Table 12.1 Type of challenges and Monetization Strategies for 6G-Based Modern Society.

Types of Challenges	Explanation	Monetization Strategies
Infrastructure Investment	One of the primary challenges in deploying 6G networks is the huge investment required to build the necessary infrastructure, including base stations, antennas, and fiber optic cables.	Telecom operators can discuss innovative financing models such as public-private partnerships, infrastructure-sharing agreements, and revenue-sharing arrangements with technology providers. Additionally, they can use network slicing and software-defined networking (SDN) to optimize resource utilization and reduce operational costs.
Regulatory Difficulties	Regulatory constraints, including spectrum allocation, licensing requirements, and compliance with data privacy regulations, face major challenges for 6G deployment.	Companies can engage with regulatory authorities to advocate for favorable policies and regulations that support 6G innovation and investment. They can also invest in compliance and cybersecurity measures to build trust with consumers and regulators, thereby mitigating regulatory risks.
Interoperability and Standards	The heterogeneous nature of 6G technologies and the lack of standardized protocols can hinder interoperability and continuous integration across different networks and devices.	Businesses can collaborate with industry consortia, standards bodies, and academic institutions to develop interoperable solutions and establish common standards for 6G technology. By promoting open innovation and interoperability, companies can unlock new revenue opportunities through ecosystem development and cross-platform compatibility.
Security and Privacy Issues	As 6G networks become increasingly interconnected and ubiquitous, they also become more vulnerable to cyber threats and privacy breaches.	Companies can invest in robust cybersecurity measures, including encryption, authentication, and threat detection technologies, to protect their networks and protect sensitive data. They can also provide cybersecurity solutions and services to customers, generating additional revenue streams while enhancing trust and confidence in their offerings.
Digital Divide and Inequality	The rapid rise of 6G technology risks exacerbating existing inequalities, as underserved communities may lack access to affordable broadband services and digital skills training.	Businesses can partner with government agencies, nonprofit organizations, and community users to bridge the digital divide through initiatives such as subsidized broadband access, digital literacy programs, and community broadband projects. By expanding access to 6G technology and promoting digital inclusion, companies can tap into new markets and foster long-term customer loyalty.
Monetization of Emerging Applications	While 6G technology enables a wide range of innovative applications and services, monetizing these offerings may pose challenges due to uncertainty around consumer demand, pricing models, and revenue-sharing arrangements.	Companies can adopt agile business models that allow for experimentation and adaptation to changing market dynamics. They can use data analytics and market insights to identify emerging trends and customer preferences, thereby tailoring their offerings to meet evolving needs and preferences.

challenges and propose monetization strategies to overcome them, as shown in Table 12.1.

In summary, addressing the business challenges associated with 6G deployment requires a combination of strategic foresight, technological innovation, and collaborative partnerships. By adopting a proactive approach and using monetization strategies tailored to their specific context and objectives, businesses can navigate the complexities of the 6G landscape and unlock new sources of value creation in modern society.

12.5 Industry Collaboration and Partnerships Required for 6G-Based Modern Society

The successful deployment and realization of the full potential of 6G technology require extensive collaboration and partnerships across various industries [12, 13]. Here are some key areas where industry collaboration and partnerships are essential for building a 6G-enabled modern society:

- **Telecommunications and Technology Providers**: Telecom companies and technology providers play a central role in developing and deploying 6G infrastructure and services. Collaboration among these entities is important for driving innovation, standardization, and interoperability in 6G networks. Partnerships can focus on joint R&D initiatives, technology trials, and ecosystem development to accelerate the commercialization of 6G technology.
- **Government and Regulatory Agencies**: Government agencies and regulatory bodies play an important role in shaping the regulatory framework and policy environment for 6G deployment. Collaboration between industry users and policymakers is essential for addressing regulatory difficulties, spectrum allocation, and privacy issues. Partnerships can involve policy advocacy, regulatory consultations, and public–private partnerships to foster a conducive environment for 6G innovation and investment.
- **Academic and Research Institutions**: Academic and research institutions are at the forefront of 6G technology development, conducting cutting-edge research, and advancing state-of-the-art technologies. Collaboration between academia and industry facilitates knowledge exchange, talent development, and technology transfer. Partnerships can involve joint research projects, technology incubators, and academic-industry consortia to drive innovation and skill development in 6G-related fields.
- **Vertical Industries and End-User Sectors**: Collaboration with vertical industries and end-user sectors is essential for understanding specific use cases and requirements for 6G technology. Industries such as healthcare, automotive,

manufacturing, and entertainment can provide valuable insights into their unique challenges and opportunities. Partnerships can involve co-innovation workshops, pilot projects, and industry alliances to cocreate 6G-enabled solutions tailored to specific verticals.

- **Startups and Innovation Ecosystems**: Startups and innovation ecosystems play a vital role in driving entrepreneurship and innovation in the 6G space. Collaboration with startups provides established companies access to disruptive technologies, agile development methodologies, and entrepreneurial talent. Partnerships can involve accelerator programs, venture capital investments, and technology scouting initiatives to identify and nurture promising startups in the 6G ecosystem.
- **International Collaboration and Standardization Bodies**: Collaboration at the international level is essential for harmonizing standards, promoting interoperability, and facilitating global deployment of 6G technology. Participation in standardization bodies such as the International Telecommunication Union (ITU) and the Institute of Electrical and Electronics Engineers (IEEE) enables industry users to contribute to the development of common standards and best practices. Partnerships can involve collaborative research projects, working groups, and joint conferences to facilitate knowledge sharing and alignment on 6G standards and specifications.

In summary, industry collaboration and partnerships are essential for realizing the vision of a 6G-enabled modern society. By using collaboration among various users, sharing resources and expertise, and aligning efforts toward common goals, industry users can accelerate the development and deployment of 6G technology and unlock its full potential to drive economic growth, innovation, and societal progress.

12.6 An Open Discussion for 6G-Based Modern Society to Achieve Sustainable Goals

An open discussion on how 6G technology can contribute to achieving sustainable goals in modern society. As we embark on this conversation, let us consider the intersection between 6G technology and sustainability and discuss potential avenues for using 6G to address pressing environmental, social, and economic challenges [14, 15]. Figure 12.2 shows a quintuple helix model and sustainable development of 6G.

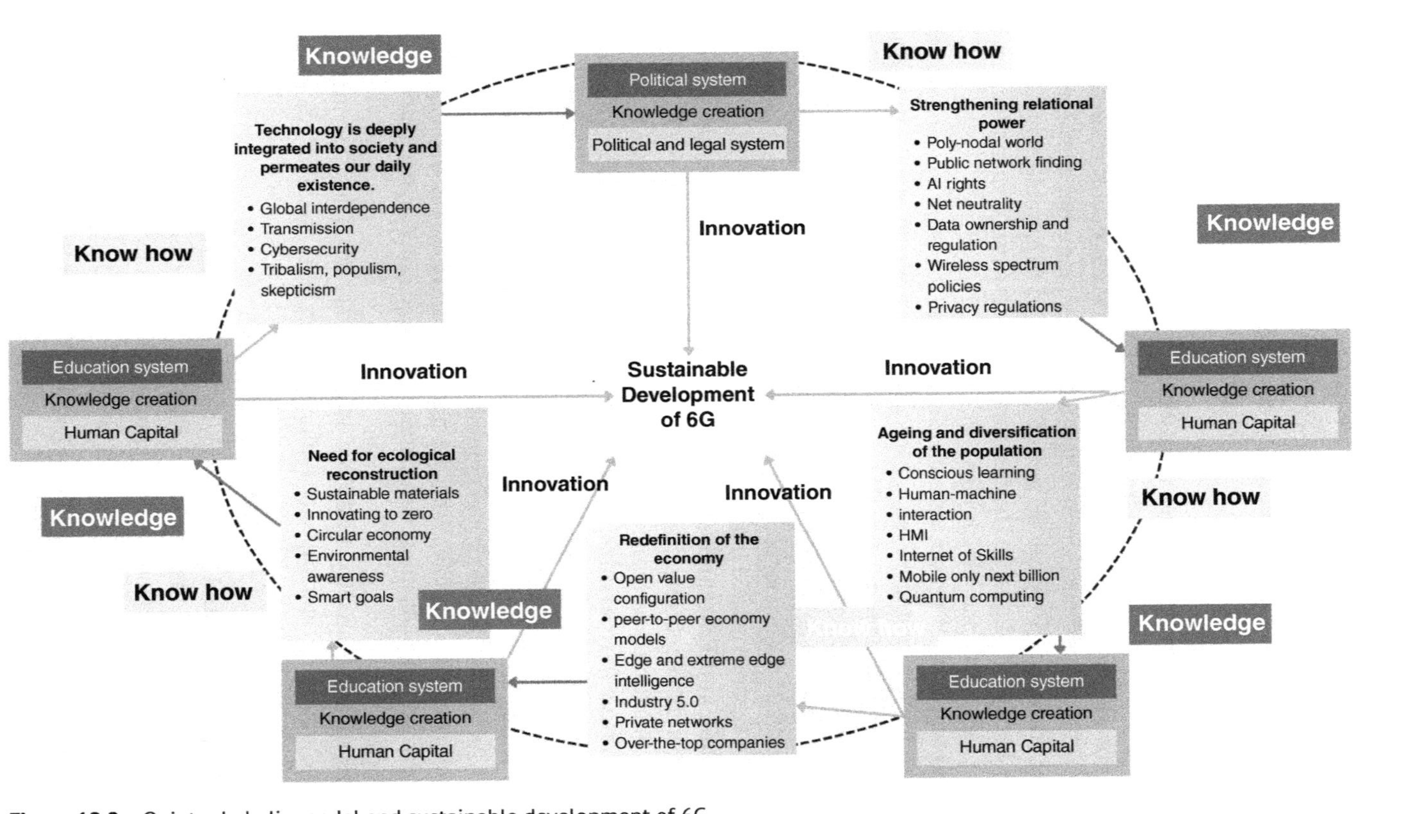

Figure 12.2 Quintuple helix model and sustainable development of 6G.

Here are some key questions to kickstart our discussion:

- **Environmental Sustainability**: How can 6G technology be designed and deployed in an environmentally sustainable manner? What are the opportunities and challenges in reducing the carbon footprint of 6G infrastructure and operations?
- **Energy Efficiency**: How can 6G networks be optimized for energy efficiency, considering the increasing energy demands associated with higher data speeds and network densification? What role can renewable energy sources and energy-efficient hardware play in mitigating the environmental impact of 6G?
- **Digital Inclusion**: How can 6G technology be used to promote digital inclusion and bridge the digital divide, ensuring that underserved communities have access to affordable and reliable broadband services?
- **Equitable Access**: What strategies can be employed to ensure equitable access to 6G technology and its benefits, particularly in rural and remote areas and among marginalized populations?
- **Privacy and Data Security**: How can privacy and data security issues associated with 6G technology be addressed to protect user privacy and protect sensitive data [15, 16]? What regulatory frameworks and technological solutions are needed to ensure trust and transparency in 6G networks?
- **Innovation and Entrepreneurship**: How can 6G technology foster innovation and entrepreneurship, particularly in addressing sustainability challenges such as climate change, resource management, and disaster response?
- **Collaborative Partnerships**: What opportunities exist for collaborative partnerships between industry users, government agencies, nonprofit organizations, and academia to use 6G technology for sustainable development?
- **Smart Cities and Communities**: How can 6G technology enable the development of smart cities and communities that are resilient, inclusive, and environmentally sustainable? What are some innovative use cases and applications of 6G in urban planning, transportation, energy management, and public services?
- **Education and Capacity Building**: How can 6G technology be used to enhance education and capacity building, particularly in STEM (science, technology, engineering, and mathematics) fields, to prepare the workforce for the jobs of the future?
- **Global Cooperation and Standards**: What role can international cooperation and standardization efforts play in ensuring the interoperability and compatibility of 6G networks on a global scale? How can global users collaborate to address shared challenges and achieve common goals?

Note that there are many questions and topics that we can discuss together in this discussion. We engage in constructive dialogue and brainstorm innovative solutions that use the transformative power of 6G for the benefit of society.

12.7 Technical, Nontechnical Issues and Challenges Toward Implementing 6G for Business/Economic Reforms for Betterment of Society

6G technology for business and economic reforms, with the aim of betterment of society, presents a range of technical and nontechnical issues and challenges [10, 12]. We discuss these in depth:

12.7.1 Technical Issues and Challenges

- **Network Infrastructure**: Building the necessary infrastructure for 6G networks, including base stations, antennas, and fiber optic cables, requires huge investment and coordination. Scaling up infrastructure to support the high data rates and low latency of 6G poses technical challenges in terms of deployment, maintenance, and scalability.
- **Spectrum Allocation**: Securing sufficient spectrum for 6G networks is important for ensuring adequate bandwidth and capacity. However, spectrum allocation processes are complex and subject to regulatory constraints, facing challenges for operators in acquiring and optimizing spectrum resources for 6G deployment.
- **Interference and Signal Propagation**: 6G networks operate at higher frequencies than previous generations, which may lead to increased susceptibility to interference and signal attenuation. Overcoming these challenges requires advancements in antenna technology, signal processing, and network optimization to ensure reliable connectivity and coverage.
- **Energy Consumption**: The energy consumption of 6G networks, particularly with the deployment of massive multiple input multiple output (MIMO) and beamforming technologies, raises issues about sustainability and environmental impact. Developing energy-efficient hardware and algorithms is essential for minimizing the carbon footprint of 6G networks.
- **Security and Privacy**: With the rapid rise of connected devices and the exponential growth of data traffic, ensuring the security and privacy of 6G networks becomes increasingly challenging. Addressing vulnerabilities such as network attacks, data breaches, and privacy infringements requires robust encryption, authentication, and cybersecurity measures [17–22].

12.7.2 Nontechnical Issues and Challenges

- **Regulatory Frameworks**: Regulatory frameworks governing spectrum allocation, licensing, data privacy, and competition play an important role in shaping the deployment and operation of 6G networks. Harmonizing regulations across jurisdictions and adapting existing frameworks to accommodate 6G technology presents challenges for policymakers and industry users.

- **Digital Divide**: The digital divide between urban and rural areas, as well as disparities in access to affordable broadband services, presents a major challenge to realizing the full potential of 6G technology. Bridging the digital divide requires concerted efforts to expand infrastructure deployment, reduce connectivity costs, and promote digital literacy and inclusion.
- **Skills Gap**: The rapid pace of technological innovation in the 6G era exacerbates the skills gap in the workforce, particularly in STEM fields and digital technologies. Addressing the skills gap through education, training, and lifelong learning initiatives is essential for preparing the workforce for the jobs of the future and using innovation and competitiveness.
- **Business Models and Monetization**: Developing sustainable business models and monetization strategies for 6G services poses challenges in terms of revenue generation, cost management, and value proposition. Identifying viable revenue streams, pricing models, and value-added services that align with consumer preferences and market dynamics requires market research, experimentation, and adaptability.
- **Ethical and Societal Implications**: The adoption of 6G technology raises ethical and societal implications related to privacy, surveillance, autonomy, and inequality. Addressing these issues requires proactive engagement with users, transparent communication, and ethical issues in the design and implementation of 6G systems.

Hence, addressing the technical and nontechnical issues and challenges associated with implementing 6G technology for business and economic reforms requires a collaborative and multidisciplinary approach. By using technological innovation, regulatory reform, education and skills development, and ethical issues, users can overcome barriers and unlock the transformative potential of 6G for the betterment of society.

12.8 Future Research Opportunities Toward Implementing 6G for Business/Economic Reforms for Betterment of Society

Future research opportunities toward implementing 6G for business and economic reforms, aimed at the betterment of society, abound in various areas. Here, Figure 12.3 shows the different visions of 6G in terms of the human, digital, and physical world.

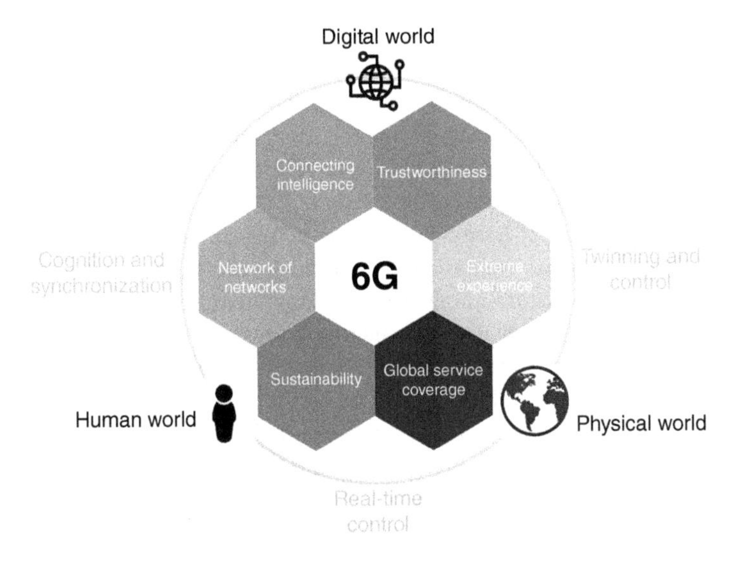

Figure 12.3 6G vision.

Here are some potential research directions:

- **Sustainable Infrastructure Deployment**: Researcher can focus on developing sustainable and cost-effective approaches to deploying 6G infrastructure, including green energy solutions, efficient resource utilization, and environmentally friendly materials.
- **Dynamic Spectrum Management**: Investigating dynamic spectrum management techniques, such as cognitive radio and spectrum sharing, can optimize spectrum utilization and accommodate diverse applications while ensuring equitable access and minimizing interference.
- **Energy-Efficient Network Design**: Researcher can discuss energy-efficient network architectures, protocols, and algorithms to minimize the energy consumption of 6G networks while maintaining high performance and reliability.
- **Security and Privacy Enhancements**: Advancing cybersecurity and privacy-preserving technologies for 6G networks, including secure authentication, encryption, and intrusion detection, can mitigate security threats and protect user data in an increasingly connected and digitized world [17, 23–25].
- **Inclusive Connectivity Solutions**: Researcher can focus on developing inclusive connectivity solutions for underserved and marginalized communities, such as rural areas, developing regions, and vulnerable populations, to bridge the digital divide and promote digital inclusion.
- **Business Model Innovation**: Investigating innovative business models and monetization strategies for 6G services, including subscription-based models,

pay-per-use schemes, and value-added services, can drive revenue generation and market growth while aligning with societal needs and preferences.

- **Regulatory and Policy Frameworks**: Researcher can discuss regulatory and policy frameworks that foster innovation, competition, and investment in 6G technology while addressing issues related to spectrum management, data privacy, consumer protection, and market competition.
- **Social and Economic Impact Assessment**: Conducting studies on the social and economic impacts of 6G technology adoption, including its effects on employment, income distribution, productivity, and quality of life, can inform evidence-based policymaking and decision-making processes.
- **Collaborative Innovation Ecosystems**: Investigating collaborative innovation ecosystems and partnership models that promote knowledge sharing, technology transfer, and ecosystem development among industry users, academic institutions, government agencies, and nonprofit organizations can accelerate 6G innovation and deployment.
- **Ethical and Societal Issues**: Researcher can discuss ethical and societal issues associated with 6G technology, including issues of privacy, surveillance, autonomy, fairness, and accountability, to ensure that technological advancements are aligned with societal values and principles.

Note that by addressing these research opportunities, scholars, policymakers, industry users/experts, and civil society can contribute to the responsible and sustainable deployment of 6G technology for the betterment of society, using economic growth, social inclusion, and environmental sustainability in the process.

12.9 Summary

The growth of 6G technology heralds a new era of connectivity and innovation, with profound implications for economic growth, business transformation, and societal progress. In this paper, we have discussed the economic and business perspectives of 6G technology and its potential to reshape industries, drive innovation, and adopt sustainable development in modern society. From infrastructure investment to regulatory frameworks, from market opportunities to revenue models, the transition to 6G presents both opportunities and challenges for businesses, policymakers, and society at large. By using the transformative power of 6G, users can unlock new sources of value creation, enhance productivity, and improve the quality of life for individuals and communities around the world.

However, realizing the full potential of 6G requires a collaborative and multidisciplinary approach, encompassing technical innovation, regulatory reform, business model innovation, and societal engagement. It requires proactive efforts to address challenges such as infrastructure deployment, spectrum management, cybersecurity, digital inclusion, and ethical issues.

As we navigate the complexities of the 6G landscape, it is essential to remain mindful of the broader societal impacts of technological advancements and ensure that the benefits of 6G are equitably distributed and sustainable over the long term. By using inclusive growth, protecting privacy and security, and promoting responsible innovation, we can use the transformative potential of 6G technology to create a more prosperous, equitable, and resilient society for future generations.

In summary, the economic and business perspectives of 6G technology indicate a future of endless possibilities, where connectivity knows no bounds, and innovation knows no limits. By using the opportunities and addressing the challenges of the 6G era, we can make the way for a brighter, more connected, and more wealthy future for all.

References

1 Lee, J.H. and Kim, S.H. (2021). 6G technology: future directions and business opportunities. *Journal of Industrial Information Integration* 23: 100234. https://doi.org/10.1016/j.jii.2021.100234.

2 Ahokangas, P. and Yrjölä, S. (2021). Scenarios and business models for 6G. In: *2021 IEEE 93rd Vehicular Technology Conference (VTC2021-Spring)*, 1–6. IEEE https://doi.org/10.1109/VTC2021-Spring51267.2021.9448899.

3 Ghosh, A. and Das, A. (2021). Economics of 6G wireless networks: A comprehensive overview. arXiv preprint arXiv:2107.09952.

4 Taibi, D., Park, S., and Kim, D.I. (2021). Building 6G economy: analysis of business model and economic potential. *IEEE Transactions on Network and Service Management* https://doi.org/10.1109/TNSM.2021.3111370.

5 Galinina, O., Andreev, S., Samuylov, A., and Koucheryavy, Y. (2021). Potential of 6G wireless systems: key features and enabling technologies. *IEEE Network* 35 (4): 6–13. https://doi.org/10.1109/MNET.011.2100351.

6 Leppänen, T. and Andreev, S. (2021). 6G business opportunities: spectrum perspective. In: *2021 IEEE 93rd Vehicular Technology Conference (VTC2021-Spring)*, 1–5. IEEE https://doi.org/10.1109/VTC2021-Spring51267.2021.9448869.

7 Samie, F., Chen, J., Saad, W., and Bennis, M. (2021). 6G wireless networks: vision, research challenges, and prospects. *IEEE Transactions on Wireless Communications* 20 (3): 1711–1726. https://doi.org/10.1109/TWC.2021.3040845.

8 Tyagi, A.K. and Aswathy, S.U. (2021). Autonomous intelligent vehicles (AIV): research statements, open issues, challenges and road for future, international journal of intelligent. *Networks* 2: 83–102, ISSN 2666-6030. https://doi.org/10.1016/j.ijin.2021.07.002.

9 Tyagi, A.K., Fernandez, T.F., Mishra, S., and Kumari, S. (2021). Intelligent automation systems at the core of Industry 4.0. In: *Intelligent Systems Design and Applications. ISDA 2020. Advances in Intelligent Systems and Computing*, vol. 1351 (ed. A. Abraham, V. Piuri, N. Gandhi, et al.). Cham: Springer https://doi.org/10.1007/978-3-030-71187-0_1.

10 Al-Masri, S., Alsmadi, I., and Rawashdeh, H. (2021). The potential business models for 6G. In: *International Conference on Smart Cities and Green ICT Systems*, 90–100. Cham: Springer https://doi.org/10.1007/978-3-030-83147-7_8.

11 Piran, M.J., Chen, L., and Khan, M.S. (2021). 6G communication systems: a comprehensive review. *IEEE Access* 9: 55313–55336. https://doi.org/10.1109/ACCESS.2021.3065697.

12 Chen, H., Wang, M., Liu, X. et al. (2021). New opportunities for the industry in the 6G era: a business model perspective. *Wireless Communications and Mobile Computing* 2021: https://doi.org/10.1155/2021/9923473.

13 Yang, C., Cheng, N., Liang, Y. et al. (2021). 6G: opportunities and challenges from industrial perspective. *China Communications* 18 (8): 1–14. https://doi.org/10.1109/CC.2021.9481424.

14 Majumder, A., Hasan, S.M.M., and Khan, M.S. (2021). 6G wireless communication networks: technologies, challenges, and future trends. *IEEE Internet of Things Journal* https://doi.org/10.1109/JIOT.2021.3106578.

15 Hu, R.Q. and Qian, Y. (2021). 6G: a vision of intelligent wireless communication systems. *China Communications* 18 (4): 1–18. https://doi.org/10.1109/CC.2021.9397455.

16 Saad, W., Bennis, M., and Chen, M. (2021). A vision of 6G wireless systems: applications, trends, technologies, and open research problems. *IEEE Network* 35 (4): 9–16. https://doi.org/10.1109/MNET.011.2100394.

17 Nair, M.M. and Tyagi, A.K. (2023). Blockchain technology for next-generation society: current trends and future opportunities for smart era. In: Blockchain Technology for Secure Social Media Computing https://doi.org/10.1049/PBSE019E_ch11.

18 Tyagi, A.K., Kumari, S., Chidambaram, N., and Sharma, A. (2024). Engineering applications of blockchain in this smart era. In: *Enhancing Medical Imaging with Emerging Technologies*. https://doi.org/10.4018/979-8-3693-5261-8.ch011.

19 Ajanthaa, L., Seranmadevi, R., Sree, P.H., and Tyagi, A.K. (2024). Engineering applications of artificial intelligence. In: *Enhancing Medical Imaging with Emerging Technologies*. https://doi.org/10.4018/979-8-3693-5261-8.ch010.

20 Tyagi, A.K., Kukreja, S., Richa, and Sivakumar, P. (2024). Role of blockchain technology in smart era: a review on possible smart applications. *Journal of Information & Knowledge Management* https://doi.org/10.1142/S0219649224500321.

21 Tyagi, A.K. and Tiwari, S. (2024). The future of artificial intelligence in blockchain applications. In: *Machine Learning Algorithms Using Scikit and TensorFlow Environments*. IGI Global https://doi.org/10.4018/978-1-6684-8531-6.ch018.

22 Kumari, S., Thompson, A., and Tiwari, S. (2024). 6G-enabled internet of things-artificial intelligence-based digital twins: cybersecurity and resilience. In: *Emerging Technologies and Security in Cloud Computing*. IGI Global https://doi.org/10.4018/979-8-3693-2081-5.ch016.

23 Nair, M.M. and Tyagi, A.K. (2023). 6G: technology, advancement, barriers, and the future. In: *6G-Enabled IoT and AI for Smart Healthcare*. CRC Press.

24 Nair, M.M. and Tyagi, A.K. (2021). Privacy: history, statistics, policy, laws, preservation and threat analysis. *Journal of Information Assurance & Security* 16 (1): 24–34. 11p.

25 Kumar, S., Singh, A., Sharma, R., and Singh, D. (2021). 6G technology: potential features, challenges, and research directions. In: *Advances in Wireless Communications*, 13–32. Singapore: Springer https://doi.org/10.1007/978-981-16-2817-6_2.

26 Petri, A., Marja, M.-B., and Seppo, Y. (2023). Envisioning a future-proof global 6G from business, regulation, and technology perspectives. *IEEE Communications Magazine* https://doi.org/10.1109/MCOM.001.2200310.

27 Schwenteck, P., Nguyen, G.T., Boche, H. et al. (2023). 6G perspective of mobile network operators, manufacturers, and verticals. *IEEE Networking Letters* https://doi.org/10.1109/lnet.2023.3266863.

28 Peter, L. (2022). 6G technologies - how can it help future green business model innovation. *Journal of ICT Standardisation* https://doi.org/10.13052/jicts2245-800x.1012.

29 Ahokangas, P., Matinmikko-Blue, M., and Yrjölä, S. (2022). Envisioning a future-proof global 6G from business, regulation, and technology perspectives. *IEEE Communications Magazine* 61 (2): 72–78.

30 Yrjölä, S.S., Ahokangas, P., and Matinmikko-Blue, M. (2022). Value creation and capture from technology innovation in the 6G era. *IEEE Access* 10: 16299–16319.

13

Ethical and Social Implications of Using Artificial Intelligence in 6G Networks

13.1 Introduction to AI-Based 6G Networks, Systems, and Communication

The emergence of sixth-generation (6G) networks marks a major milestone in the evolution of telecommunications, introducing a new level of connectivity, speed, and reliability. At the front of this technological revolution lies the integration of artificial intelligence (AI), which is used to revolutionize the way networks, systems, and communications operate [1]. It provides an overview of the transformative potential of AI-based 6G networks, with explaining the key concepts and implications driving this major change.

Evolution of Communication Networks: This begins by tracing the evolutionary trajectory of communication networks, from the growth of 1G to the forthcoming era of 6G. Each successive generation has brought about major changes in connectivity, introducing new capabilities and enabling transformative applications. The transition to 6G represents an essential forward in terms of speed, latency, and capacity, laying the foundation for futuristic use cases like augmented reality, autonomous vehicles, and the Internet of Things (IoT).

AI Integration in 6G Networks: Central to the advancements of 6G networks is the integration of AI technologies, which hold the promise of optimizing network performance, enhancing user experience, and enabling innovative applications [2]. AI algorithms can dynamically allocate resources, predict network congestion, and optimize energy consumption, thereby ensuring efficient and reliable communication services. AI-driven functionalities like intelligent routing, predictive maintenance, and anomaly detection are used to revolutionize network management and operation.

Systems and Communication in AI-Based 6G Networks: It discusses the details of AI-based systems and communication within the context of 6G networks. It explains how AI algorithms are deployed across various network components, including base stations, antennas, edge computing nodes, and user

6G-Enabled Technologies for Next Generation: Fundamentals, Applications, Analysis and Challenges, First Edition. Amit Kumar Tyagi, Shrikant Tiwari, Shivani Gupta, and Anand Kumar Mishra.

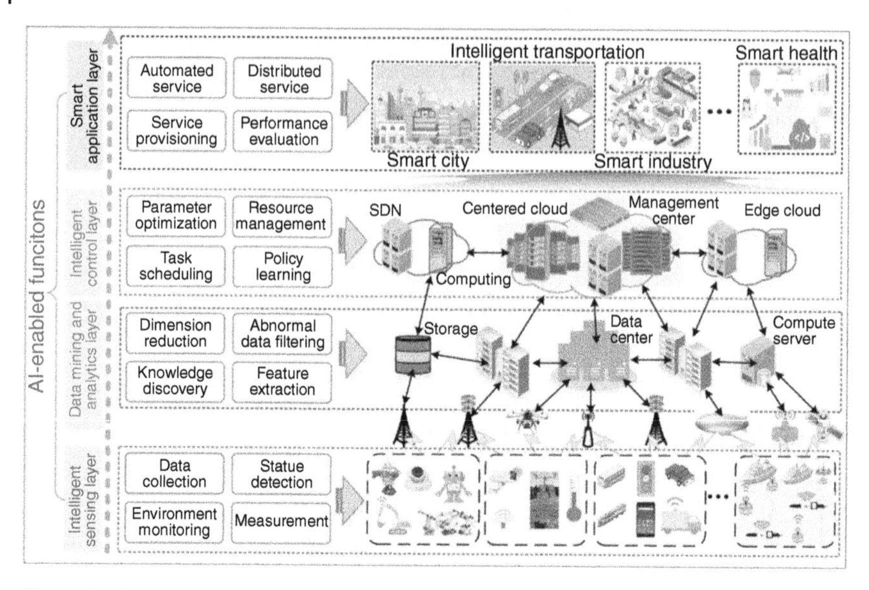

Figure 13.1 The architecture of AI-enabled 6G networks.

devices, to orchestrate continuous communication experiences. Furthermore, it highlights the role of machine learning techniques in enabling adaptive modulation and coding schemes, beamforming, and spectrum management, thereby optimizing spectral efficiency and network performance. Now the architecture of AI-enabled 6G networks can be depicted in Figure 13.1. Also, further Figure 13.2 discusses about 6G architecture for vertical applications in near future.

Implications and Challenges: While the integration of AI in 6G networks provides transformative potential, it also presents several issues and challenges. Ethical issues regarding data privacy, algorithmic bias, and societal impacts necessitate careful scrutiny and proactive measures. Moreover, technical challenges like interoperability, security vulnerabilities, and regulatory compliance must be addressed to ensure the robustness and reliability of AI-based 6G networks.

In summary, this section provides a stage for a complete exploration of AI-based 6G networks, laying the groundwork for understanding the transformative potential, implications, and challenges associated with this convergence of cutting-edge technologies like AI, blockchain technology, etc. By using the power of AI to augment communication networks, 6G has power to reshape the digital landscape and unlock new stages/possibilities of innovation and connectivity.

13.2 Literature Review

In [3], for 6G autonomous systems to be successful, they must exhibit both reliability and moral accountability. The decision-making process in 6G networks requires very little, if any, human intervention. A groundwork for 6G autonomous

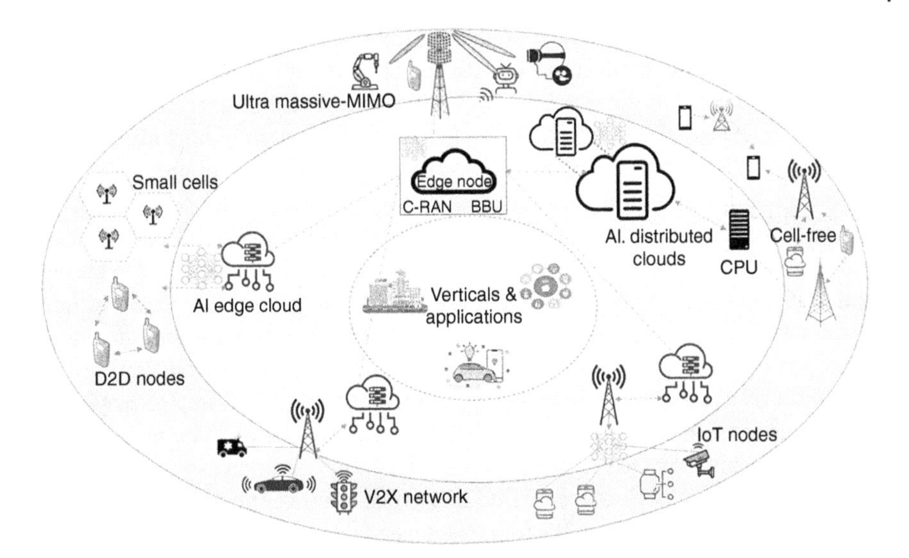

Figure 13.2 6G architecture for vertical applications in near future.

systems that are reliable and socially conscious and a closed control loop whose objective is to lessen the role of humans in decision-making. 6G autonomous systems must prioritize trust and ethical responsibility. Less human intervention is required in the decision-making process for 6G networks compared to earlier generations. A major limitation of 6G autonomous systems is their unreliability and lack of ethical responsibility.

In [4], reliability and moral accountability are two important qualities that 6G autonomous systems should possess. We anticipate a highly automated decision-making process with minimal human involvement. The first step toward trustworthy and ethically aware 6G autonomous systems is the closed control loop with the goal of reducing human involvement in decision-making. People with persistent moral compass and honesty are needed to build 6G autonomous systems. The biggest difficulties in building 6G autonomous systems are dependability and ethical responsibility.

In [5], the authors discussed the ethical issues faced by IoT devices in 6G networks. In this, researchers from different streams talk about how important their knowledge is and also analyzes the 6G network development paradigms. Also, ethical issues in 6G networks, and in particular IoT devices, are discussed in this work. Dealing with technological breakthroughs requires an interdisciplinary strategy, which is discussed in this work.

In [6], electronic healthcare provided by blockchain in 6G networks presents certain ethical challenges. An individual's right to be forgotten and the need for accurate information are fundamental ethical standards. A comprehensive examination of the moral dilemmas raised by electronic healthcare systems

(EHS) built on the blockchain. It provides a holistic view of the ethical concerns raised by healthcare applications enabled by blockchain technology. There were no moral dilemmas in the section on data ownership and accountability.

In [7], 6G networks accomplish all of these goals and more, including social responsibility, environmental reform, and autonomous operation. Using 6G open-source software, fully autonomous networks may be operated and maintained from start to finish. A comprehensive overview of the 6G open-source software's design, core technologies, and goals. An evolutionary road map and technical possibilities for open-source software in the 5G–6G transition. Open-source software for 6G networks should have the ability to automate and intelligently twin digital networks at a high level. Along with facilitating the execution of standard 6G business scenarios, the corporation is also playing a bigger role in the ESG (environment, society, and governance) domains.

In [8], technology, legislation, ethics, and administration are all touched by the problems of privacy, security, and trust. Connecting the physical and digital worlds with 6G networks requires dependable systems. Issues of privacy, security, and trust arise while building a reliable 6G network. Things like privacy solutions, security automation, and trust modeling are necessary. For 6G to be reliable, trust must be built in so that data can be better protected. To guarantee national security, 6G networks must undergo thorough security architectural planning. Regarding cloud trust services, make sure that all users are treated fairly. The challenge of determining the point at which de-identified datasets can be directly associated with persons.

In [9], politics, technology, and personal data protection are all ethical factors to think about. A shift in wireless technology's trajectory toward connected intelligence has social implications. 6G has several benefits for cellular and satellite communications. Several problems and possible solutions are brought about by new networks and technology. Note that 6G is changing the face of wireless networks with its capacity to provide infinitely rapid connectivity. Questions of ethics, new technologies, politics, and national security are all challenges, for example, the creation of an intelligent and secure network is one of the constraints.

13.3 Privacy, Surveillance, and Data Ethics for 6G-Based Systems, and Communication: From Society's Perspective

In the innovative era of 6G networks, the convergence of advanced communication systems and AI presents huge opportunities for connectivity and innovation [10, 11]. However, the rapid growth of 6G-based systems also raises several issues regarding privacy, surveillance, and data ethics from society's perspective.

It provides an overview of several implications of 6G-based systems and communication, which discuss the societal dimensions of privacy, surveillance, and data ethics.

Privacy in the Digital Age: This begins by acknowledging the fundamental importance of privacy in the digital age, wherein individuals increasingly rely on interconnected technologies for communication, commerce, and social interaction. With the growth of 6G networks, which provide enhanced capabilities like ultra-fast speeds, low latency, and ubiquitous connectivity, the volume and granularity of personal data generated and exchanged are poised to escalate exponentially. Consequently, safeguarding individuals' privacy rights against unauthorized access, data breaches, and unidentified/untraceable surveillance emerges as an essential issue of 6G.

Surveillance and Control: The proliferation of 6G-based systems raises issues about the potential for persistent surveillance and control, both by state actors and corporate entities. Advanced technologies like AI-powered surveillance cameras, facial recognition systems, and geolocation tracking face risks of undiscriminating monitoring and profiling, attacking on individuals' rights to autonomy, anonymity, and freedom of expression. Moreover, the deployment of surveillance technologies in public spaces, workplaces, and private domains amplifies issues regarding mass surveillance, social control, and the erosion of civil liberties.

Data Ethics and Governance: This discusses the importance of ethical issues and governance frameworks in shaping the development and deployment of 6G-based systems. From data collection to data processing, algorithmic decision-making and user consent, ethical principles like transparency, accountability, and fairness must note down the efficient design and operation of 6G networks. Moreover, active measures are needed to address algorithmic bias, discriminatory outcomes, and the equitable distribution of benefits and risks across different societal groups.

Empowering Society: Despite the challenges faced by privacy, surveillance, and data ethics, it highlights the potential for 6G-based systems to empower society and advance collective well-being. By using transparent communication, involved decision-making, and digital literacy, 6G networks can provide informed consent, individual autonomy, and societal resilience. Furthermore, by prioritizing user-centric design, privacy-by-design principles, and inclusive governance mechanisms, 6G technologies have the potential to engender trust, use social cohesion, and uphold democratic values in the digital ecosystem [11, 12].

In summary, this section sets the stage for the study of privacy, surveillance, and data ethics in the context of 6G-based systems and communication. By examining the societal issues and ethical dilemmas inherent in the deployment of 6G networks, users can collaboratively navigate the evolving landscape of digital transformation while safeguarding fundamental rights and values.

13.4 Socioeconomic Impact and Digital Divide for 6G-Based Systems and Communication

The growth of 6G-based systems and communication moves to a new era of technological advancement with the potential to reshape societies and economies worldwide. However, alongside the promises of enhanced connectivity and innovation, the deployment of 6G networks also brings important issues regarding their socioeconomic impact and the insistent challenge of the digital divide [13]. This section provides an overview of the socioeconomic implications and the widening digital gap in the context of 6G-based systems and communication.

Socioeconomic Transformation: It acknowledges the transformative potential of 6G networks to drive socioeconomic development, use innovation, and accelerate digital transformation. By providing ultra-fast speeds, low latency, and huge connectivity, 6G technologies have the capacity to revolutionize industries, enable new business models, and enhance productivity across sectors. Moreover, the integration of IoT devices, augmented reality (AR) applications, and autonomous systems has the power to unlock new areas/sectors for economic growth and societal progress.

Digital Inclusion and the Digital Divide: Despite the promises of 6G networks, it highlights the essential challenge of the digital divide, wherein marginalized communities and underserved populations are left behind in the digital revolution. Inequalities in access to high-speed internet, digital literacy, and technological infrastructure exacerbate existing inequalities, focusing on socioeconomic participation and intensifying social exclusion. Moreover, the digital divide suspiciously impacts rural areas, low-income households, and disadvantaged groups, preserving cycles of poverty and marginalization.

Challenges of Inclusive Connectivity: It discusses the importance of addressing the digital divide and advancing digital inclusion initiatives in the deployment of 6G-based systems and communication. Bridging the gap requires rigorous efforts to expand broadband access, improve digital literacy, and promote affordability across different communities. Moreover, targeted interventions like community networks, public–private partnerships, and regulatory incentives are essential to ensure equitable access to 6G technologies and their associated benefits.

Socioeconomic Empowerment: It provides the transformative potential of 6G networks to empower individuals, communities, and economies through inclusive connectivity and equitable access to digital opportunities [14, 15]. By bridging the digital divide, using digital skills development, and promoting inclusive innovation ecosystems, 6G-based systems can catalyze socioeconomic empowerment and use resilient, complete societies. Moreover, by prioritizing ethical issues, regulatory frameworks, and user engagement, users can collaboratively navigate the complex socioeconomic landscape of 6G deployment while advancing shared wealth and societal well-being.

In summary, this section discusses a complete description of the socio-economic impact and digital challenges inherent in the deployment of 6G-based systems and communication. By addressing the digital divide and promoting inclusive connectivity, users can use the transformative potential of 6G technologies to build more equitable, resilient, and wealthy societies.

13.5 Legal and Ethical Issues for Implementing AI in 6G Networks, Systems, and Communication in Next Decade

As the telecommunications landscape evolves toward 6G networks, the integration of AI used to be a transformative force, revolutionizing network management, system optimization, and communication capabilities [15–17]. However, the deployment of AI in 6G networks presents several legal and ethical challenges that must be carefully navigated in the coming decade. It provides an overview of the complex legal and ethical issues surrounding the implementation of AI in 6G networks, systems, and communication.

Legal Frameworks and Regulatory Compliance: It acknowledges the importance of robust legal frameworks and regulatory compliance in governing the deployment of AI in 6G networks. Given the sensitive nature of telecommunications data and the potential for AI algorithms to impact user privacy and security, adherence to data protection laws, like GDPR in Europe and CCPA in California, is important. Moreover, regulatory bodies must contend with evolving challenges related to AI transparency, accountability, and liability, ensuring that AI-driven functionalities comply with existing regulations and ethical standards.

Data Privacy and Security: The integration of AI in 6G networks raises major issues regarding data privacy and security. AI algorithms depend on a large amount of data to train and operate effectively, leading to potential privacy risks associated with data collection, processing, and storage [18, 19]. Moreover, the use of AI in network security and threat detection introduces complexities in protection against cyber threats and ensuring the integrity of communication systems. As such, robust data privacy regulations, encryption protocols, and cybersecurity measures are essential to mitigate risks and protect user privacy in AI-driven 6G networks.

Algorithmic Bias and Fairness: It discusses the ethical issues of algorithmic bias and fairness in AI-based 6G networks. AI algorithms are vulnerable to biases inherent in training data, which potentially leads to discriminatory outcomes and exacerbates existing societal inequalities [18–20]. In important applications like healthcare diagnostics, criminal justice, etc., algorithmic bias can have far-reaching consequences, underscoring the importance of addressing bias mitigation strategies and promoting algorithmic transparency and accountability.

Ethical Decision-Making and Accountability: Ethical issues surrounding AI implementation in 6G networks extend beyond technical issues to include broader societal impacts and ethical dilemmas. This highlights the need for ethical decision-making frameworks that prioritize human values, equity, and social responsibility. Moreover, establishing mechanisms for AI accountability and transparency is essential to ensure that users can understand, scrutinize, and challenge the decisions made by AI systems in 6G networks.

In summary, this section highlights the complex interaction between legal and ethical issues in the implementation of AI in 6G networks, systems, and communication. By addressing legal challenges, protecting data privacy and security, mitigating algorithmic bias, and promoting ethical decision-making, users can navigate the evolving landscape of AI-driven 6G deployment while maintaining fundamental rights, values, and societal comfort.

13.6 Technical, Nontechnical and Legal Issues Toward Using AI in 6G Networks, Systems, and Communication in Next Decade

An overview of technical, nontechnical, and legal issues associated with the implementation of AI in 6G networks, systems, and communication over the next decade, can be listed as the following.

13.6.1 Technical Issues

- **Algorithmic Complexity**: Developing AI algorithms capable of efficiently processing the large amount of data generated by 6G networks while maintaining low latency and high reliability.
- **Resource Constraints**: Optimizing AI models to operate within the resource-constrained environments of 6G networks, including limitations in processing power, memory, and energy consumption.
- **Dynamic Environments**: Adapting AI algorithms to operate effectively in dynamic and unpredictable network conditions, including variations in traffic load, interference, and topology changes.
- **Interoperability**: Ensuring interoperability among diverse AI-driven components and technologies within 6G networks, including compatibility between hardware, software, and communication protocols.

- **Security and Privacy**: Addressing security vulnerabilities and privacy issues associated with AI-enabled functionalities, including protecting against cyber threats, data breaches, and unauthorized access.

13.6.2 Nontechnical Issues

- **Ethical Issues**: Addressing ethical issues related to the use of AI in 6G networks, including issues of transparency, accountability, bias, and fairness in algorithmic decision-making.
- **User Acceptance**: Promoting user acceptance and trust in AI-driven 6G technologies by ensuring transparency, user control, and the ethical use of data.
- **Workforce Skills**: Building the technical expertise and workforce capacity to develop, deploy, and manage AI-based systems within 6G networks, including training programs and talent acquisition strategies.
- **Digital Inclusion**: Bridging the digital divide and ensuring equitable access to AI-enabled 6G technologies, particularly for underserved communities and marginalized populations.
- **Societal Impact**: Assessing the broader societal impact of AI integration in 6G networks, including implications for employment, education, healthcare, and social equity.

13.6.3 Legal Issues

- **Data Governance**: Establishing robust data governance frameworks to govern the collection, processing, and sharing of data within AI-driven 6G networks, ensuring compliance with data protection regulations and safeguarding user privacy.
- **Regulatory Compliance**: Navigating legal and regulatory requirements governing the deployment and operation of AI-based systems in 6G networks, including spectrum licensing, network neutrality, and competition laws.
- **Liability and Accountability**: Clarifying liability and accountability mechanisms for AI-driven decisions and actions within 6G networks, including legal frameworks for attributing responsibility in the event of system failures, accidents, or harm.
- **Intellectual Property**: Addressing intellectual property rights and ownership issues related to AI algorithms, models, and datasets used in 6G networks, including patents, copyrights, and trade secrets.

In summary, addressing these technical, nontechnical, and legal challenges will be essential for the successful integration of AI in 6G networks, systems, and communication over the next decade. Collaboration among industry users/experts, policymakers, researchers, and civil society will be important in navigating these complexities and ensuring the responsible and ethical deployment of AI technologies in the telecommunications sector.

13.7 Challenges and Open Research Questions AI-Based 6G Networks, Systems, and Communication

Some challenges and open research questions related to AI-based 6G networks, systems, and communication, can be discussed in Table 13.1:

Hence, addressing these challenges and research questions will be important for advancing the development and deployment of AI-based 6G networks, systems, and communication, while ensuring they are ethically sound, reliable, and aligned with societal values and legal frameworks.

13.8 Interference and Coexistence Challenges Toward Using AI in 6G Networks, Systems, and Communication

Interference and coexistence challenges are major issues when integrating AI into 6G networks, systems, and communication. Here is an overview of these challenges as mentioned in Table 13.2.

Hence, addressing these interference and coexistence challenges through AI-driven solutions will be essential for realizing the full potential of 6G networks, enabling continuous connectivity, efficient spectrum utilization, and robust communication performance in diverse and dynamic wireless environments.

13.9 Business Models and Market Challenges with Using of AI in 6G Networks, Systems, and Communication

Integrating AI into 6G networks, systems, and communication introduces various business models and market challenges [20, 21]. We can include few of the points here as the following.

13.9.1 Business Models

- **AI as a Service (AIaaS)**: Companies may provide AI algorithms, tools, and platforms as a service, allowing network operators to access and integrate AI capabilities into their 6G networks on a subscription or pay-per-use basis.
- **Network Optimization and Management**: Service providers can use AI to optimize network performance, manage resources efficiently, and automate network operations, providing enhanced services and experiences to customers.

Table 13.1 Challenges and open research questions related to AI-based 6G networks, systems, and communication.

	Challenge	Research Questions
Scalability and efficiency	Developing scalable AI algorithms capable of handling the large amounts of data generated by 6G networks while maintaining efficiency in terms of computational resources and energy consumption.	How can AI algorithms be optimized to scale effectively with the increasing complexity and size of 6G networks? What techniques can be employed to improve the energy efficiency of AI-driven functionalities in 6G systems? How can distributed AI architectures be used to enhance scalability and efficiency in 6G networks?
Robustness and reliability	Ensuring the robustness and reliability of AI-driven functionalities in 6G networks, particularly in dynamic and unpredictable environments with varying network conditions and user demands.	What methods can be employed to enhance the resilience of AI algorithms against adversarial attacks, data anomalies, and environmental uncertainties? How can AI-driven systems be designed to adapt dynamically to changing network conditions and user requirements while maintaining reliability and performance? What techniques can be used to validate and verify the correctness and safety of AI-driven decisions in real-time communication scenarios?
Privacy and security	Addressing privacy issues and security vulnerabilities associated with the use of AI in 6G networks, including protecting sensitive user data and protecting against cyber threats.	How can AI algorithms be designed to ensure data privacy and confidentiality in 6G networks, especially in scenarios involving personal and sensitive information? What strategies can be employed to detect and mitigate security threats and attacks targeting AI-enabled components in 6G systems? How can federated learning and differential privacy techniques be utilized to protect user privacy while enabling collaborative AI training across distributed 6G networks?

(Continued)

Table 13.1 (Continued)

	Challenge	Research Questions
Ethical and societal implications	Addressing ethical issues and societal impacts of AI integration in 6G networks, including issues related to algorithmic bias, fairness, transparency, and accountability.	How can algorithmic fairness and transparency be ensured in AI-driven decision-making processes within 6G networks, particularly in important applications like healthcare and public safety?
		What approaches can be adopted to mitigate biases in AI algorithms and promote equitable outcomes across diverse user populations?
		How can users be held accountable for the ethical implications and societal consequences of AI-based decisions in 6G systems, and what mechanisms can be put in place to ensure responsible AI governance?
Regulatory and legal frameworks	Navigating the legal and regulatory landscape governing the deployment and operation of AI-based 6G networks, including compliance with data protection laws, spectrum regulations, and liability frameworks.	What legal frameworks and standards are needed to govern the collection, processing, and sharing of data in AI-driven 6G networks, and how can regulatory compliance be ensured?
		How can liability and accountability mechanisms be established to attribute responsibility for AI-driven decisions and actions in 6G systems, especially in cases of harm or adverse outcomes?
		What policies and regulations are necessary to promote innovation while protecting user rights, privacy, and security in the context of AI-based 6G networks?

Table 13.2 Interference and coexistence challenges.

	Challenges	AI Solutions	Research Question
Interference management	In dense and dynamic 6G network environments, interference between neighboring cells and devices can degrade signal quality and impair communication performance.	AI algorithms can be employed to dynamically manage interference by optimizing resource allocation, adjusting transmission parameters, and coordinating spectrum usage in real time.	How can AI techniques like reinforcement learning and deep reinforcement learning be used to mitigate interference in 6G networks? What approaches can be used to model and predict interference patterns in complex and heterogeneous network environments? How can AI-driven interference management techniques adapt to changing network conditions and user demands while ensuring optimal performance and reliability?
Spectrum sharing and coexistence	With the proliferation of diverse wireless technologies and applications in 6G networks, spectrum scarcity and competition become more pronounced, necessitating efficient spectrum sharing and coexistence mechanisms.	AI-based spectrum management techniques can enable dynamic spectrum access, cognitive radio capabilities, and adaptive interference mitigation strategies to facilitate coexistence among diverse wireless systems.	How can AI algorithms be used to optimize spectrum utilization and facilitate dynamic spectrum sharing among multiple users and services in 6G networks? What methods can be employed to detect and mitigate interference between coexisting wireless systems operating in adjacent or overlapping frequency bands? How can AI-driven spectrum management techniques balance the trade-offs between spectrum efficiency, interference mitigation, and quality of service requirements in 6G networks?

(Continued)

Table 13.2 (Continued)

	Challenges	AI Solutions	Research Question
Multi-RAT (radio access technology) Integration	6G networks are expected to support a multitude of Radio Access Technologies (RATs), including traditional cellular networks, Wi-Fi, mmWave, and emerging technologies like massive MIMO and beamforming, leading to increased complexity and potential interference.	AI algorithms can facilitate continuous integration and coordination among different RATs by optimizing handover decisions, managing interference, and dynamically adapting transmission strategies based on network conditions and user requirements.	How can AI-driven techniques be used to optimize the handover process between different RATs in 6G networks, minimizing disruptions and maximizing user experience? What approaches can be employed to coordinate resource allocation and interference mitigation across heterogeneous RATs operating in the same geographical area? How can AI-based algorithms adaptively select and configure the most suitable RAT for each user and application dynamically in response to changing network conditions and traffic patterns?
Cross-layer optimization	To address interference and coexistence challenges effectively, AI algorithms need to consider interactions and dependencies across multiple protocol layers, including physical, medium access control (MAC), and network layers.	AI-driven cross-layer optimization techniques can enable holistic and coordinated resource management, traffic scheduling, and interference coordination across different layers of the communication stack in 6G networks.	How can AI algorithms be designed to capture cross-layer interactions and optimize resource allocation and interference mitigation strategies across multiple protocol layers in 6G networks? What methodologies can be used to model and analyze the impact of cross-layer optimizations on network performance, energy efficiency, and QoS (Quality of Service) metrics in diverse 6G deployment scenarios? How can AI-driven cross-layer optimization techniques adapt and evolve over time to accommodate changing network conditions, user requirements, and technological advancements in 6G networks?

- **Predictive Maintenance**: AI-based predictive analytics can enable proactive maintenance and fault detection in 6G infrastructure, reducing downtime and operational costs for network operators.
- **Customized Services**: AI-driven information about user behavior, preferences, and context can enable personalized services and content delivery, creating new revenue streams through targeted advertising and premium offerings.
- **Partnerships and Ecosystem Collaboration**: Collaboration between network operators, technology providers, and AI-based startups can use innovation and co-creation of AI-enabled solutions tailored to the specific needs of 6G networks.

13.9.2 Market Challenges

- **Data Privacy and Security**: Issues about data privacy, security breaches, and regulatory compliance may delay the adoption of AI in 6G networks, requiring robust measures to protect sensitive user information and ensure compliance with privacy regulations [21–24].
- **Interoperability and Standards**: Lack of interoperability among AI solutions and heterogeneous network components can impede continuous integration and scalability, requiring industry collaboration and standardization efforts to address compatibility issues.
- **Cost and ROI**: Initial investment costs associated with deploying AI in 6G networks, including hardware, software, and training, may present challenges for network operators, requiring careful cost–benefit analysis and long-term planning to justify investments and demonstrate return on investment (ROI).
- **Skill Gap and Talent Acquisition**: A shortage of skilled professionals with expertise in AI, machine learning, and telecommunications may face challenges for organizations looking to implement AI-based solutions in 6G networks, necessitating workforce development initiatives and talent acquisition strategies.
- **Regulatory Uncertainty**: Rapid advancements in AI technology and evolving regulatory landscapes may introduce uncertainty and compliance challenges for businesses operating in the 6G ecosystem, requiring proactive monitoring of regulatory developments and adaptation to changing requirements.
- **Ethical and Societal Issues**: Ethical issues surrounding AI algorithms, like bias, transparency, and accountability, may impact consumer trust and public perception of AI-driven 6G networks, necessitating responsible and transparent deployment practices to address societal issues.

Hence, addressing these business models and market challenges requires a complete approach that considers technical feasibility, economic viability, regulatory compliance, and ethical implications. We can say that collaboration among users, innovation in business models, and proactive engagement with regulatory bodies and policymakers are essential for realizing the transformative potential of AI in 6G networks, systems, and communication while navigating market uncertainties and ensuring societal benefits.

13.10 Future Research Opportunities Toward Using AI in 6G Networks, Systems, and Communication

Future research opportunities abound in the integration of AI into 6G networks, systems, and communication. Here are some key areas for exploration reading future work.

13.10.1 AI-Driven Network Optimization

- Research can focus on developing advanced AI algorithms for optimizing the performance of 6G networks, including resource allocation, interference management, and energy efficiency.
- Techniques like reinforcement learning, meta-learning, and federated learning can be discussed to enable autonomous and adaptive network optimization in dynamic and heterogeneous environments.

13.10.2 AI-Enabled Edge Computing

- Investigate the role of AI in enhancing edge computing capabilities in 6G networks, including edge intelligence, distributed machine learning, and real-time decision-making at the network edge [22, 25–28].
- Discuss novel architectures and algorithms for efficient data processing, storage, and inference at the edge, enabling low-latency applications and services in 6G networks.

13.10.3 AI-Driven Security and Privacy

- Research opportunities exist in developing AI-based solutions for enhancing security and privacy in 6G networks, including threat detection, anomaly detection, and user authentication.
- Investigate privacy-preserving AI techniques like federated learning, homomorphic encryption, and differential privacy to protect user data while enabling AI-driven functionalities in 6G networks.

13.10.4 AI for Dynamic Spectrum Access

- Discuss AI-driven approaches for dynamic spectrum access and management in 6G networks, including cognitive radio, spectrum sharing, and interference mitigation techniques.
- Investigate machine learning algorithms for predicting spectrum availability, optimizing spectrum allocation, and adapting to changing environmental conditions in 6G communication systems.

13.10.5 Ethical and Responsible AI Deployment

- Research can focus on ethical issues and societal impacts of AI integration in 6G networks, including issues of bias, fairness, transparency, and accountability.
- Develop frameworks and guidelines for the responsible deployment of AI in 6G networks, ensuring alignment with ethical principles, regulatory requirements, and societal values.

13.10.6 AI-Enabled Services and Applications

- Discuss novel AI-driven services and applications enabled by 6G networks, including augmented reality, autonomous systems, intelligent transportation, and immersive communication experiences [24, 28, 29].
- Investigate the integration of AI techniques like natural language processing, computer vision, and reinforcement learning about different/several 6G applications and use cases.

13.10.7 Interdisciplinary Research Collaboration

- Use interdisciplinary collaboration between researchers from different fields/sectors like telecommunications, machine learning, computer science, and social sciences to address the complex challenges of AI in 6G networks.
- Promote industry–academia partnerships and international collaboration to accelerate innovation, knowledge exchange, and technology transfer in the field of AI-driven 6G communication systems.

Here, Figure 13.2 depicts a 6G architecture for vertical applications in near future. Figure 13.3 shows the importance of AI in a 6G-based communication network in near future.

Hence, by discussing these research opportunities, researchers can advance the state-of-the-art in AI-enabled 6G networks, systems, and communication,

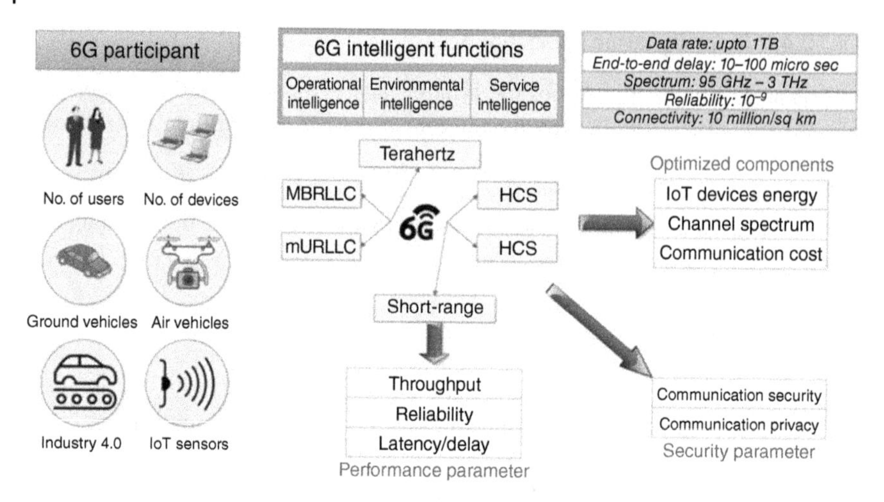

Figure 13.3 Importance of AI in 6G-based communication network in near future.

making the way for transformative advancements in connectivity, efficiency, and user experience in the next generation of wireless communication.

13.11 Summary

The integration of AI into 6G networks provide huge promise for revolutionizing communication, connectivity, and technological innovation. However, this convergence also brings forth major ethical and social issues that must be carefully considered and addressed/solved. Ethical issues surrounding privacy, fairness, transparency, and accountability are paramount in ensuring that AI-driven 6G networks provide fundamental human rights and values. Safeguarding user privacy, mitigating algorithmic biases, and promoting transparency in AI decision-making processes are essential steps toward building trust and using ethical deployment practices.

Moreover, the societal impacts of AI integration in 6G networks extend beyond technical issues, including broader socioeconomic issues. Several efforts to bridge the digital divide, promote digital inclusion, and mitigate disparities in access to AI-driven technologies are important for ensuring equitable benefits and opportunities for all segments of society.

As we navigate the complex landscape of AI-based 6G networks, it is imperative to adopt a complete approach that balances technological innovation with ethical reflection, user engagement, and proactive policy interventions. By prioritizing ethical values, promoting responsible AI deployment, and addressing societal issues, we can use the transformative potential of AI in 6G networks to create a more inclusive, equitable, and sustainable digital future for all.

References

1 Floridi, L. and Taddeo, M. (2016). What is data ethics? *Philosophical Transactions of the Royal Society A: Mathematical, Physical and Engineering Sciences* 374 (2083): 20160360. https://doi.org/10.1098/rsta.2016.0360.

2 Jobin, A., Ienca, M., and Vayena, E. (2019). The global landscape of AI ethics guidelines. *Nature Machine Intelligence* 1 (9): 389–399. https://doi.org/10.1038/s42256-019-0088-2.

3 Martine, F. and Chabalier (2022). Ethically responsible and trustworthy autonomous systems for 6G. *IEEE Network* https://doi.org/10.1109/mnet.005.2100711. Yulei, W. (2022). Ethically responsible and trustworthy autonomous systems for 6G. *IEEE Network*, https://doi.org/10.1109/MNET.005.2100711.

4 Demanboro, A.C., Bianchini, D., Iano, Y. et al. (2021). 6G networks: an innovative approach, but with many challenges and paradigms, in the development of platforms and services in the near future. In: *Brazilian Technology Symposium*, 172–187. Cham: Springer International Publishing.

5 Srivastava, V., Mahara, T., and Yadav, P. (2021). An analysis of the ethical challenges of blockchain-enabled E-healthcare applications in 6G networks. *International Journal of Cognitive Computing in Engineering* 2: 171–179.

6 Ouyang, Y., Ye, X., & Wang, X. (2023). 6G network operation support system. *arXiv preprint arXiv:2307.09045.*

7 Mika, Ylianttila, Raimo, Kantola, Andrei, Gurtov, et al. (2020). 6G white paper : research challenges for trust, security and privacy. arXiv: Cryptography and Security,

8 Arastouei, N. (2022). 6G technologies: key features, challenges, security and privacy issues. In: *International Workshop on Digital Sovereignty in Cyber Security: New Challenges in Future Vision*, 94–109. Cham: Springer Nature Switzerland.

9 Chris, Z. (2006). Social and ethical aspects of IPv6. *The Information Society* https://doi.org/10.1007/0-387-31168-8_19.

10 Mittelstadt, B.D., Allo, P., Taddeo, M. et al. (2016). The ethics of algorithms: mapping the debate. *Big Data & Society* 3 (2): 2053951716679679. https://doi.org/10.1177/2053951716679679.

11 Jobin, A., Ienca, M., and Vayena, E. (2019). Artificial intelligence: the global landscape of ethics guidelines. *Nature* 569 (7756): 131–134. https://doi.org/10.1038/d41586-019-01313-5.

12 Shaw, R. and Garlan, D. (1996). *Software Architecture: Perspectives on an Emerging Discipline*. Prentice Hall PTR.

13 Caliskan, A., Bryson, J.J., and Narayanan, A. (2017). Semantics derived automatically from language corpora contain human-like biases. *Science* 356 (6334): 183–186. https://doi.org/10.1126/science.aal4230.

14 Liu, Y., Wei, Q., and Chen, K. (2021). A survey of ethical issues in artificial intelligence. *IEEE Access* 9: 30831–30849. https://doi.org/10.1109/ACCESS.2021.3066954.

15 O'Connor, P. and Goddard, J. (2020). Ethical considerations in AI-based 5G and beyond networks. *IEEE Communications Standards Magazine* 4 (2): 48–54. https://doi.org/10.1109/MCOMSTD.001.1900457.

16 Taddeo, M. and Floridi, L. (2018). How AI can be a force for good. *Science* 361 (6404): 751–752. https://doi.org/10.1126/science.aat5991.

17 Mittelstadt, B.D., Allo, P., Taddeo, M. et al. (2016). The ethics of algorithms: mapping the debate. *Big Data & Society* 3 (2): 2053951716679679. https://doi.org/10.1177/2053951716679679.

18 Nair, M.M. and Tyagi, A.K. (2023). Blockchain technology for next-generation society: current trends and future opportunities for smart era. In: *Blockchain Technology for Secure Social Media Computing*. https://doi.org/10.1049/PBSE019E_ch11.

19 Tyagi, A.K., Kumari, S., Chidambaram, N., and Sharma, A. (2024). Engineering applications of blockchain in this smart era. In: *Enhancing Medical Imaging with Emerging Technologies*. https://doi.org/10.4018/979-8-3693-5261-8.ch011.

20 Ajanthaa, L., Seranmadevi, R., Sree, P.H., and Tyagi, A.K. (2024). Engineering applications of artificial intelligence. In: *Enhancing Medical Imaging with Emerging Technologies*. https://doi.org/10.4018/979-8-3693-5261-8.ch010.

21 Tyagi, A.K., Kukreja, S., Richa, and Sivakumar, P. (2024). Role of blockchain technology in smart era: a review on possible smart applications. *Journal of Information & Knowledge Management* https://doi.org/10.1142/S0219649224500321.

22 Tyagi, A.K. and Tiwari, S. (2024). The future of artificial intelligence in blockchain applications. In: *Machine Learning Algorithms Using Scikit and TensorFlow Environments*. IGI Global https://doi.org/10.4018/978-1-6684-8531-6.ch018.

23 Kumari, S., Thompson, A., and Tiwari, S. (2024). 6G-enabled internet of things-artificial intelligence-based digital twins: cybersecurity and resilience. In: *Emerging Technologies and Security in Cloud Computing*. IGI Global https://doi.org/10.4018/979-8-3693-2081-5.ch016.

24 Nair, M.M. and Tyagi, A.K. (2023). 6G: technology, advancement, barriers, and the future. In: *6G-Enabled IoT and AI for Smart Healthcare*. CRC Press.

25 Nair, M.M. and Tyagi, A.K. (2021). Privacy: history, statistics, policy, Laws, preservation and threat analysis. *Journal of Information Assurance & Security* 16 (1): 24–34. 11p.

26 Boyd, D. and Crawford, K. (2012). Important questions for big data: provocations for a cultural, technological, and scholarly phenomenon. *Information, Communication & Society* 15 (5): 662–679. https://doi.org/10.1080/1369118X.2012.678878.

27 Floridi, L. and Cowls, J. (2019). A unified framework of five principles for AI in society. *Harvard Data Science Review* 1 (1): https://doi.org/10.1162/99608f92.8cd550d1.

28 Nair, M.M. and Tyagi, A.K. (2023). Chapter 11 - AI, IoT, blockchain, and cloud computing: the necessity of the future. In: *Distributed Computing to Blockchain* (ed. R. Pandey, S. Goundar, and S. Fatima), 189–206. Academic Press. ISBN: 9780323961462 https://doi.org/10.1016/B978-0-323-96146-2.00001-2.

29 Winfield, A.F. (2017). Towards an ethical robot: internal models, consequences and ethical action selection. In: *2017 IEEE International Conference on Robotics and Automation (ICRA)*, 6439–6444. IEEE https://doi.org/10.1109/ICRA.2017.7989792.

14

Future Trends and Research Directions for 6G

14.1 Beyond 6G: Vision for Future Generations

This work articulates a roadmap for the evolution of wireless communication technologies beyond the anticipated sixth-generation (6G) era. It predicts a future where connectivity extends far beyond what current technologies can provide, addressing the insatiable demand for faster, more reliable, and ubiquitous communication [1, 2]. Here, Figure 14.1 shows the evolution of 6G and beyond in detail.

At the heart of this vision is the recognition of the exponential growth of connected devices and the emergence of new use cases like augmented reality (AR), autonomous vehicles, and smart cities. To meet these evolving needs, this work proposes a range of innovative concepts and technologies that could shape the future of communication networks.

One such concept is quantum communication, which uses the principles of quantum mechanics to enable secure and ultra-fast communication channels. By using the unique properties of quantum entanglement and superposition, quantum communication holds the promise of unbreakable encryption and near-instantaneous data transmission over long distances.

Another frontier is molecular communication, which draws inspiration from biological systems to enable communication between nanoscale devices. By encoding information in chemical signals, molecular communication could enable communication in environments where traditional radio-frequency-based technologies are ineffective, like inside the human body or underground.

Furthermore, the vision extends to bioinspired networking paradigm, which mimic the decentralized and adaptive nature of biological systems to improve the efficiency and resilience of communication networks. By adopting principles from swarm intelligence, neural networks, and evolutionary algorithms, future networks could self-optimize and adapt to changing conditions in real time.

6G-Enabled Technologies for Next Generation: Fundamentals, Applications, Analysis and Challenges, First Edition. Amit Kumar Tyagi, Shrikant Tiwari, Shivani Gupta, and Anand Kumar Mishra.

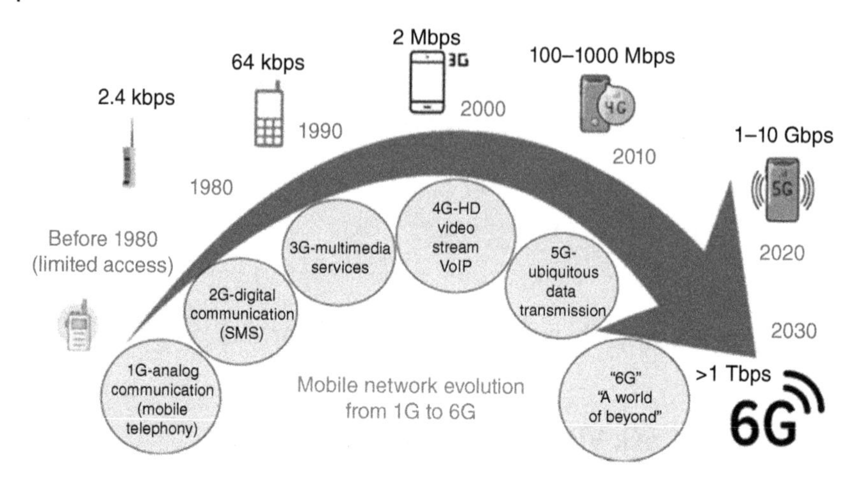

Figure 14.1 Evolution of 6G and beyond.

Hence, this work highlights the importance of ethical issues and societal implications in shaping the trajectory of future communication systems. It discusses the need for interdisciplinary collaboration and stakeholder engagement to ensure that emerging technologies are developed and deployed in a manner that is responsible, equitable, and aligned with the broader interests of society. Here, Figure 14.2 provides an overview of 6G wireless communication network.

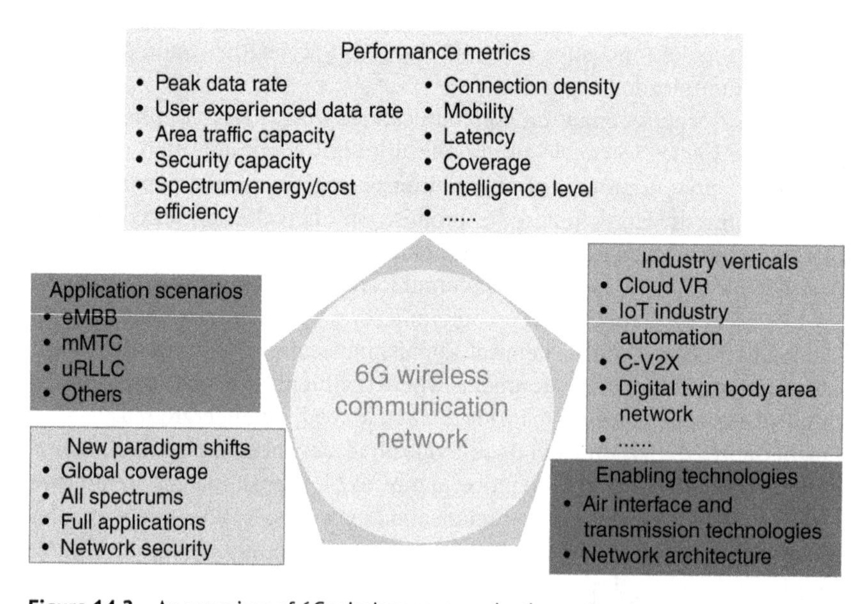

Figure 14.2 An overview of 6G wireless communication network.

In summary, this work presents a compelling vision for the future of wireless communication, one that moves the boundaries of what is possible and predicts a world where connectivity is continuous, secure, and sustainable.

14.2 Literature Work

In [3], the most important metrics to consider are peak data rate, energy per bit, traffic capacity, and latency. Internet of Things (IoT), smart systems, and autonomous networks are the main topics of emphasis. Establishing a theoretical basis for wireless communication systems of the future and creating 6G-compatible wireless communication solutions with the help of the IoTs, consumer apps and services can undergo a radical revolution. There are several limitations that come with the fifth generation (5G), including standards for latency, energy per bit, traffic capacity, and peak data rate, among others.

In [4], 5G designs, private networks, open networks, and AI-driven components are all part of the network infrastructure. In this work, a discussion on how AI and ML can facilitate better network management to enable improved service delivery, 6G fabric compiles microdomains in detail. In this, they conclude that artificial intelligence (AI) and machine learning (ML) have the potential to improve the efficiency of network administration and resource utilization.

In [5], 6G aims to improve security, sustainability, spectrum efficiency, and global coverage. Incorporating networks on land, in space, and beneath the ocean makes up the integrated network system. There have been design, technological, and problem identification changes for the 6G network. We are pointing the way for additional studies on 6G network ecosystems. As a roadmap for the implementation of 6G ecosystems in forthcoming studies found out how the 6G network is evolving in terms of architecture, technology, and challenges.

In [6], research into 6G wireless systems mainly focuses on the future of cellular technology. 6G technology is examined for its features and its uses. The development of wireless cellular technology will progress from 1G to 5G. This work goes over 6G, the following generation of technology, that is, addressing the disadvantages of 5G for future wireless networks, more and more ubiquitous 5G consequences. Even if 5G does exist, it might not last into this decade.

In [7], virtual and AR apps, as well as those that allow for three-dimensional communication, will be in high demand. 6G wireless networks will be able to handle data transfers from large devices with ease. An outline of the 6G system's goals, specifications, design, and possible implementations, study possibilities, advantages, disadvantages, and future directions for 6G wireless technology were discussed. It is used as a blueprint for 6G networks to come. Raise the bar for the amount of enthusiasm for further R&D in this area. Note that currently established performance benchmarks will be unmet by 5G capabilities.

In [8], 6G offers a lot of benefits, including minimal latency, massive connectivity, low cost, and fast data throughput. Several technologies include AI, large IoT network, video-linking, huge multiple input multiple output (MIMO), drones, and three-dimensional beam forming are useful for this chapter, that is, AI, video-on-demand (VLC), and massive MIMO are some of the hot topics in 6G based research in near future. 6G wireless technology will bridge the gap between the digital, physical, and human realms, allowing users to have an extrasensory experience.

14.3 Emerging Technologies and Innovations for Next-Generation Society with 6G

Discuss the transformative potential of cutting-edge technologies and innovations in shaping the future society empowered by 6G networks. These forward-looking predict a future where ubiquitous connectivity, enhanced by 6G, initiatives extraordinary societal advancements across various domains [9–11]. Figure 14.3 shows the explanation of the interaction between machine and human being in an era of 6G.

At the front of this vision are emerging technologies poised to revolutionize industries and redefine human experiences. One such technology is AR and virtual reality (VR), which, powered by ultra-low latency and high data rates

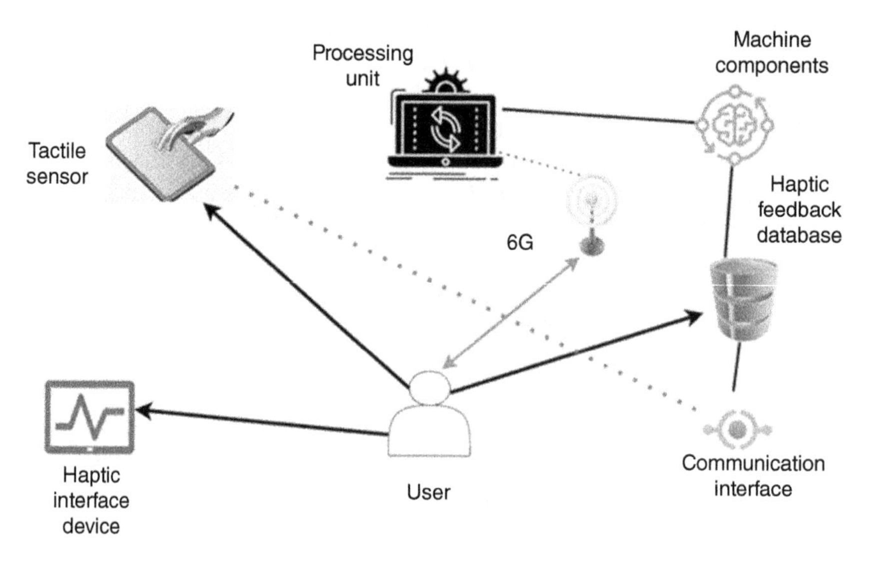

Figure 14.3 Haptic interaction with machines.

enabled by 6G, could unlock immersive experiences for education, entertainment, and collaboration. From virtual classrooms to interactive gaming and remote assistance, AR and VR applications could revolutionize how we learn, work, and interact with the world around us.

Additionally, the integration of AI and ML with 6G networks promises to move into a new era of intelligent connectivity [12–15]. AI-powered network optimization, predictive analytics, and autonomous systems could enhance the reliability, efficiency, and security of 6G networks, enabling self-healing networks, predictive maintenance, and personalized services tailored to individual user preferences.

Furthermore, the convergence of 6G with other emerging technologies like the IoT, blockchain, and edge computing could unlock new opportunities for innovation and disruption. From smart cities and connected vehicles to precision agriculture and remote healthcare, the continuous integration of these technologies with 6G networks could drive efficiency gains, cost savings, and improvements in quality of life [16–18]. Figure 14.4 shows the IoE importance with 6G mobile technology.

Importantly, this vision highlights the importance of inclusive and sustainable development, ensuring that the benefits of 6G and emerging technologies are accessible to all members of society. By addressing digital divides, using digital literacy, and promoting responsible innovation, 6G has the potential to create a more equitable and inclusive future for all.

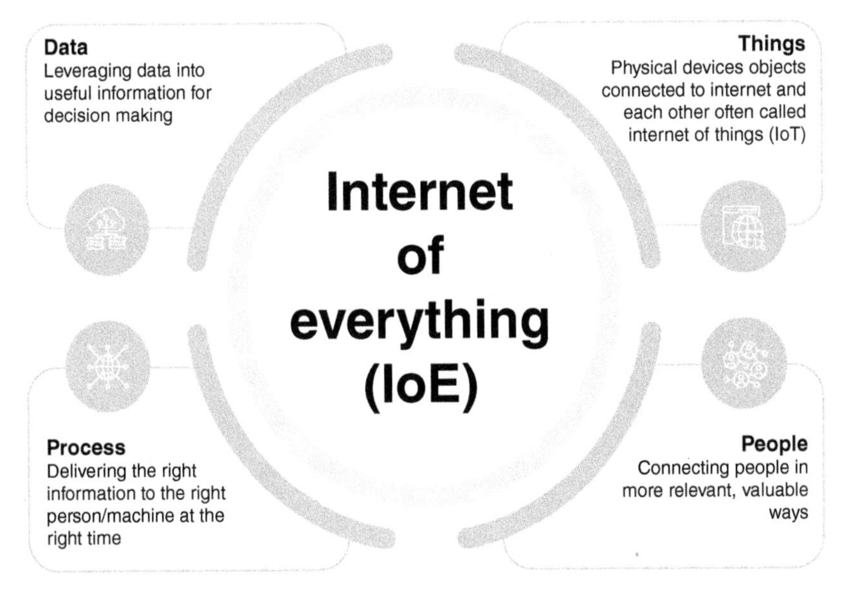

Figure 14.4 IoE with 6G mobile technology.

In summary, this work provides a vision of a future society transformed by the convergence of 6G networks and emerging technologies. By using the power of connectivity, intelligence, and innovation, this future society could unlock new opportunities for economic growth, social development, and human well-being.

14.4 Future Trends Beyond 6G

This work discusses the evolving landscape of wireless communication technologies and predicts the trajectory of advancements beyond the anticipated 6G era [19, 20].

This work anticipates several key trends and developments (refer to Figure 14.5) that may shape the future of communication networks:

Quantum Communication: Beyond 6G, the adoption of quantum communication holds promise for ultra-secure and high-speed data transmission. Using the principles of quantum mechanics, like entanglement and superposition, quantum communication could revolutionize encryption and enable unique levels of data security and privacy.

Molecular Communication: Another emerging trend is molecular communication, inspired by biological systems. This approach involves encoding and transmitting information using chemical signals, providing potential applications in environments where traditional wireless communication is limited or impractical, like within the human body or underground.

Bioinspired Networking Paradigms: Future networks may draw inspiration from biological systems to achieve greater efficiency, adaptability, and resilience.

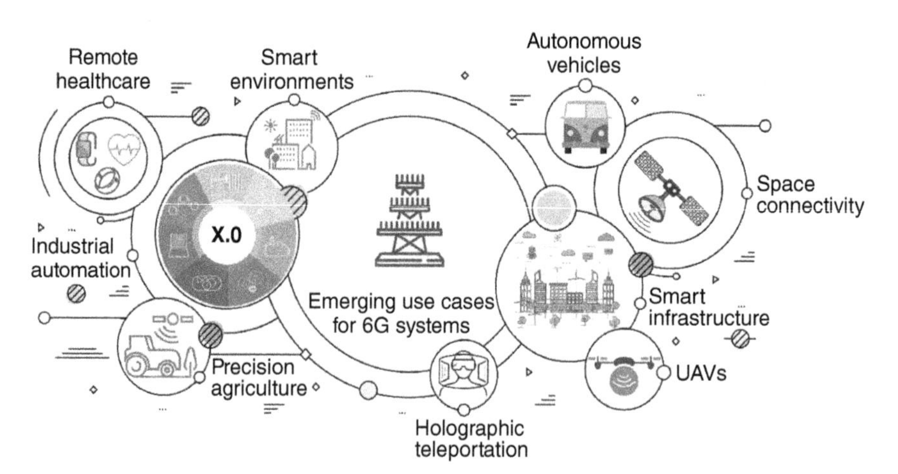

Figure 14.5 Emerging use cases: 6G and beyond.

By emulating decentralized and self-organizing behaviors observed in nature, bioinspired networking paradigms could lead to more robust and self-healing communication networks capable of autonomously adapting to dynamic conditions.

Integration with Emerging Technologies: The convergence of communication networks with other emerging technologies like AI, blockchain, and edge computing is expected to drive innovation and enable new applications and services. By using synergies between these technologies, future communication networks could support advanced functionalities like intelligent routing, distributed ledger-based authentication, and edge-based processing for low-latency applications.

Ethical and Societal Issues: As communication technologies continue to advance, it will be essential to address ethical, privacy, and societal implications. Future trends in communication beyond 6G will likely involve ongoing discussions and efforts to ensure that technological advancements are aligned with societal values, promote inclusivity, and mitigate potential risks and drawbacks.

Hence, this work provides a picture of a future where communication networks evolve to use novel technologies, biological inspirations, and ethical issues, ultimately shaping a more connected, secure, and adaptable digital landscape.

14.5 The Future of Wireless Communication

The future of wireless communication is poised for major transformation, driven by technological advancements and evolving societal needs [21–24]. Several key trends and developments are likely to shape the trajectory of wireless communication in the coming years:

5G Deployment and Optimization: While 5G networks are still being deployed globally, efforts to optimize and expand their capabilities will continue in the near term. This includes enhancing network coverage, increasing data speeds, and reducing latency to meet growing demands for connectivity and support emerging applications like autonomous vehicles, remote healthcare, and smart cities.

Beyond 5G (6G) Technologies: Research and development efforts are already underway to define the next generation of wireless communication standards, often referred to as "6G." Beyond 5G, 6G is expected to deliver even faster data speeds, lower latency, and support for massive device connectivity. Emerging technologies like terahertz communication, integrated satellite-terrestrial networks, and intelligent reflecting surfaces are being discussed to unlock the full potential of 6G networks.

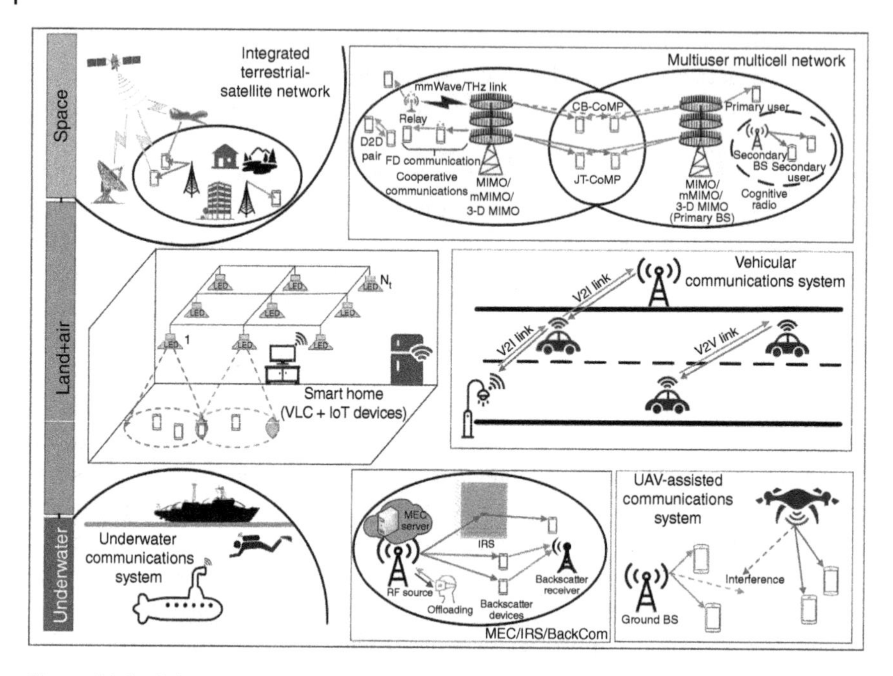

Figure 14.6 Schemes and technologies for future wireless communication network for vehicles.

Internet of Things (IoT) Expansion: The rapid growth of IoT devices and applications will continue to drive demand for wireless connectivity. Wireless communication technologies will need to evolve to support the large-scale and diverse requirements of IoT deployments, spanning industries like healthcare, agriculture, manufacturing, and smart home automation. One example of future wireless communication network for vehicles can be depicted in Figure 14.6.

Edge Computing Integration: The integration of edge computing with wireless communication networks is expected to become increasingly prevalent. By bringing computing resources closer to end-users and IoT devices, edge computing reduces latency, improves data privacy, and enables real-time processing of data-intensive applications [25, 26].

AI and Machine Learning Integration: AI and ML technologies will play an important role in optimizing wireless communication networks [27–30]. AI-powered algorithms can dynamically allocate network resources, predict traffic patterns, and optimize network performance, leading to improved efficiency, reliability, and user experience. Further, we can find another example of AI-enabled intelligent 6G network in Figure 14.7.

Security and Privacy Issues: As wireless communication becomes more pervasive and important to everyday life, ensuring security and privacy will be useful.

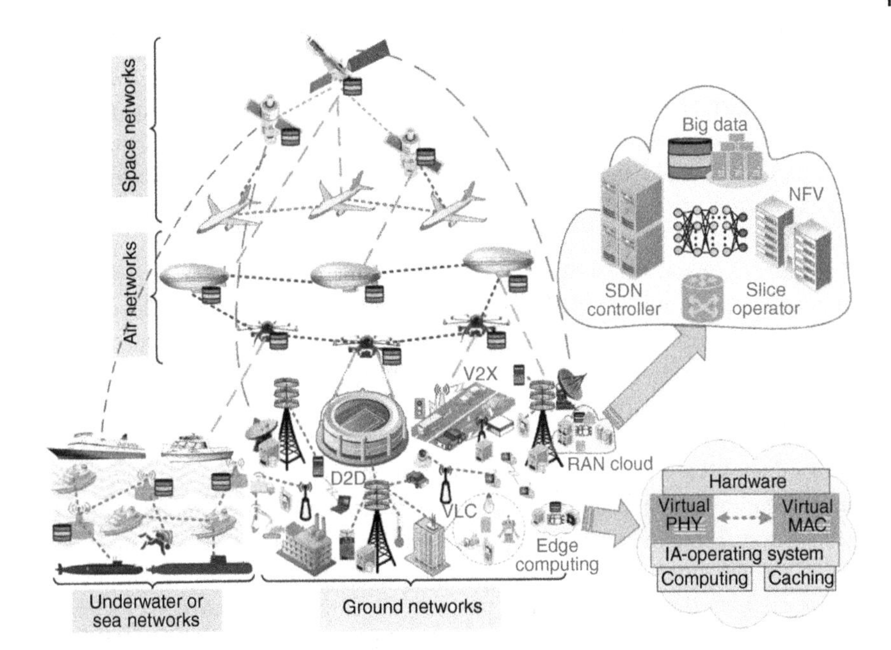

Figure 14.7 AI-enabled intelligent 6G network.

Efforts to develop robust encryption algorithms, authentication mechanisms, and secure communication protocols will be essential to protect against cyber threats and protect sensitive data.

Sustainability and Energy Efficiency: With the increasing energy consumption associated with wireless communication networks, there will be a growing highlight on sustainability and energy efficiency. Innovations in energy harvesting, power management, and network optimization will be needed to minimize the environmental impact of wireless communication infrastructure.

In summary, the future of wireless communication holds huge promise for connecting people, devices, and systems in increasingly continuous and efficient ways. By using technological innovations and addressing emerging challenges, wireless communication will continue to drive economic growth, societal advancement, and improved quality of life.

14.6 Future Potential Beyond 6G Technologies

Looking beyond 6G, the future of wireless communication holds huge potential for transformative technologies that could reshape connectivity and redefine what is possible [23, 25]. Here are some potential technologies and concepts that could emerge beyond 6G:

Quantum Communication: Quantum communication uses the principles of quantum mechanics to enable secure and ultra-fast data transmission. Quantum entanglement and cryptography provide unique levels of security, making it virtually impossible for data to be intercepted or hacked. Beyond 6G, quantum communication could become mainstream, revolutionizing cybersecurity and enabling entirely new applications in fields like finance, healthcare, and national security.

Terahertz Communication: Terahertz frequencies provide significantly higher data rates compared to current wireless communication bands. Beyond 6G, terahertz communication could enable multi-terabit-per-second data transfer speeds, unlocking new possibilities for high-bandwidth applications like real-time holographic video streaming, ultra-high-definition VR, and instantaneous file transfers.

Bioinspired Networking: Drawing inspiration from biological systems, future communication networks could adopt decentralized and self-organizing architectures. By mimicking the resilience and adaptability of biological ecosystems, these bioinspired networks could dynamically adapt to changing conditions, self-heal in the face of disruptions, and optimize resource allocation for maximum efficiency.

Holographic Communication: Holographic communication technologies could enable three-dimensional, real-time communication experiences that transcend the limitations of traditional screens and displays. Beyond 6G, holographic communication could become ubiquitous, allowing users to interact with lifelike holograms of people, objects, and environments from anywhere in the world, revolutionizing telepresence, collaboration, and entertainment.

Neuromorphic Computing: Neuromorphic computing emulates the structure and function of the human brain, enabling ultra-efficient and adaptive information processing. Beyond 6G, neuromorphic computing could be integrated into communication devices and networks, enabling cognitive capabilities like context-awareness, natural language processing, and emotion recognition. This could lead to more intuitive and personalized communication experiences.

Space-Based Communication Networks: Integrating satellite communication with terrestrial networks could extend connectivity to even the most remote and underserved regions of the world. Beyond 6G, space-based communication networks could use advancements in satellite technology, including low Earth orbit (LEO) constellations and high-capacity inter-satellite links, to deliver continuous, low-latency connectivity on a global scale.

Molecular Communication: Molecular communication involves encoding and transmitting information using chemical signals, mimicking communication processes observed in biological organisms. Beyond 6G, molecular communication could enable wireless communication in environments where traditional electromagnetic communication is impractical, like inside the human body or underwater. This could have profound implications for healthcare, environmental monitoring, and exploration.

These are few examples of the potential technologies that could emerge beyond 6G, pushing the boundaries of wireless communication and opening up new frontiers for innovation and discovery. As we continue to discuss and develop these technologies, the future of connectivity holds endless possibilities for improving lives, driving economic growth, and expanding human knowledge.

14.7 Speculation on 7G and Beyond

Speculating on 7G and beyond involves predicting the future of wireless communication technologies beyond current horizons, taking into account advancements in science, engineering, and societal needs. Here is a speculative look at what could lie ahead:

Quantum Communication Networks: In the realm of 7G and beyond, quantum communication networks may become the standard for ultra-secure and high-speed data transmission. Quantum entanglement and quantum key distribution (QKD) could enable encrypted communication channels immune to hacking or interception, revolutionizing cybersecurity and enabling entirely new levels of privacy.

Neuro-interfaced Communication: Advancements in neuroscience and brain–computer interfaces (BCIs) could lead to direct communication between brains and devices. By translating neural signals into digital data and vice versa, individuals could communicate thoughts, emotions, and sensory experiences wirelessly, opening up new possibilities for telepathic communication and immersive virtual experiences.

Teleportation-Based Communication: Speculative technologies like quantum teleportation could enable instantaneous transfer of information across vast distances. Although currently theoretical, if realized, teleportation-based communication could transcend the limitations of traditional signal propagation, enabling real-time communication with spacecraft, colonies on other planets, or even extraterrestrial civilizations.

Global Brain Networks: Building on the concept of the "global brain," future communication networks could evolve into interconnected systems that mimic the structure and function of the human brain. These networks would exhibit self-awareness, collective intelligence, and emergent behavior, facilitating continuous information exchange and decision-making on a global scale.

Exotic Communication Mediums: Beyond traditional electromagnetic waves, future communication technologies may use exotic mediums like gravitational waves, neutrinos, or even quantum entanglement. By using these phenomena, communication systems could achieve faster-than-light communication, enabling instantaneous transmission of information across vast cosmic distances.

Consciousness-Based Communication: Highly speculative but intriguing, consciousness-based communication posits that consciousness itself could serve as a medium for information exchange. If consciousness is indeed a fundamental aspect of the universe, future technologies may enable direct communication between individual consciousnesses, transcending the limitations of physical mediums altogether.

Time Travel Communication: Theoretical concepts like closed time-like curves or wormholes suggest the possibility of time travel. While highly speculative and fraught with paradoxes, if time travel were ever realized, it could open the door to communication across different points in time, allowing for interactions with past or future civilizations.

These speculative visions of 7G and beyond push the boundaries of imagination and scientific possibility, illustrating the potential for wireless communication to evolve in ways that fundamentally transform our understanding of connectivity, consciousness, and the nature of reality itself.

14.8 Open Research Gaps, Technical/Nontechnical Challenges Beyond 6G

Beyond 6G, several open research gaps and technical/non-technical challenges remain to be addressed as wireless communication technologies continue to evolve [23]. Here are some key areas of focus:

Terahertz Communication: While terahertz frequencies provide the potential for ultra-high data rates, major challenges remain in developing practical and cost-effective terahertz communication systems. Research is needed to address issues like signal attenuation, power consumption, and interference mitigation in terahertz communication links.

Quantum Communication Security: Quantum communication promises unique levels of security, but practical implementation still faces challenges. Research is needed to develop scalable QKD protocols, quantum repeaters, and quantum memory technologies to enable secure quantum communication networks.

Energy Efficiency: As wireless communication networks become increasingly energy-intensive, improving energy efficiency is an important challenge. Research is needed to develop energy-efficient communication protocols, hardware architectures, and power management techniques to reduce the environmental footprint of wireless networks.

Spectrum Management: Efficient spectrum management is essential for maximizing the capacity and performance of wireless communication systems. Research is needed to develop dynamic spectrum allocation algorithms, cognitive radio technologies, and spectrum-sharing mechanisms to optimize spectrum utilization and mitigate interference.

Interference Mitigation: As wireless networks become denser and more heterogeneous, interference mitigation becomes increasingly challenging. Research is needed to develop advanced interference cancelation techniques, beamforming algorithms, and coordination mechanisms to minimize interference and improve network performance.

Security and Privacy: With the rapid growth of connected devices and data-intensive applications, ensuring security and privacy is important. Research is needed to develop robust encryption algorithms, authentication mechanisms, and privacy-preserving protocols to protect against cyber threats and protect sensitive data [24].

Integration with Emerging Technologies: Integrating wireless communication networks with emerging technologies like AI, edge computing, and the IoT presents both technical and nontechnical challenges. Research is needed to develop interoperability standards, integration frameworks, and cross-disciplinary collaborations to use the full potential of emerging technologies in wireless communication.

Regulatory and Policy Frameworks: Developing regulatory and policy frameworks that facilitate innovation while ensuring safety, security, and equitable access to communication services is a nontechnical challenge. Research is needed to address legal, ethical, and societal implications of emerging technologies, as well as to advocate for policies that promote competition, innovation, and digital inclusion.

Hence, addressing these research gaps and challenges will be essential for realizing the full potential of wireless communication technologies beyond 6G and for building a more connected, secure, and sustainable digital future.

14.9 Technical/Nontechnical/Legal Issues Moving Beyond 6G

Moving beyond 6G will present a host of technical, nontechnical, and legal challenges that must be addressed to ensure the successful deployment and adoption of future wireless communication technologies. Here is a breakdown of some of the key issues in each category.

14.9.1 Technical Issues

Spectrum Utilization: Efficiently utilizing the radio frequency spectrum to support the growing demand for wireless connectivity while minimizing interference and maximizing spectral efficiency.

Energy Efficiency: Developing energy-efficient communication protocols, hardware components, and network architectures to reduce power consumption and environmental impact.

Interference Management: Mitigating interference in increasingly dense and heterogeneous wireless networks to ensure reliable and high-quality communication.

Security and Privacy: Enhancing security measures, encryption algorithms, and privacy-preserving protocols to protect against cyber threats and protect sensitive data transmitted over wireless networks.

Scalability and Reliability: Designing scalable and reliable communication systems capable of supporting the large influx of connected devices and applications with varying requirements.

Integration with Emerging Technologies: Integrating wireless communication networks with emerging technologies like AI, edge computing, and the IoT to enable new functionalities and applications.

14.9.2 Nontechnical Issues

Regulatory Compliance: Adhering to regulatory requirements and standards set by government agencies and international organizations to ensure compliance with spectrum allocation, privacy regulations, and safety standards.

Digital Inclusion: Addressing the digital divide by ensuring equitable access to wireless communication technologies and bridging disparities in connectivity among different regions, demographics, and socioeconomic groups.

User Acceptance and Adoption: Garnering user acceptance and adoption of new communication technologies through effective education, outreach, and user-centric design that addresses user needs and preferences.

Ethical issues: Addressing ethical implications related to the use of wireless communication technologies, including issues like data privacy, surveillance, and the impact on society and individual well-being.

Environmental Impact: Assessing and mitigating the environmental impact of wireless communication infrastructure, including energy consumption, electronic waste, and the ecological footprint of manufacturing and deployment processes.

14.9.3 Legal Issues

Spectrum Licensing and Allocation: Securing spectrum licenses and navigating spectrum allocation policies and regulations set by regulatory authorities to ensure lawful operation and compliance with frequency usage guidelines.

Intellectual Property Rights: Addressing intellectual property rights issues related to patents, trademarks, and copyrights associated with wireless communication technologies and innovations.

Liability and Legal Liability: Clarifying legal liability and accountability for issues like network failures, data breaches, and cybersecurity incidents, including liability frameworks for manufacturers, service providers, and end-users.

Privacy Regulations: Complying with data protection and privacy regulations like the General Data Protection Regulation (GDPR) and ensuring transparency, consent, and accountability in the collection, processing, and storage of user data.

International Standards and Interoperability: Adhering to international standards and interoperability requirements to facilitate global compatibility, continuous roaming, and interoperable communication across different networks and devices.

Hence, addressing these technical, nontechnical, and legal challenges will be essential for realizing the full potential of future wireless communication technologies beyond 6G and for building a connected, inclusive, and sustainable digital future.

14.10 Important Challenges Toward Beyond 6G

Moving beyond 6G will require addressing several important challenges that are essential for the development, deployment, and adoption of future wireless communication technologies.

Here are some of the key challenges:

Spectrum Availability and Efficiency: As demand for wireless connectivity continues to grow, ensuring sufficient spectrum availability and efficient spectrum utilization will be important. Future technologies must address spectrum scarcity issues and develop innovative spectrum-sharing mechanisms to accommodate diverse applications and services.

Energy Efficiency and Sustainability: Wireless communication networks consume huge amounts of energy, contributing to environmental issues and operational costs. Beyond 6G, there is a pressing need to improve energy efficiency through the development of energy-efficient hardware, communication protocols, and network management strategies to minimize energy consumption and carbon emissions.

Capacity and Throughput: Meeting the ever-increasing demand for higher data rates and throughput will be a major challenge. Future technologies must support large data transmission and provide scalable solutions to accommodate exponential growth in data traffic while maintaining quality of service and user experience.

Latency Reduction: Ultra-low latency is essential for enabling real-time applications like AR, autonomous vehicles, and telemedicine. Beyond 6G, reducing latency to imperceptible levels requires overcoming technical difficulties related to signal processing, network optimization, and propagation delays.

Security and Privacy: As wireless networks become more pervasive and interconnected, ensuring robust security and privacy protections is important [24]. Future technologies must address cybersecurity threats, vulnerabilities, and attacks by implementing advanced encryption techniques, authentication mechanisms, and intrusion detection systems to protecting sensitive data and important infrastructure.

Interference and Reliability: Wireless communication networks are susceptible to interference from various sources, including neighboring networks, environmental conditions, and electromagnetic interference. Beyond 6G, mitigating interference and ensuring reliable communication under challenging conditions require advanced signal processing techniques, interference cancelation algorithms, and adaptive resource allocation strategies.

Regulatory and Policy Frameworks: Regulatory frameworks and policies play an important role in shaping the development and deployment of wireless communication technologies. Beyond 6G, addressing regulatory barriers, spectrum licensing issues, and legal challenges related to privacy, liability, and intellectual property rights is essential for using innovation, competition, and global interoperability.

Integration with Emerging Technologies: Future wireless communication technologies must continuously integrate with emerging technologies like AI, edge computing, and the IoT to unlock new functionalities and applications. Ensuring compatibility, interoperability, and synergy between different technologies requires interdisciplinary collaboration, standardization efforts, and alignment with industry trends and market demands.

Addressing these important challenges will be essential for realizing the full potential of future wireless communication technologies beyond 6G and for building a connected, resilient, and sustainable digital future. Collaboration between industry stakeholders, government agencies, academia, and standards organizations will be key to overcoming these challenges and driving innovation in the field of wireless communications.

14.11 Future Research Opportunities Beyond 6G

Looking beyond 6G presents several exciting research opportunities that could shape the future of wireless communication [23, 25]. Few future directions and challenges toward 6G can be found in Figure 14.8.

Here are some potential areas for future research:

- **Terahertz Communication**: We investigate terahertz communication as a viable solution for achieving ultrahigh data rates and unlocking new applications like ultra-fast wireless networking, high-resolution imaging, and sensing in the terahertz frequency range.
- **Quantum Communication**: We discuss quantum communication technologies, including QKD, quantum teleportation, and quantum repeaters, to develop ultra-secure communication networks resistant to hacking and interception.
- **Bioinspired Networking**: Investigate bioinspired networking platform that mimic the decentralized and self-organizing behavior of biological systems to design more robust, adaptive, and energy-efficient communication networks.

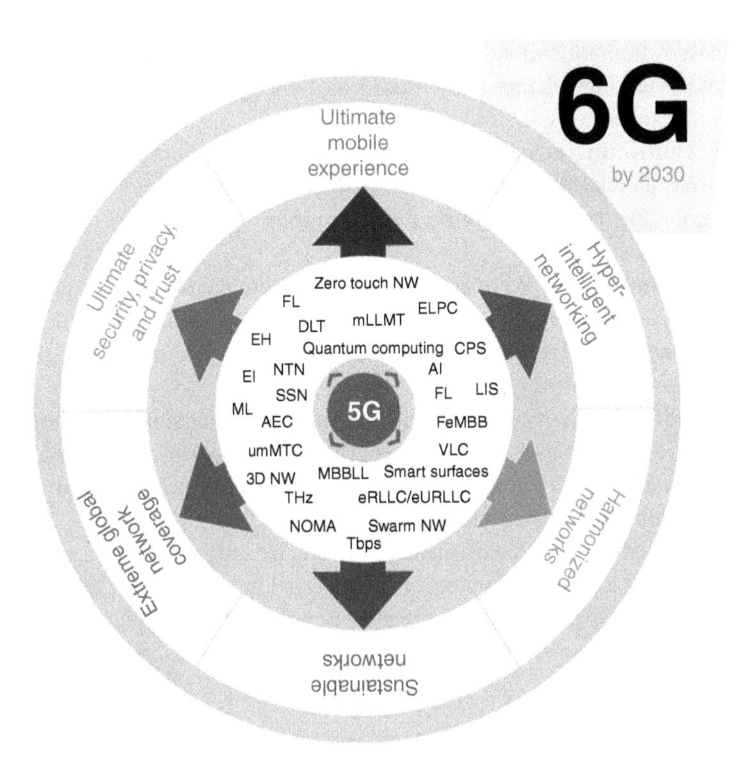

Figure 14.8 Toward 6G: future directions and challenges.

- **Molecular Communication**: We do research molecular communication as a means of enabling communication between nanoscale devices using chemical signals, with potential applications in healthcare, environmental monitoring, and biotechnology.
- **AI-Driven Network Optimization**: We discuss the integration of AI and ML techniques to optimize network performance, predict traffic patterns, and automate network management tasks, leading to more efficient and resilient communication systems.
- **Smart Antenna Technologies**: We investigate advanced antenna technologies, like massive MIMO and beamforming, to improve spectrum efficiency, increase coverage, and enhance the reliability of wireless communication networks.
- **Energy Harvesting and Wireless Power Transfer**: We discuss energy harvesting and wireless power transfer techniques to enable self-powered and energy-efficient wireless devices and networks, using ambient energy sources like solar, thermal, and kinetic energy.
- **Security and Privacy Enhancements**: We do research novel security and privacy-enhancing techniques, including post-quantum cryptography, blockchain-based authentication, and secure multiparty computation, to protect against emerging cybersecurity threats and ensure the confidentiality and integrity of transmitted data.
- **Holographic Communication**: We investigate holographic communication technologies that enable three-dimensional, real-time communication experiences, revolutionizing telepresence, virtual collaboration, and immersive entertainment applications.
- **Cross-Layer Optimization**: We discuss cross-layer optimization techniques that use interactions between different layers of the communication protocol stack to improve network performance, enhance quality of service, and optimize resource allocation in wireless communication systems.
- **Green Communication Networks**: We do research on green communication technologies and sustainable network design principles to minimize the environmental impact of wireless communication infrastructure, reduce energy consumption, and promote ecological sustainability.
- **Satellite and Space-Based Communication**: We investigate satellite and space-based communication technologies, including low Earth orbit (LEO) satellite constellations, high-altitude platforms, and inter-satellite links, to provide ubiquitous and high-speed connectivity to remote and underserved areas.

Hence, these research opportunities present new possibilities for advancing the state-of-the-art in wireless communication and shaping the future of connectivity beyond 6G. By discussing these avenues, researchers can contribute to the development of innovative solutions that address emerging challenges and unlock new capabilities in wireless communication networks.

14.12 Summary

This work provides a details explanation the evolving landscape of wireless communication technologies and outlines key features for future research on 6G. As the world moves toward the era beyond 5G, this work discusses the imperious of staying ahead of the curve and moving toward the next generation of wireless networks. This work highlights several transformative trends and technologies used to shape the future of 6G networks, including terahertz communication, intelligent reflecting surfaces, and integrated satellite-terrestrial networks. It highlights the need for interdisciplinary collaboration and exploration of emerging technologies to propel the development of 6G networks and address the evolving demands for connectivity, speed, and reliability. Furthermore, this work discusses the important role of AI and ML in optimizing network performance, predicting user behavior, and dynamically allocating resources. By using the power of AI-driven optimization, 6G networks can adapt to changing conditions in real time and deliver personalized services tailored to individual user preferences. In summary, this work serves as a roadmap for researchers, industry stakeholders, and policymakers to navigate the complex landscape of wireless communication technologies and drive innovation toward the realization of 6G networks and beyond. Through continued exploration, collaboration, and innovation, the future of wireless communication holds huge promise for transforming industries, enriching lives, and shaping the digital landscape for generations to come.

References

1 Rangan, S., Rappaport, T.S., and Erkip, E. (2014). Millimeter-wave cellular wireless networks: potentials and challenges. *Proceedings of the IEEE* 102 (3): 366–385. https://doi.org/10.1109/JPROC.2014.2299398.

2 Andrews, J.G., Buzzi, S., Choi, W. et al. (2014). What will 5G be? *IEEE Journal on Selected Areas in Communications* 32 (6): 1065–1082. https://doi.org/10.1109/JSAC.2014.2328098.

3 Swetha, M.S., Muneshwara, M.S., Murali Manohara Hegde, A.S., and Lu, Z. (2023). 6G wireless communication systems and its applications. In: *Machine Learning and Mechanics Based Soft Computing Applications*, 271–288. Springer Nature Singapore: Singapore.

4 Klaus, D., Anwer, A.-D., Harald, H. et al. (2023). 6G fabric compiles microdomains for sophisticated service delivery [from the guest editors]. *IEEE Vehicular Technology Magazine* https://doi.org/10.1109/mvt.2023.3238481.

5 Quy, V.K., Chehri, A., Quy, N.M. et al. (2023). Innovative trends in the 6G era: a comprehensive survey of architecture, applications, technologies, and challenges. *IEEE Access* 11: 39824–39844.

6 Sikiru, A.A., Olukayode, A.A., Lateef, O.A., and Sobowale, S.O. (2023). 6G wireless system: the emerging trend in cellular technology. *Journal of Electrical Engineering* 74 (3): 240–245.

7 Banafaa, M., Shayea, I., Din, J. et al. (2022). 6G mobile communication technology: requirements, targets. *Applications, Challenges, Advantages, and Opportunities. Alexandria Engineering Journal* https://doi.org/10.1016/j.aej.2022.08.017.

8 Bourbah, A., Meliani, B., Madini, Z., and Zouine, Y. (2022). The next-generation 6g: trends, applications, technologies, challenges, and use cases. In: *Proceedings of Seventh International Congress on Information and Communication Technology: ICICT 2022, London, Volume 3*, 761–770. Singapore: Springer Nature Singapore.

9 Akyildiz, I.F., Wang, P., and Wang, X. (2015). Continuous wireless connectivity for internet of things in 5G: definition, roadmap, and challenges. *IEEE Communications Magazine* 53 (5): 6–14. https://doi.org/10.1109/MCOM.2015.7105636.

10 Yang, Y., Xiao, Y., Gao, Y. et al. (2021). A survey on 6G networks: vision, requirements, enabling technologies, and emerging trends. *IEEE Access* 9: 55004–55027. https://doi.org/10.1109/ACCESS.2021.3063584.

11 Andrews, J.G. and Buzzi, S. (2020). A perspective on 6G research. *IEEE Open Journal of the Communications Society* 1: 111–125. https://doi.org/10.1109/OJCOMS.2020.3031005.

12 Chung, T.Y., So, J.W., Kim, T., and Sung, D. (2021). Evolution and future directions of mobile communications towards 6G. *IEEE Access* 9: 112224–112240. https://doi.org/10.1109/ACCESS.2021.3101372.

13 Ajanthaa, L., Seranmadevi, R., Sree, P.H., and Tyagi, A.K. (2024). Engineering applications of artificial intelligence. In: *Enhancing Medical Imaging with Emerging Technologies*. https://doi.org/10.4018/979-8-3693-5261-8.ch010.

14 Tyagi, A.K., Kukreja, S., Richa, and Sivakumar, P. (2024). Role of blockchain technology in smart era: a review on possible smart applications. *Journal of Information & Knowledge Management* https://doi.org/10.1142/S0219649224500321.

15 Tyagi, A.K. and Tiwari, S. (2024). The future of artificial intelligence in blockchain applications. In: *Machine Learning Algorithms Using Scikit and TensorFlow Environments*. IGI Global https://doi.org/10.4018/978-1-6684-8531-6.ch018.

16 Saad, W., Bennis, M., Chen, M., and Vasilakos, A.V. (2019). A vision of 6G wireless systems: applications, trends, technologies, and open research problems. *IEEE Network* 35 (4): 6–13. https://doi.org/10.1109/MNET.011.2100394.

17 Sharma, S., Kumar, N., Leung, V.C.M., and Chatzinotas, S. (2021). 6G wireless communication systems: applications, challenges, and future perspective. *IEEE Access* 9: 48731–48760. https://doi.org/10.1109/ACCESS.2020.2980502.

18 Kumari, S., Thompson, A., and Tiwari, S. (2024). 6G-enabled internet of things-artificial intelligence-based digital twins: cybersecurity and resilience. In: *Emerging Technologies and Security in Cloud Computing*. IGI Global https://doi.org/10.4018/979-8-3693-2081-5.ch016.

19 Yilmaz, T., Ahmed, S.H., Mohamed, A., and Alkhawaja, A.M. (2021). 6G technology: potential features, challenges, and research directions. In: *Advances in Wireless Communications*, 1–21. Singapore: Springer https://doi.org/10.1007/978-981-16-2817-6_1.

20 Polese, M., Giordani, M., Pizzi, S., and Zorzi, M. (2021). How will 6G be different from 5G? *IEEE Communications Magazine* 59 (3): 14–20. https://doi.org/10.1109/MCOM.001.1900009.

21 Goyal, R., Sengupta, A., Dey, S., and Chakraborty, D. (2021). A comprehensive review on 6G technology. *Journal of Network and Computer Applications* 179: 103090. https://doi.org/10.1016/j.jnca.2021.103090.

22 Galinina, O., Andreev, S., Samuylov, A., and Koucheryavy, Y. (2021). Potential of 6G wireless systems: key features and enabling technologies. *IEEE Network* 35 (4): 6–13. https://doi.org/10.1109/MNET.011.2100351.

23 Nair, M.M. and Tyagi, A.K. (2023). 6G: technology, advancement, barriers, and the future. In: *6G-Enabled IoT and AI for Smart Healthcare*. CRC Press.

24 Nair, M.M. and Tyagi, A.K. (2021). Privacy: history, statistics, policy, laws, preservation and threat analysis. *Journal of Information Assurance & Security* 16 (1): 24–34. 11p.

25 Letaief, K.B., Zhang, W., Yuan, Y., and Wang, Q. (2019). The roadmap to 6G: AI empowered wireless networks. *IEEE Communications Magazine* 57 (8): 84–90. https://doi.org/10.1109/MCOM.001.1900351.

26 Nair, M.M., Mishra, A.K., and Tyagi, A.K. (2023). Fog computing and edge computing: open issues, critical challenges and the road ahead for future. In: *Proceedings of the 2023 Fifteenth International Conference on Contemporary Computing (IC3–2023). Association for Computing Machinery, New York, NY, USA*, 66–76. https://doi.org/10.1145/3607947.3607962.

27 Chen, H., Wang, M., Liu, X. et al. (2021). New opportunities for the industry in the 6G era: a business model perspective. *Wireless Communications and Mobile Computing* 2021: 9923473. https://doi.org/10.1155/2021/9923473.

28 Xia, F., Xiao, X., Tao, X., and Yang, L.T. (2021). Toward 6G networks: use cases and technologies. *IEEE Communications Magazine* 58 (3): 55–61. https://doi.org/10.1109/MCOM.001.1900323.

29 Nair, M.M. and Tyagi, A.K. (2023). Blockchain technology for next-generation society: current trends and future opportunities for smart era. In: *Blockchain Technology for Secure Social Media Computing*. https://doi.org/10.1049/PBSE019E_ch11.

30 Tyagi, A.K., Kumari, S., Chidambaram, N., and Sharma, A. (2024). Engineering applications of blockchain in this smart era. In: *Enhancing Medical Imaging with Emerging Technologies*. https://doi.org/10.4018/979-8-3693-5261-8.ch011.

15

Evolution of Hybrid Li-Fi–Wi-Fi Networks: Technology, Barriers, Advancement, and Future

15.1 Introduction

15.1.1 Background

The concept of hybrid Light Fidelity Li-Fi–Wi-Fi Wireless Fidelity (Li-Fi–Wi-Fi) networks represents a fusion of two distinct wireless communication technologies: Li-Fi and Wi-Fi [1]. To understand the background of hybrid Li-Fi–Wi-Fi networks, it is essential to explore the individual histories and characteristics of these technologies.

Wi-Fi: Wi-Fi, based on the Institute of Electrical and Electronics Engineers (IEEE) 802.11 standards, has been a cornerstone of wireless communication for decades. It was first introduced in the late 1990s and rapidly gained popularity for providing wireless internet access in homes, businesses, and public spaces. Wi-Fi primarily operates in the radio frequency (RF) spectrum, with variants like 2.4 and 5 GHz bands. It provides the convenience of wireless connectivity, allowing multiple devices to connect to a network, making it an integral part of our modern digital lifestyle.

Li-Fi: Li-Fi, on the other hand, is a relatively newer technology that was conceptualized in the early 2000s by Professor Harald Haas. Li-Fi utilizes visible light for data transmission, relying on light-emitting diodes (LEDs) to modulate data signals. This technology uses the properties of light to transmit data, providing several unique advantages. Li-Fi provides significantly higher data transfer rates compared to traditional Wi-Fi. It is also more secure as it operates through line-of-sight communication, making it less susceptible to interference and eavesdropping. Moreover, Li-Fi has the potential to reduce electromagnetic interference in environments where RF communication is an issue [2, 3]. The emergence of Li-Fi can be traced back to the growing need for

6G-Enabled Technologies for Next Generation: Fundamentals, Applications, Analysis and Challenges, First Edition. Amit Kumar Tyagi, Shrikant Tiwari, Shivani Gupta, and Anand Kumar Mishra.

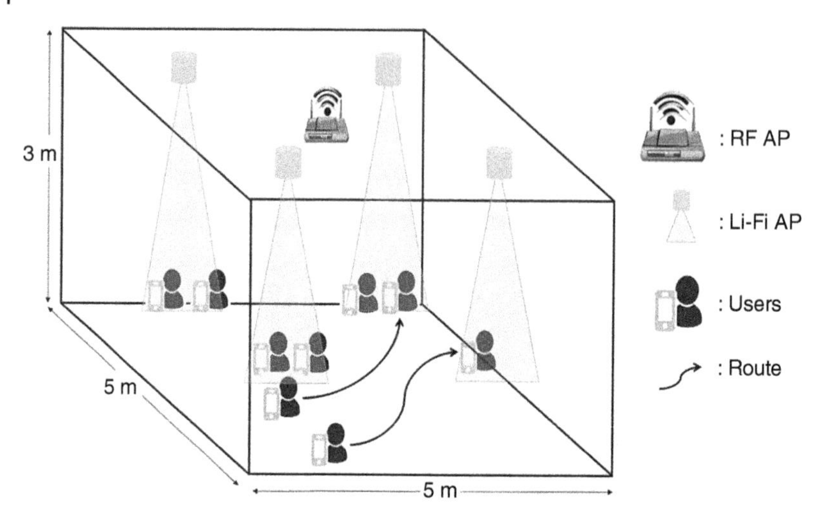

Figure 15.1 Schematic diagram of a hybrid Li-Fi–Wi-Fi network.

increased data bandwidth and more efficient use of the electromagnetic spectrum. As Wi-Fi networks became more congested and data demands continued to rise, researchers began discussing alternative technologies. Li-Fi's use of the light spectrum presented a novel solution.

The concept of hybrid Li-Fi–Wi-Fi networks arose as researchers and engineers recognized the complementary nature of these two technologies. Figure 15.1 shows a schematic diagram of a hybrid Li-Fi–Wi-Fi network. By combining Li-Fi's strengths in high-speed, secure data transmission through visible light with Wi-Fi's ubiquitous coverage and ability to penetrate walls and obstacles, the potential for creating hybrid networks became evident. The hybridization of Li-Fi and Wi-Fi networks looks to use the advantages of both technologies while mitigating their respective limitations. This innovation opens the door to a future where wireless communication systems are more robust, flexible, and capable of addressing the diverse connectivity needs of our increasingly interconnected world. As research and development in this field continue, we can anticipate the evolution of hybrid Li-Fi–Wi-Fi networks and their integration into various aspects of our daily lives, from homes and offices to smart cities and beyond.

15.1.2 Evolution of Hybrid Li-Fi–Wi-Fi Networks

The evolution of hybrid Li-Fi–Wi-Fi networks represents major advancement in wireless communication technology [3, 4]. These networks combine the strengths of two distinct technologies, Li-Fi and Wi-Fi, to create a more versatile and efficient wireless communication ecosystem. The journey of the evolution of hybrid Li-Fi–Wi-Fi networks can be summarized in several key stages:

- **Conceptualization and Early Research**: The idea of utilizing visible light for data transmission, which led to Li-Fi, was first introduced by Professor Harald Haas in a TED Talk in 2011. This ignited interest in the possibilities of Li-Fi technology. Researchers and engineers began to discuss its potential and limitations in comparison to traditional Wi-Fi.
- **Li-Fi Technology Development**: Over the years, Li-Fi technology has shown rapid advancements. Researchers developed specialized hardware and communication protocols to enable high-speed data transmission through LED lighting. Li-Fi provided the promise of extremely high data transfer rates, security through line-of-sight communication, and the potential to alleviate RF interference issues in densely populated areas.
- **Complementary Nature of Li-Fi and Wi-Fi**: It became increasingly evident that Li-Fi and Wi-Fi could complement each other. While Li-Fi excelled in high-speed, secure data transmission, Wi-Fi provided broader coverage and penetration through walls. Researchers recognized the potential for hybrid networks that could switch between these technologies based on the specific needs of a given environment. Advancements in Hybrid Network Infrastructure: As interest in hybrid Li-Fi–Wi-Fi networks grew, infrastructure development followed suit. Prototypes of hybrid routers and access points that could continuously switch between Li-Fi and Wi-Fi modes were developed. These devices aimed to create a cohesive network that provided both high-speed Li-Fi connections and broader Wi-Fi coverage.
- **Overcoming Technical Challenges**: Engineers and researchers worked on addressing technical challenges in the integration of Li-Fi and Wi-Fi. These challenges included continuous handovers between Li-Fi and Wi-Fi, synchronization of devices, and interference management when operating both technologies in the same space.
- **Regulatory and Standardization Efforts**: The development of hybrid networks also required regulatory and standardization support. As with any new technology, regulations and standards had to adapt to accommodate Li-Fi and its integration with Wi-Fi.
- **Use Cases and Applications**: Hybrid Li-Fi–Wi-Fi networks found applications in various domains. These networks could enhance wireless communication in environments where RF interference was an issue, like hospitals and industrial settings. They also played a role in improving security, as Li-Fi's line-of-sight nature made eavesdropping more difficult.
- **Ongoing Research and Future Prospects**: Research in this field continues to evolve, with ongoing efforts to refine hybrid Li-Fi–Wi-Fi technologies and discuss new applications. As Internet of Things (IoT) devices become more prevalent, the role of hybrid networks in supporting these devices and ensuring secure, high-speed connectivity is expected to expand.

The evolution of hybrid Li-Fi–Wi-Fi networks represents a promising direction in wireless communication, providing a solution to the increasing demand for faster, more reliable, and secure connectivity. As technology and infrastructure continue to advance, we can anticipate these networks becoming an integral part of our connected world, enhancing the way we interact with information and devices in our homes, businesses, and beyond.

15.1.3 Organization of the Work

This work is summarized in eight sections.

15.2 Literature Review

In [5], the sixth-generation (6G) features that are driven by AI make it easier to process communication data. Enhancements, both internal and external, driven by AI, features that facilitate collaborative communication and sensing, and optimizations in these areas. Making better use of data for a range of applications to enhance the rollout of 6G technology. Using AI to enhance communication and include the third-party apps in this work.

In [6], it is discussed that design of the waveform, critical features, and technical considerations for integration methods are all crucial. 6G networks are now exploring new waveforms for monostatic sensing. Waveform design issues and future research directions are discussed. Several benefits include better spectrum utilization, lower costs, lower latency, less area required, and lighter weight. Potential novel waveforms for monostatic sensing have been identified in this work.

In [7], future 6G wireless communication network construction plans are discussed. Wireless optical technology and terahertz communication are two examples of the new technologies under development. Enhanced security, improved quality of service, reduced latency, and greater data rate are just a few advantages. Note that quality of service, capacity, data speed, latency, and security are all issues beyond fifth generation (5G).

In [8], mobile networks that use integrated sensing and communication (ISAC), such as 5G-A and 6G are discussed. A prototype outfitted with 5G technology for radar monitoring and high-data-rate communications. ISAC uses scenarios, technical components, and sensing algorithm proof-of-concept, etc. It was proven that orthogonal frequency division multiplexing (OFDM) radar sensing and high data rate communications are compatible. It is demonstrated that ISAC proof of concept (PoC) can communicate and detect radar at the same time using 5G. The approach seems achievable for future cellular networks, according to the preliminary results.

In [9], it was shown that advanced sensing and communication systems, known as 6G networks, will make use of ISAC technology. In the context of IoT networks, open intelligent sensor networks (OISNs), communication, and computing over the air were discussed. Sensing is being integrated into wireless networks to enhance them for intelligent applications.

In [10], it was discussed that more sophisticated services are now possible because of 6G RAN's incorporation of radio sensing. Collaboration on improving sensing and communication capacities for the benefit of both parties. And making the most of what is available through the integration of sensing and communication and better location-aware services available at every level, from the physical to the application.

In [11], a system that connects the sky, the ground, and the water and utilization of tools such as network slicing, big data, and artificial intelligence (AI) are discussed. 6G networks are more efficient, have faster data rates, and have more connections per square inch. An overview of futuristic enabling technologies and paradigm which uses for 6G wireless networks to improve the potential users of 6G wireless communication networks across several sectors.

In [12], it was shown that distributed communication, control, computation, sensing, energy, and the tactile web are just a few examples of possible uses in 6G. Infrastructure technologies enabling 6G networks and the seven most essential technologies that will enable 6G networks are identified, together with their potential disruptions and probable applications. This chapter discusses potential applications and great areas for future research.

15.2.1 Overview of Li-Fi, Wi-Fi Technologies, and Hybrid Li-Fi–Wi-Fi Networks

Li-Fi is a wireless communication technology that utilizes visible light to transmit data [13, 14]. Here are the key aspects of Li-Fi technology:

- **Light-Based Data Transmission**: Li-Fi uses LEDs to transmit data. These LEDs rapidly modulate the intensity of light, which is imperceptible to the human eye, to carry digital information.
- **High Data Transfer Rates**: Li-Fi can achieve exceptionally high data transfer rates, often surpassing the capabilities of traditional Wi-Fi. Speeds of several gigabits per second are possible.
- **Line-of-Sight Communication**: Li-Fi relies on line-of-sight communication, meaning the receiver must be within the direct view of the transmitting LED. This inherent property enhances security, as it limits signal leakage and makes eavesdropping more challenging.
- **Limited Coverage**: Li-Fi has a relatively limited coverage area, as it relies on visible light. Signals do not penetrate through walls or obstacles, making them suitable for specific use cases like indoor environments with controlled lighting.

- **Low Interference**: Since Li-Fi operates in the visible light spectrum, it is immune to electromagnetic interference, making it an attractive solution for settings where RF interference is an issue.

15.2.1.1 Wi-Fi (Wireless Fidelity) Technology

- Wi-Fi is a widely adopted wireless communication technology that operates in the RF spectrum. Here are the key aspects of Wi-Fi technology:
- **RF-Based Data Transmission**: Wi-Fi uses radio waves for data transmission. It operates in different frequency bands, like 2.4 GHz and 5 GHz, to connect devices to a network.
- **Ubiquitous Coverage**: Wi-Fi provides broader network coverage and can penetrate walls and obstacles. It is the most common technology for wireless connectivity in homes, businesses, and public spaces.
- **Moderate to High Data Transfer Rates**: Wi-Fi can provide moderate to high data transfer rates, with the latest standards providing multi-gigabit speeds, although often lower than Li-Fi's potential.
- **Susceptible to Interference**: Wi-Fi signals can be affected by electromagnetic interference from other electronic devices or networks operating in the same frequency band.

15.2.1.2 Hybrid Li-Fi–Wi-Fi Networks

- Hybrid Li-Fi–Wi-Fi networks combine the strengths of Li-Fi and Wi-Fi to create a more versatile and efficient wireless communication system. Here is an overview of these hybrid networks:
- **Complementary Technologies**: Li-Fi and Wi-Fi are inherently complementary. Li-Fi excels in high-speed, secure data transmission, while Wi-Fi provides broader coverage and can extend connectivity through walls and obstacles.
- **Continuous Transition**: In hybrid networks, devices can continuously transition between Li-Fi and Wi-Fi based on their location and connectivity needs. This transition allows for an optimized wireless experience.
- **Enhanced Security**: The line-of-sight nature of Li-Fi enhances network security, as data transmission is limited to areas within the direct view of the transmitting light source. This feature is particularly valuable in environments where data privacy is important.
- **Applications**: Hybrid Li-Fi–Wi-Fi networks find applications in various settings, including hospitals, industrial facilities, smart homes, and smart cities. They address the demand for high-speed, secure, and reliable wireless connectivity in diverse environments.

Hence, the development of hybrid Li-Fi–Wi-Fi networks represents an exciting evolution in wireless communication technology, providing solutions to the challenges and demands of our increasingly interconnected world. These networks are poised to play a pivotal role in ensuring robust, efficient, and secure wireless connectivity across various sectors and applications.

15.2.2 Existing Research and Technologies Toward Hybrid Li-Fi–Wi-Fi Networks

A few of the existing research and technologies that were relevant to the evolution of these networks at that, time can be listed here:

- **Hybrid Network Protocols**: Researchers were working on protocols that allow continuous switching between Li-Fi and Wi-Fi to create a cohesive hybrid network. These protocols aimed to optimize network performance and user experience by dynamically selecting the best technology based on factors like device location, network load, and signal quality.
- **LED and Photodetector Technologies**: Advancements in LED and photodetector technology were important for improving the performance of Li-Fi. Researchers were discussing ways to enhance the efficiency and data-carrying capacity of LEDs while also improving the sensitivity and speed of photodetectors.
- **Synchronization and Handover Solutions**: To enable smooth transitions between Li-Fi and Wi-Fi, synchronization and handover mechanisms were under development. These technologies would ensure that devices could switch between the two modes without disruption, maintaining a consistent connection.
- **Hybrid Routers and Access Points**: Hardware development was a key focus, with the creation of hybrid routers and access points that could operate in both Li-Fi and Wi-Fi modes. These devices were designed to simplify the deployment of hybrid networks in various settings.
- **Regulatory issues**: Regulatory frameworks needed to adapt to accommodate Li-Fi technology and its integration with Wi-Fi. Researchers and industry users were working with regulatory bodies to define standards and guidelines for Li-Fi and hybrid networks.
- **Security Enhancements**: Researchers were addressing security issues specific to Li-Fi–Wi-Fi networks. As Li-Fi operates in a more confined space, security measures were designed to protect data within line-of-sight. This included encryption techniques and authentication protocols.
- **Real-World Applications**: Various industries, including healthcare, manufacturing, and smart cities, were discussing the practical applications of hybrid networks. Hospitals, for example, were considering using Li-Fi for secure data transmission in sensitive areas, while smart cities were investigating Li-Fi for enhanced connectivity in urban environments.

- **IoT Integration**: The integration of the IoT into hybrid Li-Fi–Wi-Fi networks was a focus area. These networks were seen as valuable for supporting the growing number of IoT devices, which require reliable and secure connections.
- **Commercialization and Market Adoption**: Companies were increasingly investing in the commercialization of Li-Fi and hybrid network solutions. While not yet as globally as Wi-Fi, Li-Fi technology was starting to appear in certain commercial applications.

Note that this is an evolving field, and ongoing research is likely to lead to further innovations and practical implementations.

15.2.3 Advantages and Challenges of Hybrid Li-Fi–Wi-Fi Networks

Hybrid Li-Fi–Wi-Fi networks provide a unique combination of advantages and challenges [15], which are essential to consider when assessing the potential of this technology. Here, we discuss both aspects:

15.2.3.1 Advantages

- **High-Speed Data Transfer**: Li-Fi provides extremely high data transfer rates, often surpassing the capabilities of traditional Wi-Fi. By incorporating Li-Fi into the hybrid network, users can access ultra-fast data connectivity in areas with Li-Fi coverage.
- **Improved Network Security**: Li-Fi's line-of-sight communication enhances network security. Data transmission occurs within the direct view of the transmitting light source, making it less susceptible to eavesdropping and interference, which is valuable in sensitive environments.
- **Reduced Electromagnetic Interference**: Li-Fi operates in the visible light spectrum, which is separate from the RF spectrum used by Wi-Fi. This segregation reduces electromagnetic interference, making hybrid networks suitable for locations with high RF interference issues.
- **Enhanced Connectivity in Specific Environments**: Hybrid networks are well suited for environments where the characteristics of both Li-Fi and Wi-Fi are beneficial. For instance, in hospitals, Li-Fi can provide secure communication in operating rooms, while Wi-Fi can provide broader coverage in patient areas.
- **IoT Support**: The growing IoT ecosystem benefits from hybrid networks. Li-Fi's secure and high-speed data transmission is ideal for connecting and controlling IoT devices, contributing to the development of smart homes and smart cities.
- **Dynamic Load Balancing**: Hybrid networks can dynamically balance the network load between Li-Fi and Wi-Fi to optimize performance. This ensures that devices receive the best possible connectivity based on their location and network conditions.

15.2.3.2 Challenges

- **Limited Coverage of Li-Fi**: Li-Fi's coverage is limited to the areas illuminated by the light source. It does not penetrate walls or obstacles, restricting its use to specific areas and requiring careful infrastructure planning.
- **Device Synchronization**: Continuous handovers between Li-Fi and Wi-Fi can be technically challenging. Devices need to transition smoothly as they move between areas covered by different technologies, and ensuring synchronization is essential.
- **Hardware and Infrastructure Costs**: Deploying hybrid networks may require investment in specialized hardware, including Li-Fi-enabled LEDs, photodetectors, and hybrid routers. The initial infrastructure setup can be cost-intensive.
- **Complex Network Management**: Managing a hybrid Li-Fi–Wi-Fi network can be complex, as administrators need to coordinate and optimize the operation of both technologies. Efficient network management tools are necessary.
- **Integration Challenges**: Integrating Li-Fi technology into existing environments can be challenging, especially in retrofitting scenarios. Ensuring continuous coexistence with Wi-Fi and other technologies is vital.
- **Limited Ecosystem**: While Wi-Fi has a well-established ecosystem with a wide range of devices, Li-Fi is still emerging. As a result, the number of Li-Fi-enabled devices and products is more limited.

Note that hybrid Li-Fi–Wi-Fi networks represent a promising evolution in wireless communication, providing solutions to address the challenges and demands of an increasingly connected world. While there are challenges to overcome, the advantages, particularly in terms of speed, security, and IoT support, make hybrid networks a compelling option for specific applications and environments. The development of this technology is ongoing, and as it matures, it is expected to find broader adoption and further innovation.

15.3 Li-Fi Technology: Definition, Principles, and Transmission Techniques (in Li-Fi)

Li-Fi is a wireless communication technology that employs visible light to transmit data. It is based on the principle of modulating the intensity of light emitted by LEDs to encode digital information [15–17]. Li-Fi provides a means of wireless data transfer using visible light, including the light emitted by LED bulbs, as a carrier for communication. Here are the key principles and transmission techniques in Li-Fi:

- **Data Transmission Through Light Modulation**: Li-Fi relies on the principle of light modulation, where the intensity of light is altered at a rapid rate to carry digital data. This modulation is typically imperceptible to the human eye, as it occurs at speeds far beyond the capability of human vision.

- **LEDs as Transmitters**: LEDs serve as the primary transmitters in Li-Fi systems. These LEDs emit light, which is modulated to encode data. The modulation can be achieved through various techniques, including amplitude modulation (AM) and pulse-width modulation (PWM).
- **Photodetectors as Receivers**: On the receiving end, photodetectors, like photodiodes or image sensors, capture the modulated light signals. These detectors convert the variations in light intensity into electrical signals that can be decoded to retrieve the transmitted data.
- **Line-of-Sight Communication**: Li-Fi operates using line-of-sight communication. This means that for data transmission to occur, the receiver must be within the direct view of the transmitting LED. Obstacles like walls can block the signal, limiting its range and enabling spatial reuse of the spectrum.
- **Rapid Data Transfer**: Li-Fi technology can achieve remarkably high data transfer rates, with speeds ranging from hundreds of megabits per second to several gigabits per second. The fast modulation of light allows for quick data exchange.
- **Multidimensional Communication**: Li-Fi systems can use multiple dimensions of light for data communication. These dimensions include intensity modulation (changing the brightness of the light), color modulation (changing the color of the light), and spatial modulation (using different LEDs in a room for simultaneous communication).
- **Secure and Resistant to Interference**: Li-Fi is inherently more secure and less susceptible to interference compared to traditional RF-based technologies. The line-of-sight nature of Li-Fi means that data signals are contained within a specific area, reducing the risk of unauthorized interception and minimizing electromagnetic interference.
- **Application Diversity**: Li-Fi has found applications in various settings, including indoor environments, smart lighting, vehicular communication, and secure data transfer in sensitive areas where RF interference might be an issue.

Note that Li-Fi technology provides a promising solution for high-speed, secure wireless communication, especially in environments where RF-based communication faces challenges. Its unique principles and transmission techniques make it a valuable addition to the wireless communication landscape, providing an alternative and complementary technology to traditional Wi-Fi and other RF-based systems.

15.3.1 Li-Fi Applications and Use Cases in Modern Era

Li-Fi technology has a range of applications and use cases in the modern era, providing innovative solutions for high-speed, secure wireless communication in various settings [17, 18]. Here are some of the key applications and use cases of Li-Fi:

- **Indoor Wireless Connectivity**: Li-Fi is well suited for providing high-speed wireless internet access in indoor environments like offices, homes, and public buildings. It can complement traditional Wi-Fi to provide faster data transfer rates.
- **Hospitals and Healthcare**: In healthcare settings, Li-Fi can be used to establish secure communication in areas where electromagnetic interference must be minimized, like operating rooms and intensive care units.
- **Smart Lighting Systems**: Li-Fi can be integrated into smart lighting systems, where LED bulbs serve a dual purpose as both illumination sources and data transmitters. This is particularly useful for smart homes and offices.
- **Retail and Indoor Navigation**: Li-Fi technology can be used to provide location-based services and indoor navigation in retail stores and shopping malls, enhancing the shopping experience.
- **Transportation and Vehicular Communication**: Li-Fi can be integrated into vehicles to enable high-speed data transfer for infotainment systems, autonomous vehicles, and vehicle-to-vehicle (V2V) communication.
- **Aviation and Aerospace**: Li-Fi can provide secure and high-speed data communication for in-flight entertainment systems, cockpit communications, and aircraft-to-ground communication.
- **Underwater Communication**: Li-Fi's use of visible light is suitable for underwater communication, where traditional RF signals are ineffective. It can be employed in underwater vehicles, research, and offshore industries.
- **Defense and Military Applications**: Li-Fi's resistance to RF interference and secure communication characteristics make it valuable in military and defense applications, where data security is paramount.
- **Data Centers and Secure Facilities**: Li-Fi can provide secure communication within data centers and other sensitive facilities where electromagnetic interference and data breaches must be minimized.
- **IoT Networks**: Li-Fi can support IoT devices in various settings by providing reliable and high-speed connectivity, ensuring efficient data transfer for smart devices and sensors.
- **Educational Environments**: Li-Fi technology can enhance digital learning experiences in classrooms by providing high-speed wireless connectivity to students and teachers.
- **Environmental Monitoring and Smart Cities**: Li-Fi can be used for environmental monitoring systems and smart city applications, enabling real-time data collection and communication in urban and remote areas.
- **Energy Efficiency and Sustainability**: Li-Fi contributes to energy-efficient lighting and communication systems, reducing energy consumption and supporting sustainability goals.

- **High-Security Environments**: Li-Fi's line-of-sight communication and resistance to eavesdropping make it suitable for applications where data security is a top priority, like military bases and government installations.

Hence, these applications and use cases highlight the versatility and potential of Li-Fi technology in addressing the increasing demand for high-speed, secure, and reliable wireless communication in various sectors and settings. As Li-Fi technology continues to develop and mature, it is expected to find even more innovative applications in the modern era.

15.3.2 Current and Future Advancements in Li-Fi Technology

We highlight some of the trends and potential future advancements in Li-Fi technology, such as:

- **Higher Data Transfer Rates**: Researchers and engineers were working on pushing the limits of Li-Fi's data transfer rates even further. This included discussing advanced modulation techniques, multi-beam communication, and improved signal processing to achieve gigabit and multi-gigabit speeds.
- **Integration with Emerging Technologies**: Li-Fi was being integrated with other emerging technologies like 5G and IoT. This integration aimed to create continuous connectivity across different communication platforms and enhance the capabilities of these technologies.
- **Li-Fi-Enabled Devices**: The development of more Li-Fi-enabled devices, including smartphones, laptops, and IoT devices, was anticipated. This would make Li-Fi technology more accessible to consumers and businesses.
- **Enhanced Security Measures**: Researchers were focusing on enhancing the security features of Li-Fi networks, including implementing advanced encryption methods and improving authentication protocols to protect data from potential threats.
- **Standardization and Regulatory Support**: Efforts were ongoing to establish standardized practices and regulations for Li-Fi technology. This would help ensure interoperability, safety, and global adoption of Li-Fi networks.
- **Commercialization and Deployment**: The commercialization of Li-Fi technology was expected to expand. Companies were working on implementing Li-Fi in various commercial and industrial applications, like retail, healthcare, and manufacturing.
- **Hybrid Li-Fi–Wi-Fi Networks**: Research continued in the development and deployment of hybrid Li-Fi–Wi-Fi networks, allowing for continuous transitions between the two technologies. This integration would provide flexibility and optimization in wireless connectivity.
- **Smart Cities and IoT**: Li-Fi was anticipated to play a role in smart city initiatives, providing reliable and high-speed connectivity for smart infrastructure, autonomous vehicles, and IoT devices.

- **Li-Fi for Underwater and Harsh Environments**: Li-Fi's ability to function in environments with electromagnetic interference and where traditional RF communication is ineffective, like underwater applications, was a focus area for research and development.
- **Miniaturization and Embedded Li-Fi**: Efforts were being made to miniaturize Li-Fi transmitters and receivers, making it easier to integrate Li-Fi technology into various devices and equipment, including wearables and small IoT sensors.

Note that as Li-Fi technology matures and becomes more widely adopted, it is expected to bring huge advancements to the field of wireless communication. These advancements will cater to the growing demand for faster, more secure, and more reliable connectivity in a variety of environments and applications. To stay up-to-date with the latest developments in Li-Fi, it is advisable to follow recent research papers, industry news, and technology announcements in the field.

15.4 Wi-Fi Technology: Definition, Principles, Standards, and Protocols (of Wi-Fi)

Wi-Fi, short for "Wireless Fidelity," is a popular and widely used wireless communication technology that enables devices to connect to the internet and local area networks (LANs) without the need for physical cables [19–21]. Wi-Fi technology uses RF signals to transmit data over the airwaves. Here are the key aspects of Wi-Fi technology, including its principles, standards, and protocols:

15.4.1 Principles of Wi-Fi

- **RF Communication**: Wi-Fi relies on radio waves to establish wireless connections between devices and network access points. Devices with Wi-Fi capabilities have built-in radios for transmitting and receiving data.
- **Wireless Access Points (APs)**: Wi-Fi networks typically include one or more APs such as routers or wireless hotspots. These access points broadcast Wi-Fi signals, allowing nearby devices to connect to the network.
- **Wireless Standards**: Wi-Fi operates according to a set of standards that define the rules for wireless communication. These standards ensure compatibility between different devices and manufacturers.
- **SSID (Service Set Identifier)**: Each Wi-Fi network is identified by an SSID, which is a unique name. Devices scan for available networks based on their SSID and can connect to a specific network using the appropriate SSID and password.
- **Security Protocols**: Wi-Fi networks can be secured using encryption methods like Wi-Fi Protected Access (WPA) 2 or WPA3 to protect data from unauthorized access and eavesdropping. Passwords or passphrases are typically used to authenticate users.

15.4.2 Wi-Fi Standards and Protocols

Wi-Fi standards and protocols are established by the IEEE and are designated by numerical identifiers like 802.11. Different generations of Wi-Fi technology have been developed over the years, each providing improvements in data transfer rates, range, and other features. Some notable Wi-Fi standards and protocols include:

- **802.11a**: This was one of the first Wi-Fi standards, operating in the 5 GHz frequency band. It provided data rates of up to 54 Mbps.
- **802.11b**: Operating in the 2.4 GHz frequency band, 802.11b provided data rates of up to 11 Mbps. It was one of the early global Wi-Fi standards.
- **802.11g**: This standard improved upon 802.11b by providing data rates of up to 54 Mbps in the 2.4 GHz band. It provided greater speed and backward compatibility.
- **802.11n**: Also known as "Wi-Fi 4," 802.11n introduced multiple-input multiple-output (MIMO) technology, which significantly improved data rates, range, and reliability. It operated in both 2.4 and 5 GHz bands.
- **802.11ac**: "Wi-Fi 5" or 802.11ac provided data rates of up to several gigabits per second and operated exclusively in the 5 GHz band. It introduced beamforming and multiuser MIMO (MU-MIMO) for better performance in crowded environments.
- **802.11ax**: "Wi-Fi 6" or 802.11ax is the latest Wi-Fi standard as of my last knowledge update in January 2022. It focuses on improving network efficiency, capacity, and performance, especially in high-density environments. It supports both 2.4 and 5 GHz bands.
- **802.11ay**: This is a standard that operates in the 60 GHz frequency band and is designed for very high-speed and short-range communication. It is suitable for applications like wireless docking and augmented reality (AR).

These Wi-Fi standards and protocols have evolved to keep pace with the increasing demand for wireless connectivity, providing faster data transfer rates and enhanced features in each new iteration. Users can choose the appropriate Wi-Fi standard to suit their specific needs, with backward compatibility ensuring that older devices can still connect to newer networks, albeit potentially at reduced speeds.

15.4.2.1 Wi-Fi Applications and Use Cases

Wi-Fi technology has a wide range of applications and use cases across various sectors, providing wireless connectivity for data communication [22, 23]. Here are some of the key Wi-Fi applications and use cases (refer Figure 15.2):

Figure 15.2 Wi-Fi use cases.

- **Internet Access at Home and Offices**: Wi-Fi is commonly used for providing internet access to computers, smartphones, tablets, and other devices in homes, offices, and small businesses.
- **Public Wi-Fi Hotspots**: Public places such as airports, coffee shops, libraries, and hotels provide free or paid Wi-Fi access to provide internet connectivity to visitors and travelers.
- **Education**: Wi-Fi is essential in educational institutions, enabling students and teachers to access online resources, conduct research, and participate in remote learning.
- **Enterprise Networks**: Large organizations and corporations deploy Wi-Fi networks to provide connectivity for employees, guests, and various devices within their facilities.
- **Smartphones and Mobile Devices**: Wi-Fi is integrated into smartphones, tablets, and other mobile devices, allowing them to connect to the internet and other devices without the need for cellular data.

- **IoT Connectivity**: IoT devices use Wi-Fi to connect to the internet and interact with other devices in applications like smart homes, smart cities, and industrial IoT.
- **Wireless Printing**: Wi-Fi-enabled printers allow users to print documents wirelessly from their computers, smartphones, and other devices.
- **Video and Audio Streaming**: Wi-Fi is important for streaming services, such as Netflix, YouTube, and music streaming platforms, delivering high-quality video and audio content to users.
- **Gaming**: Online gaming relies on low-latency, high-speed Wi-Fi connections to provide real-time gaming experiences for players.
- **Retail and E-commerce**: Retailers use Wi-Fi for inventory management, point-of-sale systems, and providing in-store Wi-Fi access for customers. It is also used for e-commerce order fulfillment in warehouses.
- **Telehealth and Remote Healthcare**: Wi-Fi enables telehealth services, allowing patients and healthcare providers to conduct remote consultations and monitor health data.
- **Manufacturing and Industry 4.0**: Wi-Fi is used in smart manufacturing for real-time data collection, monitoring, and control of industrial processes, contributing to the Industry 4.0 revolution.
- **Logistics and Warehousing**: Wi-Fi is vital for managing inventory, tracking shipments, and optimizing logistics operations in warehouses and distribution centers.
- **Agriculture and Precision Farming**: In precision agriculture, Wi-Fi is used for data collection, remote monitoring, and automation of farming processes.
- **Tourism and Hospitality**: The hospitality industry relies on Wi-Fi to provide guests with internet access, room automation, and personalized services.
- **Energy Management**: Wi-Fi networks are employed for monitoring and controlling smart energy grids, optimizing power consumption, and enhancing energy efficiency.
- **Transportation**: In transportation, Wi-Fi is used for passenger connectivity on trains, buses, and subways, as well as for real-time tracking and communication in logistics and fleet management.
- **Security and Surveillance**: Wi-Fi cameras and sensors are used for surveillance and security applications, allowing for remote monitoring and video streaming.
- **Outdoor and Public Safety Networks**: Public safety agencies use Wi-Fi for data exchange and connectivity in emergency response situations, and cities deploy Wi-Fi networks for public safety monitoring.
- **Cultural and Entertainment Events**: Wi-Fi is set up at sports stadiums, music festivals, and other events to provide attendees with internet access, ticket scanning, and event-related information.

Wi-Fi technology's versatility and global adoption make it an integral part of modern life, supporting a broad spectrum of applications and enabling continuous connectivity in various environments and industries.

15.4.2.2 Current and Future Advancements in Wi-Fi Technology

Wi-Fi technology is continuously evolving to meet the growing demand for faster, more reliable, and more secure wireless connectivity [20, 21]. Here are some of the current advancements in Wi-Fi technology, as well as future trends to look out for.

Current Advancements in Wi-Fi Technology

- **Wi-Fi 6 (802.11ax)**: Wi-Fi 6 is the latest standard as of my last update in January 2022. It brings major improvements in data transfer rates, network efficiency, and capacity, making it well suited for crowded environments. Wi-Fi 6-enabled devices and routers have become more widely available.
- **Mesh Networking**: Mesh Wi-Fi systems have gained popularity for their ability to provide continuous coverage in larger spaces, eliminating dead zones and ensuring consistent connectivity throughout a home or office.
- **WPA3 Security**: WPA3 is the latest security protocol for Wi-Fi networks. It provides stronger encryption and protection against common security vulnerabilities, enhancing network security.
- **MU-MIMO**: MU-MIMO technology allows routers to communicate with multiple devices simultaneously, improving network efficiency and reducing latency for connected devices.
- **Improved Beamforming**: Beamforming technology, used in Wi-Fi 6, allows routers to focus signals directly at connected devices, optimizing signal strength and coverage.
- **IoT-Focused Wi-Fi Standards**: Wi-Fi standards are being tailored to better support IoT devices, which often have low-power and low-data requirements. Technologies like Wi-Fi HaLow (802.11ah) are designed for extended range and efficiency in IoT applications.

Future Advancements and Trends in Wi-Fi Technology

- **Wi-Fi 7 (802.11be)**: The development of Wi-Fi 7 is expected to continue, bringing even higher data transfer rates, lower latency, and improved network performance. This will further support emerging technologies and applications.
- **6 GHz Band Expansion**: The opening of the 6 GHz band for unlicensed use in some regions is anticipated to provide additional spectrum for Wi-Fi, allowing for faster and less congested networks.
- **Enhanced Security Measures**: As cybersecurity threats evolve, Wi-Fi technology is likely to implement advanced security features, potentially integrating features like enhanced encryption and authentication.

- **Better Integration with 5G**: The synergy between Wi-Fi and 5G networks will become more continuous, enabling devices to continuously switch between cellular and Wi-Fi connections for optimized performance and coverage.
- **Quantum-Safe Encryption**: As quantum computing becomes a reality, the development of quantum-safe encryption methods for Wi-Fi networks will be important to ensure long-term security.
- **In-Band Full Duplex Communication**: Full-duplex Wi-Fi technology, allowing simultaneous transmission and reception, is being discussed to further increase network efficiency and capacity.
- **Enhanced Location Services**: Improved Wi-Fi location services will provide more accurate indoor and outdoor positioning, benefiting applications like AR and asset tracking.
- **Energy-Efficient Wi-Fi**: Energy-efficient Wi-Fi standards will be essential for extending the battery life of IoT devices and reducing the overall energy consumption of networks.
- **Smart and Autonomous Wi-Fi Networks**: Wi-Fi networks will become smarter, capable of autonomously optimizing network performance and adapting to changing conditions without human intervention.
- **Enhanced Device-to-Device Communication**: Technologies like Wi-Fi direct will continue to evolve, enabling devices to communicate with each other directly without the need for a traditional access point.

Hence, these advancements in Wi-Fi technology are driven by the ever-increasing demand for connectivity, the emergence of new applications, and the need for more efficient and secure networks. While Wi-Fi 6 and Wi-Fi 6E are already making a major impact, the ongoing development of Wi-Fi 7 and related technologies will shape the future of wireless communication.

15.5 Hybrid Li-Fi–Wi-Fi Networks: Introduction

15.5.1 Integration of Li-Fi and Wi-Fi Technologies

The integration of Li-Fi and Wi-Fi technologies, often referred to as hybrid Li-Fi–Wi-Fi networks, provides a powerful solution to address various connectivity challenges and provide a more versatile and efficient wireless communication system (refer Figure 15.3).

Here is how these two technologies can be integrated:

- **Continuous Handover**: Hybrid networks enable continuous handovers between Li-Fi and Wi-Fi as devices move through an environment. When a device is within the range of a Li-Fi transmitter (e.g., an LED bulb), it can connect to the Li-Fi network for high-speed data transfer. As the device moves out of the Li-Fi coverage area, it can transition to the Wi-Fi network, ensuring uninterrupted connectivity.

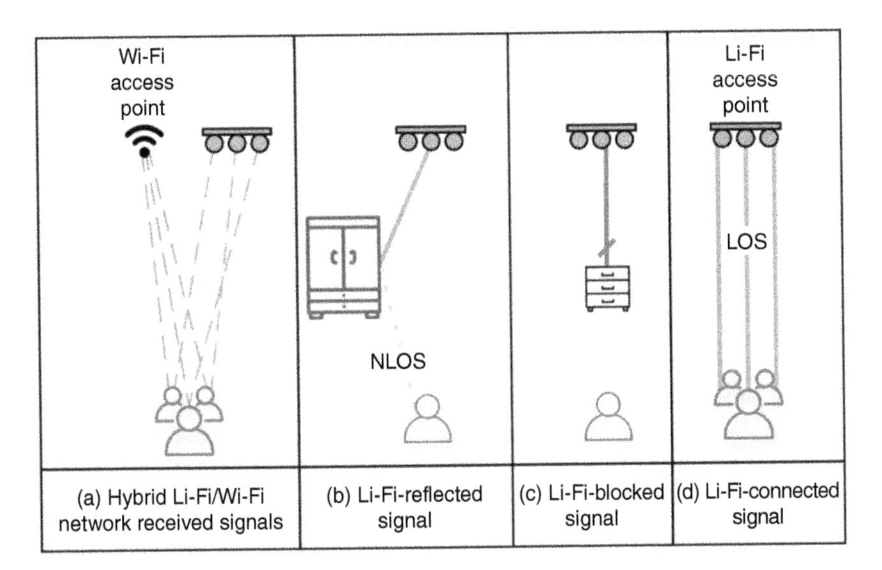

Figure 15.3 Hybrid Li-Fi.

- **Complementary Technologies**: Li-Fi and Wi-Fi are complementary technologies. Li-Fi excels in high-speed data transmission and secure communication, while Wi-Fi provides broader coverage and the ability to penetrate walls and obstacles. By integrating both technologies, users benefit from the strengths of each, ensuring efficient connectivity in different scenarios.
- **Network Optimization**: Hybrid networks can optimize network performance by dynamically selecting the best technology based on factors like device location, network load, and signal quality. For example, in an environment with many devices, Li-Fi can offload some of the data traffic, reducing congestion on the Wi-Fi network and improving overall performance.
- **Improved Security**: Li-Fi's line-of-sight communication enhances network security. Data transmission is limited to areas within the direct view of the transmitting light source, reducing the risk of eavesdropping and unauthorized access. In scenarios where data privacy and security are important [24–30], Li-Fi can be used to transmit sensitive information.
- **Diverse Applications**: Hybrid Li-Fi–Wi-Fi networks find applications in a wide range of settings, including healthcare facilities, industrial environments, smart homes, and smart cities. In hospitals, for example, Li-Fi can be used in operating rooms and patient areas for secure data transmission, while Wi-Fi provides coverage in common areas.
- **IoT Support**: The integration of Li-Fi and Wi-Fi can benefit the IoT. Li-Fi provides a high-speed and secure data transfer solution for IoT devices, while Wi-Fi can connect devices that require broader coverage and connectivity in areas outside the line of sight.

- **Enhanced User Experience**: Users can enjoy an improved wireless experience with hybrid networks. Devices automatically switch between Li-Fi and Wi-Fi based on their location, ensuring they are connected to the best available network. This continuous transition results in higher data transfer rates and better network reliability.
- **Compatibility and Standards**: To facilitate the integration of these technologies, industry stakeholders and standardization bodies may work on defining protocols and guidelines for hybrid Li-Fi–Wi-Fi networks. This will ensure interoperability and ease of deployment.

Here, Figure 15.4 provides information about heterogenous Li-Fi–Wi-Fi with multipath transmission protocol for effective access point selection and load balancing. Note that the integration of Li-Fi and Wi-Fi technologies represents an exciting evolution in wireless communication. By combining the unique strengths of each technology, hybrid networks provide solutions to the challenges and demands of our increasingly interconnected world. They are poised to play a pivotal role in ensuring robust, efficient, and secure wireless connectivity across various sectors and applications.

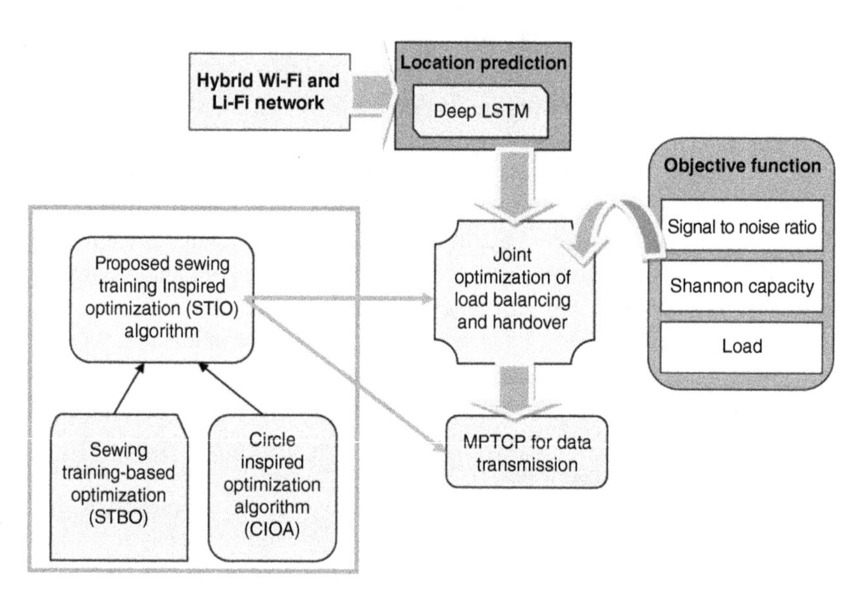

Figure 15.4 Heterogenous Li-Fi–Wi-Fi with multipath transmission protocol for effective access point selection and load balancing.

15.5.2 Benefits and Features of Hybrid Li-Fi–Wi-Fi Networks

Hybrid Li-Fi–Wi-Fi networks provide a combination of benefits and features that make them a compelling solution for various wireless communication scenarios. Here are some of the key advantages and features of hybrid Li-Fi–Wi-Fi networks:

15.5.2.1 Benefits

- **High-Speed Data Transfer**: Hybrid networks provide high-speed data transfer by using the speed of Li-Fi in areas with line-of-sight communication. This is particularly valuable for applications requiring ultra-fast data exchange.
- **Enhanced Security**: Li-Fi's line-of-sight nature enhances network security. Data transmission occurs within the direct view of the transmitting light source, reducing the risk of eavesdropping and unauthorized access, making it suitable for secure communication.
- **Reduced Electromagnetic Interference**: Li-Fi operates in the visible light spectrum, which is separate from the RF spectrum used by Wi-Fi. This separation reduces electromagnetic interference, making hybrid networks ideal for locations with high RF interference issues.
- **Improved Connectivity in Specific Environments**: Hybrid networks are well suited for environments where both Li-Fi and Wi-Fi characteristics are beneficial. For example, in healthcare facilities, Li-Fi can provide secure communication in operating rooms, while Wi-Fi can provide broader coverage in patient areas.
- **IoT Support**: The high-speed and secure data transmission capabilities of hybrid networks make them suitable for supporting the growing number of IoT devices, ensuring reliable connectivity for smart devices and sensors.
- **Dynamic Load Balancing**: Hybrid networks can dynamically balance the network load between Li-Fi and Wi-Fi to optimize performance. This ensures that devices receive the best possible connectivity based on their location and network conditions.
- **Improved Network Reliability**: The continuous handover between Li-Fi and Wi-Fi minimizes network disruptions, enhancing the reliability of wireless connectivity.
- **Multimodal Communication**: Hybrid networks enable multimodal communication by combining Li-Fi, Wi-Fi, and potentially other wireless technologies. This flexibility supports various use cases and applications.

15.5.2.2 Features

- **Continuous Handover**: Devices can transition continuously between Li-Fi and Wi-Fi based on their location and signal quality, ensuring uninterrupted connectivity.
- **Optimized Performance**: Hybrid networks optimize network performance by selecting the most suitable technology for specific situations, resulting in efficient and reliable data transfer.
- **Compatibility**: Hybrid networks are designed to be compatible with existing Wi-Fi infrastructure, making it easier to integrate Li-Fi technology into various environments and applications.
- **Network Management**: Administrators can manage hybrid networks to ensure optimal operation, load balancing, and user experience. Management tools and software can be used to monitor and control the network.
- **Privacy and Security**: The line-of-sight communication in Li-Fi enhances privacy and security, making it a valuable feature for applications where data confidentiality is important [30, 31].
- **Application Diversity**: Hybrid networks have a wide range of applications, including healthcare, manufacturing, smart cities, transportation, education, and more. Their flexibility makes them suitable for various use cases.
- **Energy Efficiency**: Li-Fi technology can also be energy efficient as it operates using LED lighting, making it suitable for environments with a focus on energy conservation.
- **Scalability**: Hybrid networks can be scaled to accommodate a growing number of devices and applications, making them suitable for environments with evolving connectivity needs.

Hybrid Li-Fi–Wi-Fi networks provide a versatile and powerful solution for addressing the challenges and demands of a connected world. They combine the strengths of Li-Fi and Wi-Fi technologies to provide reliable, high-speed, and secure wireless communication, enhancing connectivity in various sectors and applications.

15.5.3 Current and Future Advancements in Hybrid Li-Fi–Wi-Fi Networks

Hybrid Li-Fi–Wi-Fi networks, which combine the capabilities of both Li-Fi and Wi-Fi technologies, represent a promising and evolving field of wireless communication. While I cannot provide the very latest advancements beyond my last knowledge update in January 2022, I can highlight some of the current and potential future advancements in this field:

15.5.3.1 Current Advancements

- **Standardization and Protocols**: Ongoing work in standardization bodies to establish protocols and guidelines for hybrid Li-Fi–Wi-Fi networks, ensuring interoperability and compatibility.
- **Commercial Deployments**: The deployment of hybrid Li-Fi–Wi-Fi networks in commercial and industrial settings is expanding. This includes applications in healthcare, manufacturing, smart cities, and retail environments.
- **IoT Integration**: Integration with the IoT is a current trend. Hybrid networks are being used to provide secure and high-speed connectivity for IoT devices, contributing to the growth of smart cities and industrial IoT applications.
- **Improved Device Ecosystem**: An increasing number of devices with integrated Li-Fi technology are becoming available. This includes Li-Fi-enabled smartphones, laptops, and IoT devices, making hybrid networks more accessible.

15.5.3.2 Future Advancements and Trends

- **Enhanced Continuous Handovers**: Future advancements may focus on making handovers between Li-Fi and Wi-Fi even more continuous, ensuring a smoother transition as devices move between coverage areas.
- **Enhanced Network Management**: The development of advanced network management and optimization tools to dynamically balance network load and maximize performance is expected.
- **Extended Coverage**: Researchers are discussing ways to extend Li-Fi coverage, potentially using reflectors or repeaters to bounce signals into areas not directly illuminated by Li-Fi transmitters.
- **Energy Efficiency**: Efforts to make Li-Fi technology more energy efficient by optimizing LED lighting and reducing power consumption will continue to align with sustainability goals.
- **Hybrid Standards**: The development of hybrid communication standards that allow continuous switching between Li-Fi and Wi-Fi based on factors like signal quality, network load, and energy efficiency.
- **Application Diversity**: The identification of new and diverse applications for hybrid networks is expected. Potential areas of growth include automotive communication, AR, and energy-efficient lighting solutions.
- **Li-Fi for Specific Verticals**: Tailoring hybrid Li-Fi–Wi-Fi solutions to specific verticals, like healthcare, aviation, and smart manufacturing, to address their unique requirements and challenges.
- **Government and Regulatory Support**: Collaboration with regulatory bodies to define and establish standards and guidelines for hybrid networks, ensuring global adoption and compliance with regulations.

Hence, hybrid Li-Fi–Wi-Fi networks have the potential to provide high-speed, secure, and reliable wireless communication in a wide range of environments and applications. As these networks continue to mature and evolve, it is expected that they will find broader adoption and innovation in various sectors, providing solutions to the challenges and demands of a connected world. To stay up-to-date with the latest advancements in hybrid Li-Fi–Wi-Fi networks, it is advisable to follow recent research papers, industry news, and technology announcements in the field.

15.5.4 Simulators Existed for Implementing Hybrid Li-Fi–Wi-Fi Networks

As this is a relatively specialized and emerging field, few researchers and developers in the wireless communication and networking domains often depend on general network simulation tools and platforms to model and study hybrid networks. These simulators can be adapted to simulate the behavior of both Li-Fi and Wi-Fi technologies within a hybrid network. Some widely used simulation tools include:

- **ns-3 (Network Simulator 3)**: ns-3 is a popular open-source network simulator used for modeling and simulating various networking technologies. Researchers can create custom modules to simulate both Li-Fi and Wi-Fi technologies within a hybrid network.
- **OPNET (Riverbed Modeler)**: OPNET, now part of Riverbed Modeler, is a widely used commercial network simulation tool. It allows for modeling and simulating complex network scenarios, making it suitable for studying hybrid networks.
- **OMNeT++**: OMNeT++ is an open-source discrete event simulation framework that can be adapted for simulating hybrid Li-Fi–Wi-Fi networks by creating custom models for both technologies.
- **MATLAB and Simulink**: Matrix Laboratory (MATLAB) and Simulink provide simulation capabilities for various communication protocols, and researchers can build custom models for Li-Fi and Wi-Fi within these environments.
- **QualNet**: QualNet is a commercial network simulation tool that can be adapted to simulate hybrid networks. It provides a wide range of modeling capabilities for different network technologies.
- **Custom Simulation Platforms**: In some cases, researchers may develop custom simulation platforms tailored to their specific needs for simulating hybrid Li-Fi–Wi-Fi networks.
- While these simulators may not have specific pre-built modules for hybrid Li-Fi–Wi-Fi networks, researchers and developers have the flexibility to create custom models and scenarios to study the behavior, performance, and optimization of such networks.

Note that the availability of simulators and tools may evolve over time, so it is a good practice to check for updates and specialized tools that may emerge to support the simulation and analysis of hybrid Li-Fi–Wi-Fi networks in the future. Additionally, academic institutions and research organizations may develop in-house simulators to support their research in this field.

15.6 Current Barriers and Challenges Toward Hybrid Li-Fi–Wi-Fi Networks

The development and deployment of hybrid Li-Fi–Wi-Fi networks, which combine Li-Fi and Wi-Fi technologies, face several current barriers and challenges. These challenges need to be addressed to enable the global adoption of these networks in various applications and environments. Some of the key barriers and challenges include:

- **Limited Coverage Range**: Li-Fi technology relies on visible light communication, which has a limited coverage range compared to Wi-Fi. Ensuring consistent and reliable Li-Fi coverage across larger areas can be challenging, particularly in outdoor or open spaces.
- **Line-of-Sight Requirement**: Li-Fi communication requires a direct line of sight between the light source (e.g., LED bulb) and the receiving device. This can be restrictive in environments with obstacles, and mobility may be limited.
- **Interference**: Hybrid networks must manage potential interference between Li-Fi and Wi-Fi signals. Coexistence mechanisms are needed to ensure that Li-Fi and Wi-Fi devices do not disrupt each other's operation.
- **Device Compatibility**: The availability of Li-Fi-enabled devices, like smartphones, laptops, and IoT devices, is currently limited compared to Wi-Fi-enabled devices. Achieving an important mass of compatible devices is essential for global adoption.
- **Standardization**: Standardization for hybrid Li-Fi–Wi-Fi networks is still evolving. The lack of widely accepted standards can hinder interoperability and compatibility between different equipment and vendors.
- **Infrastructure Investment**: Implementing Li-Fi technology often requires the installation of LED light fixtures with integrated Li-Fi capabilities. This infrastructure investment can be a barrier to adoption, particularly in existing buildings.
- **Cost**: Li-Fi transmitters and receivers, while becoming more affordable, may still be more expensive than traditional Wi-Fi equipment. The initial cost of deployment can be a challenge for some organizations.
- **Security**: While Li-Fi provides enhanced security due to its line-of-sight nature, the security of hybrid networks must be carefully managed. Ensuring data privacy, authentication, and encryption across both Li-Fi and Wi-Fi is important.

- **Scalability**: Hybrid networks need to be designed for scalability to accommodate a growing number of devices and users. The challenge lies in managing increased traffic and ensuring network efficiency.
- **Network Management**: The management of hybrid networks, including load balancing between Li-Fi and Wi-Fi, as well as network optimization, can be complex and requires advanced management tools.
- **Regulatory Compliance**: Compliance with regulations regarding both Li-Fi and Wi-Fi spectrum usage is important. Regulatory frameworks may vary by region, and adherence to these regulations is essential for deployment.
- **Education and Awareness**: Building awareness and understanding of the benefits and limitations of hybrid Li-Fi–Wi-Fi networks among potential users and decision-makers is essential for adoption.
- **Adaptation to Specific Verticals**: Tailoring hybrid network solutions to specific industries and verticals, like healthcare, manufacturing, and transportation, requires specialized knowledge and customization.
- **Energy Efficiency**: While Li-Fi can be energy efficient when integrated with LED lighting, optimizing energy consumption remains a challenge, especially in environments where energy conservation is a priority.

Hence, addressing these barriers and challenges is essential for the successful integration of Li-Fi and Wi-Fi technologies into hybrid networks. As technology advances and standards mature, many of these challenges are expected to be overcome, making hybrid networks more practical and beneficial for a wide range of applications.

15.7 Future Directions and Research Opportunities Toward Hybrid Li-Fi–Wi-Fi Networks

The field of hybrid Li-Fi–Wi-Fi networks is rich with research opportunities and future directions that can drive innovation and advance the adoption of this technology. Here are some key research areas and opportunities for the future of hybrid Li-Fi–Wi-Fi networks:

- **Standardization and Interoperability**: Research efforts should continue to focus on standardizing protocols and interfaces for hybrid networks. Developing open standards and ensuring interoperability among different vendors' equipment is important.
- **Continuous Handover**: Improving the continuous handover mechanism between Li-Fi and Wi-Fi networks as devices move between coverage areas remains an essential research area. This includes minimizing disruption during the transition and optimizing handover decisions.

- **Energy Efficiency**: Research can discuss ways to enhance the energy efficiency of Li-Fi technology, especially issuing LED lighting. Developing techniques for power optimization in Li-Fi transmitters and receivers can make Li-Fi more sustainable.
- **Scalability and Load Balancing**: As the number of connected devices continues to grow, research can focus on scalable solutions for hybrid networks. This includes load balancing mechanisms and efficient resource allocation to manage network congestion and ensure high-quality service.
- **Security and Privacy**: Future research can discuss about advanced security measures for hybrid networks. This includes encryption, authentication, and privacy enhancements that protect data transmission in both Li-Fi and Wi-Fi modes.
- **Advanced Network Management**: Developing intelligent and self-optimizing network management solutions is essential. AI-driven network management can adapt to changing conditions, user behaviors, and network load, improving overall performance.
- **5G and Beyond**: Integration with 5G networks represents a promising research area. Investigating how hybrid networks can work in harmony with 5G technologies to provide a continuous and efficient user experience is important.
- **Localization and Positioning**: Research can discuss advanced localization techniques that use Li-Fi signals for precise indoor positioning, benefiting applications in healthcare, retail, and asset tracking.
- **Industry-Specific Applications**: Tailoring hybrid Li-Fi–Wi-Fi solutions to specific industries and verticals provides several research opportunities. Customizing network solutions for healthcare, manufacturing, transportation, and smart cities can address industry-specific challenges.
- **Multimodal Communication**: Investigating how hybrid networks can support multimodal communication, including integration with other wireless technologies like Bluetooth, Zigbee, and 5G, provides diverse research opportunities.
- **Integration with Emerging Technologies**: Discussing the integration of hybrid networks with emerging technologies like AR, virtual reality (VR), and blockchain can unlock new applications and use cases.
- **Quantum-Safe Security**: As quantum computing advances, research into quantum-safe encryption and security measures for hybrid networks is vital to protect against future threats.
- **Lighting Innovations**: Research into advanced lighting technologies that enhance Li-Fi capabilities, including new modulation techniques and improved LED transmitters, can drive the evolution of Li-Fi.
- **Government and Regulatory Engagement**: Collaboration with government bodies and regulatory authorities is essential for establishing standardized

practices and regulations for hybrid Li-Fi–Wi-Fi networks, ensuring global adoption and compliance.

Note that these research areas and opportunities have the potential to shape the future of hybrid Li-Fi–Wi-Fi networks and contribute to their adoption in a wide range of applications and industries. As technology advances and research progresses, hybrid networks are expected to become more practical, efficient, and beneficial for various connectivity needs

15.7.1 Integration of Hybrid Li-Fi–Wi-Fi Networks with 5G and Beyond

The integration of hybrid Li-Fi–Wi-Fi networks with 5G and beyond represents an important step in advancing wireless communication technologies to meet the demands of a highly connected and data-intensive future. Here are some key aspects of the integration of hybrid networks with 5G and beyond:

- **Continuous Connectivity**: Continuous handover and connectivity management: Integration efforts should focus on enabling devices to continuously transition between Li-Fi, Wi-Fi, and 5G networks, based on factors like signal strength, network load, and device capabilities. Smooth handovers will ensure uninterrupted and efficient connectivity.
- **Enhanced User Experience**: Improved quality of service: The combination of Li-Fi, Wi-Fi, and 5G technologies can enhance the overall user experience. Li-Fi's high-speed, low-latency capabilities can supplement 5G connectivity, providing users with consistent, high-quality connections, especially in crowded or challenging RF environments.
- **Coverage and Capacity**: Optimized network capacity: hybrid networks can be designed to balance the network load between Li-Fi, Wi-Fi, and 5G based on the number of connected devices and data traffic. This dynamic load balancing ensures efficient use of available resources.
- **Extending 5G Coverage with Li-Fi**: In areas where 5G signals may not penetrate, Li-Fi can be used to extend coverage. For example, Li-Fi can be deployed in indoor environments, providing high-speed connectivity where 5G signals are weaker.
- **Security and Privacy**: Enhanced security: the integration can provide enhanced security through both Li-Fi and 5G technologies. Li-Fi's inherent security benefits, like reduced interference and eavesdropping risks, complement the secure features of 5G networks.
- **IoT Connectivity**: Support for IoT devices: 5G networks are expected to play a pivotal role in connecting IoT devices. The integration with Li-Fi can provide high-speed and secure connectivity for IoT devices, especially in scenarios where data security is important.

- **Edge Computing**: Edge computing integration: 5G networks often involve edge computing for low-latency processing of data. The combination of edge computing with Li-Fi and Wi-Fi in hybrid networks can support real-time applications and services.
- **Multimodal Communication**: Supporting multiple wireless technologies: hybrid networks can continuously integrate Li-Fi, Wi-Fi, and 5G with other wireless technologies like Bluetooth, Zigbee, and LoRa, providing diverse communication options and use cases.
- **Emerging Applications**: Advanced applications: integration with 5G and beyond can drive the development of advanced applications, like augmented and VR, autonomous vehicles, smart cities, and industrial IoT. Li-Fi can serve as a reliable and high-speed connectivity solution in these contexts.
- **Regulatory and Spectrum Coordination**: Collaboration with regulatory authorities: Ensuring compliance with regulatory guidelines and spectrum coordination is important for the integration of different wireless technologies. Engaging with government bodies can help establish standards and practices for coexistence.
- **Network Management**: Advanced network management: utilizing advanced network management tools, including AI-driven management, can help optimize hybrid networks, ensuring that devices connect to the best available technology based on their location and network conditions.

Hence, the integration of hybrid Li-Fi–Wi-Fi networks with 5G and beyond represents a step toward a more interconnected and efficient wireless communication ecosystem. This integration uses the unique strengths of each technology to address the evolving needs of industries, users, and emerging applications in a highly connected world. As these technologies continue to evolve and mature, they have the potential to shape the future of wireless communication and connectivity.

15.7.2 Integration of Hybrid Li-Fi–Wi-Fi Networks with Internet of Things (IoT), AI/ML, and Blockchain

The integration of hybrid Li-Fi–Wi-Fi networks with the IoT, AI/machine learning (AI/ML), and blockchain technologies can create a powerful synergy, enabling new capabilities and use cases. Here's how these technologies can be integrated:

15.7.2.1 Integration with IoT

- **IoT Connectivity**: Hybrid networks can provide high-speed, reliable, and secure connectivity for IoT devices. Li-Fi can be particularly valuable for IoT applications that require low latency and high bandwidth, like industrial automation, smart cities, and healthcare.

- **Real-Time Data Transfer**: Li-Fi's low-latency communication can support real-time data transfer from IoT sensors and devices. This is important for applications like remote monitoring, predictive maintenance, and autonomous systems.
- **Location Services**: Li-Fi technology can provide precise indoor positioning and location services for IoT devices, enabling asset tracking, navigation, and contextual awareness in smart environments.

15.7.2.2 Integration with AI/ML

- **Data Processing**: AI/ML algorithms can be deployed at the edge of the network, where data is collected. The high-speed data transfer of Li-Fi can support real-time analytics and decision-making.
- **Network Optimization**: AI-powered network management can dynamically optimize the operation of hybrid networks. For example, AI algorithms can predict network congestion and direct devices to the most suitable connectivity option (Li-Fi, Wi-Fi, or 5G).
- **Resource Allocation**: AI can assist in the efficient allocation of resources in hybrid networks, balancing the load between Li-Fi and Wi-Fi based on device requirements, energy efficiency, and network congestion.

15.7.2.3 Integration with Blockchain

- **Security and Trust**: Blockchain technology can enhance the security and trustworthiness of hybrid networks. Smart contracts and decentralized identity solutions can be used to ensure secure access and authentication [24–31].
- **Payment and Micropayments**: Blockchain can enable secure and transparent micropayments for network access. Users can pay for connectivity in a secure and automated manner using cryptocurrencies or blockchain-based tokens.
- **Data Integrity**: Blockchain can be used to ensure the integrity and immutability of data transmitted over the network. This is valuable for applications that require tamperproof data, like supply chain management and healthcare.
- **Device Identity and Management**: Blockchain can provide a decentralized and secure way to manage device identities and access permissions in IoT networks.

15.7.2.4 Synergy of Technologies

- **Edge Computing**: The integration of IoT, AI/ML, blockchain, and hybrid networks can benefit from edge computing, where data processing and decision-making occur at the network's edge. This minimizes latency and conserves bandwidth.
- **Smart Contracts**: Smart contracts on a blockchain can automate and enforce agreements related to IoT device interactions and network access. For instance,

an IoT device can automatically negotiate the best available connectivity and pay for services through a smart contract.

- **Security and Privacy**: Blockchain's decentralized and immutable ledger can enhance the security and privacy of IoT data transmitted over hybrid networks. Users have more control over their data and can provide or revoke access as needed.
- **Data Monetization**: Users can monetize their IoT data securely through blockchain-based systems, creating new revenue opportunities.

Note that the integration of hybrid Li-Fi–Wi-Fi networks with IoT, AI/ML, and blockchain technologies creates a robust ecosystem capable of supporting advanced applications in various sectors, including smart cities, healthcare, manufacturing, and transportation. These technologies work together to provide high-speed, secure, and efficient data transfer and processing, enabling innovative solutions and improving the overall quality of connected experiences.

15.7.3 Integration of Hybrid Li-Fi–Wi-Fi Networks with Quantum Computing

The integration of hybrid Li-Fi–Wi-Fi networks with quantum computing represents a convergence of cutting-edge technologies that can potentially revolutionize the field of wireless communication. Here is how the integration can be projected.

15.7.3.1 Quantum-Safe Encryption

- **Enhanced Security**: Quantum computing threatens current encryption methods, as it has the potential to break traditional encryption algorithms. The integration of hybrid networks with quantum-safe encryption techniques ensures the long-term security of data transmitted over Li-Fi, Wi-Fi, and other technologies.
- **Secure Data Transfer**: Quantum-safe encryption can be applied to all aspects of hybrid networks, securing data transfer, authentication, and device management. This is important for maintaining data confidentiality and privacy.

15.7.3.2 Quantum Key Distribution (QKD)

- **Secure Key Exchange**: QKD can be used to securely exchange encryption keys between devices in hybrid networks. QKD ensures that keys are transmitted in a quantum-secure manner, making them resistant to eavesdropping.
- **Unbreakable Communication**: The use of QKD guarantees that communication between devices is theoretically unbreakable, providing a high level of security for important applications like military, healthcare, and financial services.

15.7.3.3 Quantum Networking

- **Quantum Communication Nodes**: Hybrid networks can serve as a foundation for quantum communication nodes. These nodes can relay quantum-secured data, expanding the reach of quantum-secured communication.
- **Quantum Repeaters**: Quantum repeaters integrated with hybrid networks can extend the distance over which quantum-secured communication can take place. This is especially important for long-distance quantum networks.

15.7.3.4 Quantum Machine Learning

- **Quantum ML Algorithms**: Integration with quantum ML can enable advanced data analytics and pattern recognition in hybrid networks. Quantum ML algorithms can process large datasets efficiently and uncover hidden insights.
- **Real-Time Analytics**: Quantum ML can be employed for real-time analytics in applications like IoT, optimizing network management, and identifying anomalies or security threats.

15.7.3.5 Quantum Sensors

- **Quantum Sensors Integration**: Quantum sensors can be integrated into hybrid networks for various applications, including environmental monitoring, healthcare, and navigation. These sensors provide unparalleled precision and sensitivity.
- **Data Fusion**: Quantum sensors can collect high-quality data, which can be fused with data from Li-Fi, Wi-Fi, and other sensors, creating a detailed and accurate dataset for analysis.

15.7.3.6 Quantum Computing as a Service

- **Quantum Computing Providers**: Quantum computing providers can provide their services over hybrid networks, allowing users to use the computational power of quantum computers for complex simulations, optimization problems, and cryptography.
- **Federated Learning**: Quantum computing can be used to enhance federated learning, allowing devices in hybrid networks to collaboratively train AI models without sharing sensitive data.

15.7.3.7 Quantum-Secured Blockchain

- **Enhanced Blockchain Security**: Hybrid networks can support blockchain applications with quantum-safe features, ensuring the security and integrity of blockchain transactions.
- **Tamperproof Transactions**: Transactions recorded on a quantum-secured blockchain are resistant to tampering and hacking, making them ideal for applications like supply chain management and secure financial transactions.

Hence, the integration of hybrid Li-Fi–Wi-Fi networks with quantum computing enhances security, data processing, and communication capabilities. It creates a foundation for quantum-secured communication and opens up new possibilities for secure, high-speed data transmission, making it suitable for applications where data privacy and security are paramount. As quantum computing continues to advance, the integration of these technologies will become increasingly important for industries, governments, and organizations that require the highest level of security and performance in their wireless communication systems.

15.7.4 Integration of Hybrid Li-Fi–Wi-Fi Networks with Digital Twin

The integration of hybrid Li-Fi–Wi-Fi networks with digital twin technology presents exciting opportunities to enhance the capabilities of digital twins and improve various applications in diverse industries. Here is how the integration can be realized:

15.7.4.1 Real-Time Data Feeds

- **High-Speed Data Transfer**: Li-Fi technology, known for its high-speed data transfer capabilities, can provide real-time data feeds to digital twins. This is particularly valuable for applications where instant feedback and updates are important.
- **IoT Device Connectivity**: Li-Fi can support the connectivity of IoT devices within the physical environment represented by the digital twin. This enables data collection from various sensors and devices in real time.

15.7.4.2 Augmented Reality (AR) and Virtual Reality (VR)

- **Immersive Visualization**: Integration with AR and VR technologies allows users to interact with digital twins in immersive ways. Li-Fi can provide high-bandwidth connectivity for AR and VR headsets, ensuring a continuous and realistic experience.
- **Remote Collaboration**: Teams working with digital twins can collaborate in real time from different locations, thanks to Li-Fi's low-latency capabilities. This is beneficial for design, engineering, and maintenance tasks.

15.7.4.3 Smart Manufacturing and Industry 4.0

- **IoT Integration**: In smart manufacturing environments, Li-Fi–Wi-Fi networks can connect IoT devices and sensors to digital twins, providing real-time monitoring and control of machinery, processes, and quality control.
- **Predictive Maintenance**: Data from digital twins can be analyzed using AI and ML algorithms to predict maintenance needs. Li-Fi can enable quick updates to the digital twin based on real-time equipment data.

15.7.4.4 Smart Cities and Infrastructure

- **Urban Planning**: Digital twins of cities and infrastructure can benefit from Li-Fi–Wi-Fi networks to gather real-time data from cameras, sensors, and traffic management systems for urban planning and public safety.
- **Emergency Response**: Li-Fi can provide fast and reliable communication for emergency responders in smart city environments, enhancing public safety.

15.7.4.5 Healthcare and Telemedicine

- **Medical IoT Devices**: Li-Fi–Wi-Fi networks can support the connectivity of medical IoT devices, enabling remote monitoring and telemedicine applications. Real-time patient data can be integrated into healthcare digital twins.
- **High-Quality Video**: Li-Fi can ensure high-quality video streaming for telemedicine consultations, enabling detailed visual examinations and remote diagnostics.

15.7.4.6 Building Automation and Smart Environments

- Energy Efficiency: Integration with digital twins can optimize building automation systems. Li-Fi can control lighting and HVAC systems based on real-time occupancy data and user preferences.
- Security and Access Control: Li-Fi–Wi-Fi networks can enhance security and access control systems within smart buildings, integrating data from sensors and cameras into the digital twin.

15.7.4.7 Education and Training

- **Virtual Learning Environments**: Integration with digital twins can create virtual learning environments for educational institutions. Li-Fi can ensure a high-quality connection for remote students accessing these environments.
- **Hands-On Training**: Digital twins can be used for hands-on training in various industries, with Li-Fi supporting the real-time interaction of trainees with the virtual environment.

15.7.4.8 Retail and Customer Experience

- **Personalized Shopping**: Digital twins of stores can provide personalized shopping experiences. Li-Fi–Wi-Fi networks can guide customers to products, provide product information, and provide special deals through their smartphones.
- **Inventory Management**: Real-time data from digital twins can optimize inventory management and restocking processes in retail environments.

Hence, the integration of hybrid Li-Fi–Wi-Fi networks with digital twin technology enhances real-time data exchange, interactive experiences, and data-driven decision-making across various sectors. By combining the capabilities of these technologies, businesses and organizations can improve efficiency, increase productivity, and provide innovative solutions to their customers and users.

15.8 Conclusion

The evolution of hybrid Li-Fi–Wi-Fi networks represents a captivating frontier in the realm of wireless communication. As we have discussed in this chapter, the integration of Li-Fi and traditional Wi-Fi technologies holds the promise of transforming how we connect to the digital world. This concluding section summarizes the key takeaways and highlights the importance of this evolution. Li-Fi's utilization of visible light for data transmission brings unique advantages, including high data transfer rates, reduced electromagnetic interference, and improved security through line-of-sight communication. These attributes make it a compelling complement to Wi-Fi, which still underpins the majority of wireless connectivity today. However, it is important to acknowledge that the journey toward the continuous integration of Li-Fi and Wi-Fi is not without its challenges. The technological, practical, and regulatory barriers are substantial, necessitating innovative solutions and cooperation across industries. Note that, the advancements in hardware development, communication protocols, and deployment strategies provide a glimpse into a future where these difficulties are surmountable.

Looking forward, the potential applications of hybrid Li-Fi–Wi-Fi networks are diverse and impactful. They can ensure reliable connectivity in environments with high electromagnetic interference, bolster security, and play a pivotal role in supporting the growing IoT ecosystem. As the IoT continues to expand, these hybrid networks are poised to become indispensable for accommodating the evolving demands of our interconnected world. In summary, the evolution of hybrid Li-Fi–Wi-Fi networks is a testament to the ongoing innovation in the field of networking. This technology promises to deliver faster, more secure, and more reliable wireless connectivity. As researchers and engineers continue to push the boundaries of what is possible, we can anticipate a future where Li-Fi–Wi-Fi hybrid networks become an integral part of our daily lives, facilitating the digital transformation of society and industry. While challenges remain, the prospects for this technology are undeniably exciting, and the journey toward their realization continues with enthusiasm and determination.

References

1 Haas, H., Yin, L., Wang, C., and Chen, C. (2018). LiFi: transforming fibre to the edge. *Journal of Lightwave Technology* 36 (15): 3085–3091. https://doi.org/10.1109/JLT.2018.2848486.

2 Elgala, H., Mesleh, R., and Haas, H. (2011). Indoor optical wireless communication: potential and state-of-the-art. *IEEE Communications Magazine* 49 (9): 56–62. https://doi.org/10.1109/MCOM.2011.6012008.

3 Kavehrad, M. and Chizari, M. (2019). Hybrid RF/optical wireless networks for indoor connectivity: evolution and potential. *IEEE Transactions on Communications* 67 (6): 4043–4056. https://doi.org/10.1109/TCOMM.2019.2903099.

4 Jungnickel, V., Fischer, M., Goleva, M. et al. (2016). Performance of hybrid RF-optical wireless networks using FSO links. *IEEE Transactions on Wireless Communications* 15 (2): 1370–1383. https://doi.org/10.1109/TWC.2015.2487066.

5 Christoph, F., Dennis, K., Maximilian, B., and Schotten, H.D. (2022). Future integrated mobile communication systems — an outlook towards 6G. *International Journal of Future Computer and Communication*. https://doi.org/10.18178/ijfcc.2022.11.2.584.

6 Wang, Q., Kakkavas, A., Gong, X., and Stirling-Gallacher, R.A. (2022). Towards integrated sensing and communications for 6G. In: *2022 2nd IEEE International Symposium on Joint Communications & Sensing (JC&S)*, 1–6. IEEE.

7 Swamy, A. (2023). Advance cellular networks (4G, 5G, 6G). *International Journal of Health Sciences* II: 10955–10966.

8 Wild, T., Grudnitsky, A., Mandelli, S. et al. (2023, September). 6G integrated sensing and communication: from vision to realization. In: *2023 20th European Radar Conference (EuRAD)*, 355–358. IEEE.

9 Christos, M., Zhang, J.A., Liu, F. et al. (2023). Guest editorial: integrated sensing and communications for 6G. *IEEE Wireless Communications*. https://doi.org/10.1109/mwc.2023.10077115.

10 Fan, L., Christos, M., Jie, X. et al. (2022). Guest editorial special issue on integrated sensing and communication - part I. *IEEE Journal on Selected Areas in Communications*. https://doi.org/10.1109/jsac.2022.3157507.

11 Xiaohu, Y., Wang, C.X., Huang, J. et al. (2021). Towards 6G wireless communication networks: vision, enabling technologies, and new paradigm shifts. *Science in China Series F: Information Sciences*. https://doi.org/10.1007/S11432-020-2955-6.

12 Lina, B., Lina, M., Sami, M. et al. (2020). A prospective look: key enabling technologies, applications and open research topics in 6G networks. *IEEE Access*. https://doi.org/10.1109/ACCESS.2020.3019590.

13 Shi, H. and Elgala, H. (2017). High-capacity optical-wireless networks for indoor communication: a review of potential architectures and challenges. *IEEE Communications Surveys & Tutorials* 19 (2): 1360–1393. https://doi.org/10.1109/COMST.2016.2632857.

14 Cossu, G. and Haas, H. (2018). Energy efficiency of LiFi in comparison to WiFi in dense scenarios. *IEEE Access* 6: 24037–24050. https://doi.org/10.1109/ACCESS.2018.2838498.

15 Karmakar, S. and Wu, Z. (2020). Enhanced internet connectivity using hybrid LiFi-WiFi systems. *IEEE Internet of Things Journal* 7 (3): 2564–2571. https://doi.org/10.1109/JIOT.2019.2956818.

16 Afgani, M., Haas, H., Uysal, M., and Yuan, J. (2006). Optical spatial modulation. *IEEE Transactions on Communications* 53 (2): 207–215. https://doi.org/10.1109/TCOMM.2005.863105.

17 Wu, Z., Karmakar, S., and Haas, H. (2019). Hybrid RF-LiFi systems: a comprehensive survey. *IEEE Communications Surveys & Tutorials* 21 (2): 1711–1734. https://doi.org/10.1109/COMST.2019.2902462.

18 Arnon, S. (2019). The potential of LiFi technology for IoT connectivity in smart buildings. *Sensors* 19 (20): 4386. https://doi.org/10.3390/s19204386.

19 Jovicic, A., Li, J., and Richardson, T.H. (2009). Visible light communication: opportunities, challenges and the path to market. *IEEE Communications Magazine* 47 (6): 92–96. https://doi.org/10.1109/MCOM.2009.5165397.

20 Li, J. and Leith, D. (2019). A survey of MAC protocols for visible light communication. *IEEE Communications Surveys & Tutorials* 21 (2): 1746–1770. https://doi.org/10.1109/COMST.2018.2884261.

21 Vucic, J., Kottke, C., and Nerreter, S. (2016). 2.3 Gbit/s visible light communications link based on DMT-modulation of a white LED. *Journal of Lightwave Technology* 34 (6): 1533–1543. https://doi.org/10.1109/JLT.2016.2518502.

22 Mesleh, R., Elgala, H., and Haas, H. (2011). Indoor optical wireless communication: potential and state-of-the-art. *IEEE Communications Magazine* 49 (9): 56–62. https://doi.org/10.1109/MCOM.2011.6012008.

23 Yin, L. and Haas, H. (2016). Light fidelity (LiFi): towards all-optical networking. *Photonic Network Communications* 31 (3): 351–361. https://doi.org/10.1007/s11107-015-0575-0.

24 Nair, M.M. and Tyagi, A.K. (2023). Blockchain technology for next-generation society: current trends and future opportunities for smart era. In: *Blockchain Technology for Secure Social Media Computing*. https://doi.org/10.1049/PBSE019E_ch11.

25 Tyagi, A.K., Kumari, S., Chidambaram, N., and Sharma, A. (2024). Engineering applications of blockchain in this smart era. In: *Enhancing Medical Imaging with Emerging Technologies*. https://doi.org/10.4018/979-8-3693-5261-8.ch011.

26 Ajanthaa, L., Seranmadevi, R., Sree, P.H., and Tyagi, A.K. (2024). Engineering applications of artificial intelligence. In: *Enhancing Medical Imaging with Emerging Technologies*. https://doi.org/10.4018/979-8-3693-5261-8.ch010.

27 Tyagi, A.K., Kukreja, S., Richa, and Sivakumar, P. (2024). Role of blockchain technology in smart era: a review on possible smart applications. *Journal of Information & Knowledge Management*. https://doi.org/10.1142/S0219649224500321.

28 Tyagi, A.K. and Tiwari, S. (2024). The future of artificial intelligence in blockchain applications. In: *Machine Learning Algorithms Using Scikit and TensorFlow Environments*. IGI Global. https://doi.org/10.4018/978-1-6684-8531-6.ch018.

29 Kumari, S., Thompson, A., and Tiwari, S. (2024). 6G-enabled internet of things-artificial intelligence-based digital twins: cybersecurity and resilience. In: *Emerging Technologies and Security in Cloud Computing*. IGI Global. https://doi.org/10.4018/979-8-3693-2081-5.ch016.

30 Nair, M.M. and Tyagi, A.K. (2023). 6G: technology, advancement, barriers, and the future. In: *6G-Enabled IoT and AI for Smart Healthcare*. CRC Press.

31 Nair, M.M. and Tyagi, A.K. (2021). Privacy: history, statistics, policy, Laws, preservation and threat analysis. *Journal of Information Assurance & Security* 16 (1): 24–34. 11p.

16

6G-Enabled Emerging Technologies for Next-Generation Society: Challenges and Opportunities

16.1 Introduction to 6G-Enabled Technologies

With the rapid evolution of wireless communication systems, fifth-generation (5G) networks are currently being deployed worldwide, providing huge advancements in terms of speed, capacity, and latency. However, the relentless growth of data-driven applications, emerging technologies, and the increasing demands of various industries have already sparked the exploration of the next generation of wireless networks – 6G [1, 2].

- 6G is envisioned as a transformative leap forward in wireless communication, surpassing the capabilities of its predecessors. It aims to provide unique levels of performance, enabling applications and services that were previously unimaginable. As the successor to 5G, 6G will be designed to tackle the ever-increasing connectivity needs of the digital era, including a wide range of sectors including telecommunications, healthcare, transportation, manufacturing, entertainment, and beyond.
- The development of 6G will be guided by a set of key objectives and requirements. These include ultrahigh data rates, exceeding the multi-gigabit-per-second speeds provided by 5G, to support bandwidth-intensive applications like virtual reality (VR), augmented reality (AR), and holography. Ultralow latency, in the sub-millisecond range, will be essential to enable real-time communication for mission-important applications like autonomous vehicles, remote surgery, and industrial automation. Additionally, 6G will aim to support large connectivity, allowing billions of devices to connect continuously, and it will use advanced network intelligence to enable efficient resource allocation, dynamic network orchestration, and intelligent decision-making.
- To achieve these ambitious goals, 6G will depend on several technologies. These technologies may include new radio (NR) frequency bands, like terahertz (THz) frequencies, as well as innovative modulation schemes, beamforming

6G-Enabled Technologies for Next Generation: Fundamentals, Applications, Analysis and Challenges, First Edition. Amit Kumar Tyagi, Shrikant Tiwari, Shivani Gupta, and Anand Kumar Mishra.

techniques, and antenna architectures. Large multiple-input, multiple-output (MIMO) systems with thousands of antenna elements will enhance spectral efficiency and spatial multiplexing capabilities. Moreover, artificial intelligence (AI) and machine learning (ML) will play an important role in optimizing network performance, managing resources, and enabling intelligent and autonomous network operations.
- The development and deployment of 6G also face major challenges. Spectrum availability is one of the foremost issues, as finding suitable frequency bands for THz communication and accommodating the vast number of devices and services will require careful planning and coordination. Energy efficiency will be another important aspect, considering the large scale of 6G networks and the need for sustainable operations. Additionally, ensuring robust security and privacy in an increasingly connected world will be a paramount challenge.

In summary, 6G is poised to redefine the boundaries of wireless communication and enable a host of revolutionary applications and services. This introduction sets the stage for exploring the various facets of 6G-enabled technologies, including their fundamentals, applications, analysis, and challenges. By understanding the potential of 6G and addressing its associated hurdles, researchers, industry professionals, and policymakers can collectively shape the future of wireless networks, empowering a hyper-connected society with unparalleled connectivity, intelligence, and innovation.

16.1.1 Definition and Overview

6G-enabled technologies include innovative communication systems, protocols, and infrastructure that will power the 6G of wireless networks. Building upon the achievements of previous generations, 6G aims to revolutionize connectivity by delivering unparalleled speeds, ultralow latency, large connectivity, and intelligent networking capabilities. At its core, 6G is envisioned as a platform that goes beyond traditional mobile communications, catering to a wide range of sectors and applications. It aims to provide connectivity and support services for emerging technologies like AR, VR, Internet of Things (IoT) devices, autonomous systems, smart cities, advanced healthcare, industrial automation, and beyond [3, 4].

The defining features of 6G-enabled technologies can be summarized as follows:

- **Ultrahigh Data Rates**: 6G networks aim to achieve data rates beyond what is currently achievable in 5G. This involves using new frequency bands, advanced modulation techniques, and spectral efficiency improvements to provide multi-gigabit-per-second or even terabit-per-second speeds.
- **Ultralow Latency**: In 6G, latency will be reduced to unique levels, reaching sub-millisecond ranges. This will enable real-time applications, like remote

surgery, autonomous vehicles, and immersive AR experiences, where even the slightest delay can have major consequences.

- **Large Connectivity**: 6G is designed to support large connectivity, allowing billions of devices to connect simultaneously. It will enable continuous communication between a vast array of IoT devices, sensors, wearables, and other connected devices, creating a highly interconnected and intelligent ecosystem.
- **Intelligent Networking**: AI and ML will play an important role in 6G-enabled technologies. AI-driven network management, resource allocation, and optimization will enhance the efficiency and adaptability of the network, enabling autonomous decision-making and proactive network operations.

To realize these goals, 6G-enabled technologies will include various innovative approaches and advancements. These may include the utilization of higher frequency bands, like THz frequencies, to enable faster data transmission, as well as the development of advanced antenna technologies like massive MIMO and holographic beamforming to improve spectral efficiency and coverage. Moreover, 6G will likely integrate edge computing and distributed computing architectures to reduce latency and enhance real-time processing capabilities. It may also use advanced security measures, like quantum encryption and privacy-enhancing technologies, to address the growing issues regarding cybersecurity and privacy in an increasingly connected world.

In summary, 6G-enabled technologies represent the next frontier of wireless communication, promising unique speeds, ultralow latency, large connectivity, and intelligent networking capabilities. By using advanced technologies and addressing the challenges associated with spectrum, energy efficiency, and security, 6G aims to unlock new possibilities and drive the digital transformation across industries, enabling innovative applications and services for a hyper-connected future.

16.1.2 Evolution from 5G to 6G

The evolution from 5G to 6G represents a huge leap in wireless communication technology, aiming to address the limitations of previous generations and unlock new capabilities [5, 6]. Here is an overview of the key aspects of the evolution:

- **Performance Enhancements**: While 5G networks provide substantial improvements in speed, capacity, and latency compared to 4G, 6G aims to push these boundaries even further. It looks to deliver ultrahigh data rates, surpassing the multi-gigabit-per-second speeds of 5G, to support bandwidth-intensive applications like VR, AR, and holography. Additionally, ultralow latency in the sub-millisecond range will enable real-time communication for mission-important applications like autonomous vehicles and remote surgery.

- **Large Connectivity**: 5G supports connectivity for a large number of devices, but 6G aims to enhance this capability further. It will enable continuous connectivity for billions of devices, including IoT devices, sensors, and wearables. This expansive connectivity will make the way for advanced applications in smart cities, industrial automation, and intelligent transportation systems.
- **Terahertz Communication**: 6G is expected to use higher frequency bands, like THz frequencies, to achieve faster data transmission rates. THz waves have importantly wider bandwidth, enabling higher data rates and improved spectrum utilization. However, utilizing THz frequencies faces technical challenges, like signal attenuation and propagation limitations, which need to be overcome for practical implementation.
- **Intelligent and Autonomous Networking**: 6G will incorporate advanced network intelligence and autonomous decision-making capabilities. AI and ML algorithms will be used to optimize network resource allocation, manage network operations, and adapt to changing network conditions dynamically. This intelligence will enable efficient use of network resources, improved user experiences, and proactive network management.
- **Multidimensional Connectivity**: 6G aims to provide continuous connectivity across various dimensions, including space, air, and underwater. It will enable communication between satellites, aerial vehicles (e.g., drones), and underwater devices, facilitating applications like global connectivity, remote sensing, and underwater exploration.
- **Enhanced Security and Privacy**: With the increasing connectivity and data exchange, 6G will prioritize robust security and privacy measures. It will incorporate advanced encryption techniques, quantum-resistant algorithms, and privacy-enhancing technologies to protect user data and ensure the integrity and confidentiality of communications.
- **Sustainable and Energy-Efficient Networks**: Energy efficiency will be an important focus in 6G. With the large scale of connectivity and data processing, sustainable network architectures and energy-efficient designs will be essential to minimize the environmental impact of wireless networks.

Hence, the evolution from 5G to 6G involves not only technological advancements but also addressing regulatory, standardization, and deployment challenges (refer Figure 16.1). Collaborative efforts between industry users, researchers, and policymakers will be important to define the vision, standards, and roadmap for 6G, ensuring its successful implementation and globally adoption in the future.

16.1.3 Literature Survey

Wireless information traffic has dramatically increased due to the accelerated development of smart terminals and the emergence of new applications, and

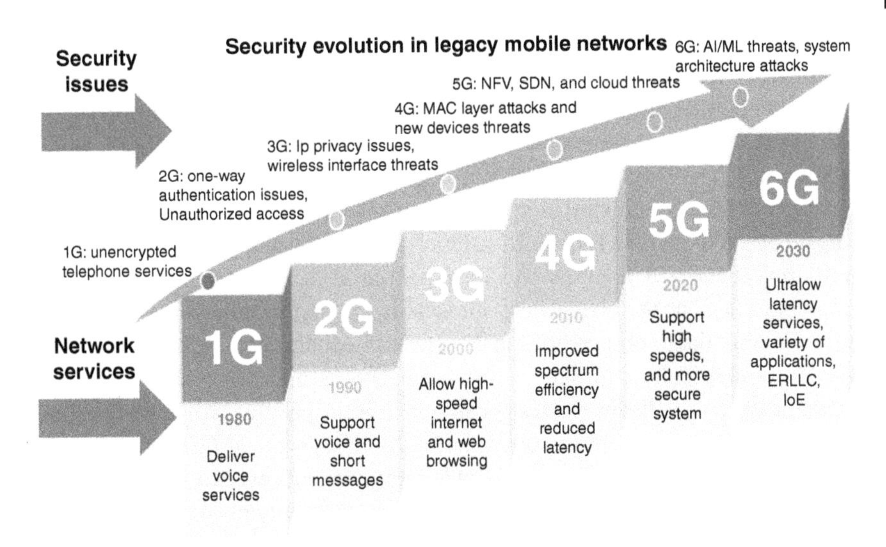

Figure 16.1 Evolution of 6G in terms of network services and security issues.

current cellular networks (even 5G) cannot fully keep up with the constantly evolving technical requirements. The sixth era (6G) of wireless communication, which would utilize floating AI, is predicted to be installed between 2027 and 2030 [7]. As 5G deployments begin, the focus of wireless research is progressively shifting toward 6G. It is important to create a vision for future communications at this point in order to provide direction for that research. In this chapter, we try to present a comprehensive picture of communication requirements and technology in the 6G era. The future of connectedness lies in the development of digital twin worlds, which will unite our experiences in the physical, biological, and digital worlds by accurately representing them in every moment of space and time. New themes that will influence 6G system requirements and technologies are likely to appear, like (i) new man–machine interfaces made by a group of several local devices acting cooperatively; (ii) ubiquitous universal computing distributed among several local devices and the cloud; (iii) multisensory data fusion to create multiverse maps and new mixed-reality experiences; and (iv) precision sensing and actuation to control the physical world. The 6G air interface and network might be built on the foundation of AI, making data, computation, and energy the new resources that can be used to achieve greater performance [8]. A quick and dependable connection between RES equipment and components is essential to provide high-quality power delivery as the use of renewable energy sources (RES) in the power system increases because of emissions reduction and sustainability laws [9]. Millions of people currently use 5G mobile communication technology, which is now widely accessible in many nations. So now is the time for industry and academia to concentrate on the next generation [10]. From 2030 onward, the entire health business will be dominated by the promised 6G communication technology. It will rule a variety of industries in addition to the health industry. Healthcare is one of the many industries that 6G is predicted to revolutionize. The future of healthcare will be totally AI-driven and reliant on 6G connectivity technologies, altering how we view lifestyle. The main obstacles to health care today are time and distance, which 6G will be able to remove. Additionally, 6G will demonstrate its potential as a game-changing medical technology [11]. For the increasingly intelligent, automated, and pervasive digital world, reliable data communication is essential. Mobile networks are the data highways and will be required to connect everyone and everything in a fully interconnected, intelligent digital society, including people with cars, sensors, data, cloud resources, and even robotic agents. Although currently being installed 5G wireless networks provide considerable improvements over LTE, it is possible that they would not be able to fully satisfy the connectivity needs of the coming digital society [12].

The development of the 6G communication system has been facilitated by the global use of new-generation information and communication technologies (ICTs), like AI, VR, AR, extended reality (XR), the IoT, and blockchain technology. The development of 6G, which builds on the foundation of 5G and consists of intelligent connection, deep connectivity, holographic connectivity, and ubiquitous connectivity, will have a major impact on the intelligence process of communication development [13]. Since the non-standalone (NSA) and standalone (SA) versions of the 5G NR are available in the 3GPP, researchers and businesses have been on pace with their 6G research. Although 6G is rumored to have far higher capabilities than 5G, its exact nature is still unknown [14]. NR NSA and NR standalone-based 5G commercial networks were introduced in substantial numbers in 2019 and will continue to do so in 2020. Industry and academia have begun to look toward the next generation of mobile networks, known as 6G, which is targeted for the 2030 timeline and aims to address challenges not easily achievable in a backwards-compatible manner with 5G NR evolution. While 3GPP is continuously working on the evolution of 5G NR with Release 16 being finalized in early 2020 and the start of Release 17, industry and academia have started looking toward 6G. 6G should address a new set of pressing issues for networks and the rapidly evolving society, issues that necessitate new architectural and technical approaches [15]. The main forces behind 6G are not just the difficulties and performance constraints that come with 5G but also the paradigm shift brought on by technology and the ongoing development of wireless networks.

Core 6G requirements are being created by technological advancements in industries like intelligent transportation and business, which will result in service classes including ubiquitous mobile ultrabroadband (uMUB), ultrahigh speed with low-latency communications (uHSLLC), and ultrahigh data density (uHDD) [16]. Following the adoption of 5G technology, 6G wireless network technology has been the subject of research in both academia and industry. Around 2030, 6G is predicted to be operational. By enabling hyper-connectivity between people and everything, it will provide a profound experience for everyone. Additionally, it is anticipated to expand mobile communication options in areas where past generations were unable to do so. It is anticipated that several technologies will form the basis of 6G networks. These include both present and emerging technologies like post-quantum cryptography, enhanced edge computing, molecular communication, THz, visible light communication (VLC), and distributed ledger (DL) systems like blockchain [17]. Worldwide, 5G networks that use large MIMO, millimeter-wave communication, and other technologies to boost system performance have been widely implemented. On the other hand, it has been suggested that forthcoming essential technologies, including THz communication, intelligent reflective surface (IRS), blockchain, and others, will improve the signal quality for upcoming 6G networks. Blockchain can increase system security across the entire communication process while IRS can intelligently regulate a large number of passive reflecting devices to program signal propagation [18].

Over the past few decades, the need for wireless connectivity has increased dramatically. Worldwide deployment of 5G communications, which has many more features than fourth-generation communications, is likely to begin. The 6G system, a new wireless communication paradigm with full AI support, is anticipated to be put into use between 2027 and 2030. Higher system capacity, higher data rate, lower latency, higher security, and improved quality of service (QoS) compared to the 5G system are some basic issues that need to be addressed beyond 5G [19]. The Internet of Everything will be made possible by 6G mobile communication technology, which we will start to see in 2030. However, Beyond 5G, innovations must be created and 5G related applications must be experienced by everyone in the world; researchers have already begun to plan for and collect needs for the 6G. Additionally, many nations have already started their 6G research. Every smart gadget, including smartphones and advanced cars, is expected to be connected to the Internet by 6G. Holographic communication, AR/VR, and many other features will be available with 6G. Additionally, it will highlight Quality of Experience (QoE) to deliver rich 6G experiences. Notably, it is important to understand the problems and difficulties associated with 6G technology [20].

Few of the application, requirement, and challenges related to 5G are listed in Tables 16.1 and 16.2 [39].

Few more literature in recent years have been listed in the following.

Table 16.1 Summary of 5G limitation and challenges.

Key Indicator	Main Industrial Application	Industry Requirement	5G KPI	Future Research Challenge
Global Coverage	Marine communication Satellite communication	Cover sea and remote areas	Ocean coverage: only 5% Land coverage: only 20%	Global coverage: space-air-ground-sea
Ultrahigh Data Rate Transmission	Ultra-high definition (HD) video, holographic image	Very high speed transmission	Transmission rate: <20 Gbps User experienced data rate: ~100 Mbps	Peak data rate at Tbps level User experienced data rate: 1–10 Gbps
Ultralow Latency	Automatic drive, high-precision industrial production	High speed and low latency transmission	Delay: <1 ms at static and low speeds, but cannot be reached at high speeds	Sub-second (<1 ms)
Ultra-dense Connection	Crowded shopping malls, stations, fully automatic production line	Super dense population, super dense equipment	10^6 devices/km^2	Connection density: up to 108 devices/km^2
High Precision Positioning	Unmanned vehicle positioning and navigation, indoor precise positioning	Any circumstances of outdoor/indoor precise positioning	Outdoor: ~10 m Indoor: ~1 m	Achieve outdoor meter level, indoor centimeter level positioning
Ultra-reliable/safe	Tactile Internet, V2X, telemedicine, wireless data center, wireless brain–machine interface	Super reliable/safe	99.999%	99.99999%
Low Power Consumption/High-Energy Efficiency	Internet of bio-nano-things, intermediate altitude communications	Reduce the power consumption and improve the energy efficiency as much as possible	Network energy efficiency: 107 bit/J	Enhance the network energy efficiency to 109 bit/J
Ubiquitous Intelligent	Digital twins, integrated sensing and communication (ISAC), AI applications	Support a series of intelligent applications	Low	High

Table 16.2 Literature review work.

Ref.	Year	Country	6G Project
[21]	2018	Finland	6G-enabled wireless smart society and ecosystem (6Genesis)
[22]	2019	China	Satellite communication technology integrated with 5G/6G
[23]	2019	EU	Artificial intelligence aided D-band network for 5G long term evolution
[24]	2020	China	Research on the theory and key technologies of ultra-wideband photonic THz wireless transmission
[25]	2020	Japan	Research and development on satellite-terrestrial integration technology for beyond 5G
[26]	2021	China	6G communication-aware-computing converged network architecture and key technologies
[26]	2021	China	6G ultralow latency ultra-reliable large-scale wireless transmission technology
[27]	2021	Germany	Project SENTINEL for flexible 6G networks consisting of non-terrestrial networks, THz, localization, etc.
[28]	2021	EU	Reconfigurable intelligent sustainable environments for 6G wireless networks (RISE-6G)
[29]	2021	EU	Project REINDEER for smart connectivity platform creating hyper-diversity
[30]	2021	EU	Project Hexa-X for 6G vision and intelligent fabric of technology enablers connecting human, physical, and digital worlds
[31]	2021	USA	Resilient and intelligent NextG systems (RINGS)
[32]	2022	China	AI-driven 6G wireless intelligent air-interface transmission technologies
[32]	2022	China	6G research on smart and simple network architecture and autonomous technologies
[32]	2022	China	Endogenous security and privacy protection technologies for 6G
[32]	2022	China	New network architecture and transmission methods for 6G smart applications
[33]	2022	Korea	Quantum cryptographic communication for aerospace and space applications
[34]	2022	Germany	Project Open6GHub for society and sustainability including adaptive 6G RAN technologies, connected intelligence, etc.
[35]	2022	EU	6GStart: Starting the Sustainable 6G SNS Initiative for Europe
[36]	2022	EU	Project 6GTandem for dual-frequency distributed MIMO technologies
[37]	2023	EU	Project Hexa-X-II launched to address challenges in sustainability, inclusion, and trustworthiness
[38]	2023	EU	TERAhertz integrated systems enabling 6G Terabit-per-second ultra-massive MIMO wireless networks (TERA6G)

16.2 Fundamentals of 6G-Enabled Technologies

The fundamentals of 6G-enabled technologies include the key principles and concepts that define the next generation of wireless communication systems. While 6G is still in the early stages of development, several fundamental characteristics and goals are being considered. Here are some of the key fundamentals of 6G:

- **Ultrahigh Data Rates**: 6G aims to provide significantly higher data rates compared to previous generations. It envisions data rates in the terabits per second (Tbps) range, enabling extremely fast and continuous data transfer for a wide range of applications.
- **Ultralow Latency**: 6G aims to achieve ultralow latency, reducing the delay between data transmission and reception. The goal is to achieve latency in the sub-millisecond (ms) range, enabling real-time applications that require instant responsiveness, like autonomous vehicles, remote surgery, and immersive experiences.
- **Large Device Connectivity**: 6G envisions supporting a large number of connected devices simultaneously. It aims to provide high-capacity connections to accommodate the exponential growth of IoT devices, smart sensors, and other connected devices, enabling continuous and efficient communication between devices.
- **Hyperconnectivity**: 6G looks to provide ubiquitous and uninterrupted connectivity, ensuring continuous communication regardless of location or mobility. It aims to achieve pervasive coverage, extending connectivity to remote and underserved areas, and enabling reliable communication during high-speed mobility scenarios.
- **Energy Efficiency**: Energy efficiency is an important issue for 6G. It aims to develop technologies and strategies to minimize energy consumption and optimize power usage. This includes energy-efficient network design, advanced power management techniques, and the integration of energy harvesting technologies.
- **Spectrum Utilization**: 6G aims to utilize a wide range of spectrum resources, including the exploitation of previously unused or underutilized frequency bands. It aims to use new spectrum bands, including millimeter-wave (mmWave) frequencies, to increase the available bandwidth and support higher data rates.
- **Intelligent Network Architecture**: 6G networks are expected to adopt intelligent and flexible network architectures. This includes the integration of AI, ML, and edge computing technologies to enable efficient data processing, real-time decision-making, and dynamic resource allocation.

- **Security and Privacy**: With the increasing connectivity and data exchange in 6G networks, robust security and privacy mechanisms are important. 6G aims to enhance security protocols, encryption techniques, and privacy-enhancing technologies to protect user data and ensure secure communication.

Note that these fundamentals are subject to ongoing research and may evolve as 6G technology progresses. The specific implementation and realization of these fundamentals will be determined through collaboration among industry players, standardization bodies, and research institutions.

16.2.1 Spectrum and Bandwidth Issues

Spectrum and bandwidth issues are of utmost importance in the development of 6G-enabled technologies. As the demand for wireless connectivity and data-intensive applications continues to grow, adequate spectrum resources and efficient utilization become important for achieving the ambitious goals of 6G. Here are some key aspects related to spectrum and bandwidth issues for 6G:

- **Spectrum Availability and Allocation**: Identifying and allocating suitable frequency bands for 6G is a fundamental requirement. The exploration of new frequency ranges, including THz frequencies, is expected to be a key focus for 6G. THz frequencies provide wider bandwidths and the potential for ultrahigh data rates. However, utilizing THz frequencies comes with challenges like higher signal attenuation and limited propagation range. Overcoming these challenges and ensuring efficient spectrum utilization will be important for 6G deployment.
- **Spectrum Sharing and Coexistence**: With the increasing demand for wireless connectivity, spectrum-sharing techniques will play an important role in 6G networks. Dynamic spectrum access, cognitive radio, and other spectrum-sharing mechanisms can enable efficient utilization of limited spectrum resources. Spectrum coexistence mechanisms will also be essential to ensure compatibility and interference management between different services and technologies operating in the same frequency bands.
- **Aggregation and Carrier Aggregation**: 6G is expected to employ advanced techniques for spectrum aggregation and carrier aggregation. By combining multiple frequency bands or carriers, 6G networks can achieve wider bandwidths and higher data rates. Carrier aggregation techniques allow devices to simultaneously use multiple frequency bands, enabling efficient use of available spectrum resources.
- **Beamforming and Spatial Multiplexing**: Beamforming technologies, like large MIMO, will continue to be important in 6G. By using a large number of antennas, beamforming focuses the transmission energy toward specific users

or locations, increasing spectral efficiency and improving coverage. Spatial multiplexing techniques enable the simultaneous transmission of multiple data streams using different spatial channels, further enhancing bandwidth utilization.

- **Dynamic Spectrum Access and Network Slicing**: 6G is expected to use dynamic spectrum access to dynamically allocate spectrum resources based on demand and quality-of-service requirements. Network slicing, which involves partitioning the network into multiple virtual networks, can also be used to allocate specific slices of spectrum to different services and applications, ensuring efficient and customized spectrum allocation.
- **Energy Efficiency and Spectrum Efficiency Trade-Offs**: Energy efficiency is a key issue in 6G networks. Efficient spectrum utilization can contribute to energy savings. However, there is often a trade-off between spectrum efficiency and energy efficiency. Striking the right balance between these factors is important to ensure sustainable operations while meeting the increasing demands for wireless connectivity.
- **Spectrum Management and Regulatory Issues**: Effective spectrum management and regulatory frameworks are important for the successful deployment of 6G. Governments and regulatory bodies need to allocate and manage spectrum resources efficiently, considering factors like harmonization, licensing frameworks, interference mitigation, and international coordination.

Note that addressing these spectrum and bandwidth issues will be important to unlock the full potential of 6G-enabled technologies. Collaborative efforts between industry users, policymakers, and regulatory bodies are necessary to ensure optimal spectrum allocation, efficient utilization, and continuous coexistence of diverse wireless services in the 6G ecosystem.

16.2.2 Massive MIMO and Beamforming

Massive MIMO and beamforming are key technologies that are expected to play a major role in 6G-enabled networks. They are designed to enhance spectral efficiency, improve coverage, and support the large connectivity requirements of the next generation of wireless communication. Here is an overview of Massive MIMO and beamforming in the context of 6G:

16.2.2.1 Massive MIMO

Massive MIMO refers to the use of a large number of antennas at the base station (BS) or access point in wireless networks. In 6G, massive MIMO systems are anticipated to scale up even further, with potentially thousands of antenna elements. By using a large array of antennas, massive MIMO provides spatial multiplexing, beamforming, and interference mitigation capabilities.

- **Spatial Multiplexing**: With multiple antennas, massive MIMO enables the simultaneous transmission of multiple data streams to multiple users in the same frequency band, improving spectral efficiency and overall system capacity.
- **Beamforming**: Massive MIMO employs beamforming techniques to focus the transmit energy toward specific users or areas of interest. By forming highly directional beams, beamforming enhances signal strength, improves coverage, and reduces interference, thereby enhancing the overall network performance and user experience.
- **Interference Mitigation**: The use of massive MIMO systems allows for efficient interference mitigation. By employing advanced signal processing algorithms and spatial filtering techniques, interference from neighboring cells or users can be mitigated, leading to improved signal quality and overall network performance.

16.2.2.2 Beamforming

Beamforming is a technique used to concentrate radio waves in a particular direction by adjusting the phase and amplitude of signals transmitted from an antenna array. It plays an important role in enhancing the coverage, capacity, and energy efficiency of wireless networks. In the context of 6G, beamforming is expected to evolve further to meet the demanding requirements of high-speed, low-latency, and large connectivity.

- **Analog Beamforming**: Analog beamforming is a cost-effective approach that adjusts the signal phase and amplitude at the radio frequency (RF) level to create focused beams. It provides beamforming capabilities with reduced complexity compared to digital beamforming, making it suitable for large-scale antenna arrays in 6G systems.
- **Digital Beamforming**: Digital beamforming involves manipulating the signals in the baseband domain, utilizing digital signal processing techniques. It provides greater flexibility and beamforming precision, allowing for adaptive and dynamic beamforming to accommodate changing channel conditions and user requirements. However, digital beamforming may require more complex processing and hardware compared to analog beamforming.
- **Holographic Beamforming**: Holographic beamforming is an advanced technique that employs principles of holography to generate and shape multiple beams simultaneously. It enables beamforming with high granularity, allowing for precise control and optimization of signal coverage and capacity. Holographic beamforming has the potential to provide adaptive and intelligent beamforming in 6G networks.

By using massive MIMO and beamforming technologies, 6G-enabled networks can achieve higher spectral efficiency, improved coverage, increased capacity, and better interference management. These advancements are important for supporting the ultrahigh data rates, ultralow latency, and large connectivity requirements of 6G applications, like AR, VR, smart cities, and IoT devices. The integration of massive MIMO and beamforming in 6G networks will enable enhanced network performance, efficient spectrum utilization, and improved user experiences in the future wireless communication landscape.

16.2.3 Terahertz (THz) Communication

THz communication is one of the promising areas of research and development for 6G-enabled technologies. THz frequencies refer to the frequency range between 100 GHz (GHz) and 10 THz, which falls between the microwave and infrared regions of the electromagnetic spectrum. THz communication is expected to bring several advantages to the sixth generation of wireless networks. Here is an overview of THz communication in the context of 6G:

- **Higher Data Rates**: One of the primary advantages of THz communication is its potential to deliver extremely high data rates. The large available bandwidth in the THz frequency range enables transmission speeds in the order of terabits per second (Tbps). This ultrahigh data rate capability is essential to support bandwidth-intensive applications in 6G, like ultra-HD video streaming, holographic communication, and immersive VR experiences.
- **Increased Spectrum Availability**: The THz frequency band provides a vast amount of spectrum resources that have been largely untapped for wireless communication. The availability of such a wide frequency range provides an opportunity to alleviate spectrum congestion in lower frequency bands and allows for the allocation of larger bandwidths to support high-speed data transmission. THz communication can potentially address the spectrum scarcity challenge faced by current wireless networks.
- **Short-Range Communication**: THz signals are highly susceptible to absorption and scattering by obstacles like walls and atmospheric gases. As a result, THz communication is expected to be primarily suitable for short-range communication links, typically within a few meters or tens of meters. This characteristic makes THz communication ideal for applications like wireless personal area networks (WPANs), in-room communication, and wireless connectivity within confined spaces.
- **Large Connectivity**: THz communication can support large connectivity requirements in 6G networks. Due to the short-range nature of THz signals, a large number of devices can be served simultaneously within a specific area, enabling dense deployments and accommodating large device connectivity. This capability is important for applications like IoT, where billions of devices need to be interconnected continuously.

- **Technical Challenges**: Deploying THz communication in practice faces several technical challenges that need to be addressed. THz signals are highly sensitive to atmospheric conditions and can experience huge signal attenuation, resulting in limited propagation range. Additionally, the design and integration of THz transceivers and antennas present technical complexities. Overcoming these challenges requires advancements in antenna technologies, signal processing techniques, and channel modeling for THz communication.
- **Applications**: THz communication holds promise for a wide range of applications in 6G. These include high-speed wireless data transmission, ultralow latency communication for real-time applications, short-range wireless links for virtual and AR systems, wireless sensing and imaging, and wireless backhaul for small cell networks.

Note that THz communication is still an area of active research, and rapid advancements are needed to overcome the technical challenges and enable practical implementations. Nevertheless, the potential of THz communication in delivering ultrahigh data rates, addressing spectrum scarcity, and supporting large connectivity makes it an intriguing area of exploration for 6G-enabled technologies.

16.2.4 Artificial Intelligence (AI) Integration

The integration of AI with 6G-enabled technologies holds huge potential to revolutionize various aspects of our lives. Here are some areas where AI integration with 6G can have a profound impact:

- **Enhanced Connectivity**: 6G is expected to provide ultrahigh-speed and low-latency connectivity, enabling continuous communication between devices and systems. AI can use this connectivity to optimize network management, enhance data transmission, and improve overall network performance.
- **Intelligent IoT**: The IoT is set to expand with 6G, connecting billions of devices and sensors. AI can play an important role in managing and analyzing the large amounts of data generated by these devices. AI algorithms can provide real-time insights, enable predictive maintenance, and optimize resource allocation in IoT systems.
- **Autonomous Systems**: AI-powered autonomous systems, like self-driving cars, drones, and robots, can benefit from the ultralow latency and high bandwidth of 6G networks. The integration of AI with 6G can enable these systems to make more informed decisions, improve perception capabilities, and enhance overall safety and efficiency.
- **Edge Computing and AI**: Edge computing, which involves processing data closer to the source, is a key component of 6G. By combining edge computing

with AI algorithms, complex data processing tasks can be performed locally, reducing latency and minimizing the need for data transmission to centralized servers. This integration enables real-time decision-making, improved privacy, and efficient resource utilization.

- **Immersive Experiences**: 6G is expected to support immersive technologies like AR and VR. AI integration can enhance these experiences by providing personalized content recommendations, real-time object recognition, and natural language understanding for more interactive and intuitive interactions.
- **Network Optimization and Security**: AI algorithms can be utilized to optimize 6G networks dynamically. They can analyze network traffic patterns, predict demand, and optimize resource allocation, leading to improved network efficiency and better QoS. Additionally, AI can enhance the security of 6G networks by detecting anomalies, identifying threats, and mitigating potential attacks.

In summary, the integration of AI with 6G-enabled technologies opens up new possibilities for advanced applications, intelligent automation, and transformative experiences across various domains, ranging from communication and transportation to healthcare and entertainment.

16.2.5 Energy Efficiency and Sustainability

Energy efficiency and sustainability are important issues in the development and deployment of 6G-enabled technologies. Here are some ways in which 6G can contribute to energy efficiency and sustainability:

- **Intelligent Resource Management**: 6G networks can use AI algorithms to optimize resource allocation and energy consumption. AI can analyze network traffic patterns, predict demand, and dynamically adjust the allocation of network resources, minimizing energy wastage and improving overall efficiency.
- **Energy-Aware Networking**: 6G can incorporate energy-aware networking protocols and techniques to reduce power consumption. For example, intelligent sleep mode mechanisms can be employed to activate or deactivate network elements based on demand, reducing energy consumption during low-traffic periods.
- **Edge Computing and Local Data Processing**: With 6G's importance on edge computing, data processing tasks can be performed closer to the source, reducing the need for data transmission to centralized data centers. This approach minimizes energy consumption associated with long-distance data transfers and enables more energy-efficient processing.
- **Green Infrastructure**: As 6G networks are deployed, there is an opportunity to incorporate environmentally friendly infrastructure. This includes using RES,

like solar or wind power, to power BS and network equipment. Additionally, the use of energy-efficient hardware components and optimized cooling systems can further contribute to reducing energy consumption.

- **Smart Grid Integration**: 6G can facilitate the integration of communication networks with smart grids, enabling better energy management and optimization. AI algorithms can analyze real-time energy data and make intelligent decisions to optimize energy distribution, reduce peak demand, and enhance the overall efficiency of the electrical grid.
- **Environmental Monitoring and Sustainability Applications**: 6G can support a wide range of environmental monitoring applications, like air quality monitoring, water management, and precision agriculture. By using AI and IoT technologies, these applications can provide real-time data and insights, enabling more effective and sustainable resource management.
- **E-waste Management**: As 6G devices and infrastructure are deployed, proper e-waste management becomes essential. Manufacturers and policymakers can implement sustainable practices, including recycling programs and responsible disposal of outdated or damaged equipment, to minimize the environmental impact of electronic waste.

Note that by prioritizing energy efficiency and sustainability in the design and deployment of 6G-enabled technologies, we can reduce energy consumption, minimize environmental impact, and contribute to building a more sustainable and resilient future.

16.3 Applications and Use Cases of 6G

6G, the next generation of wireless communication technology (refer Figure 16.2 for more detail), is expected to bring transformative advancements across various sectors [38, 40]. While the full range of applications and use cases of 6G is yet to be realized, here are some potential areas where 6G can make a huge impact:

- **Immersive Extended Reality (XR)**: 6G can revolutionize immersive experiences, including AR, VR, and mixed reality (MR). It can provide ultrahigh-speed, low-latency connectivity to enable continuous and immersive XR applications, like realistic gaming, remote collaboration, immersive education, and virtual tourism.
- **Smart Cities and Infrastructure**: 6G can support the development of smart cities by connecting and integrating various urban systems. It can enable intelligent transportation systems, smart grids, efficient waste management, real-time monitoring of air and water quality, and enhance public safety through advanced surveillance and emergency response systems.

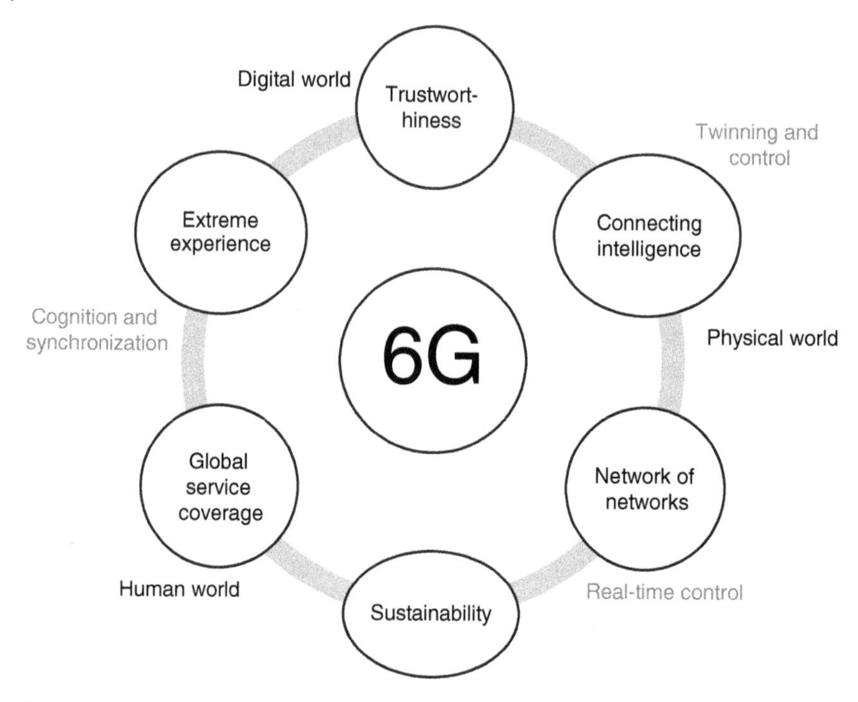

Figure 16.2 6G network.

- **Autonomous Systems**: 6G can play an important role in advancing autonomous systems, including self-driving cars, drones, robots, and unmanned aerial vehicles (UAVs). With its ultralow latency and high reliability, 6G can enable real-time communication and decision-making capabilities for autonomous vehicles, leading to safer transportation and optimized logistics operations.
- **Healthcare and Telemedicine**: 6G can transform healthcare delivery by enabling remote diagnosis, telemedicine, and real-time monitoring of patients. It can facilitate the transmission of high-resolution medical imaging, support robotic surgery, and enable reliable and instantaneous communication between medical professionals for collaborative decision-making.
- **Industrial Automation**: 6G can enhance industrial automation by enabling real-time communication and control in manufacturing processes. It can support advanced robotics, machine-to-machine (M2M) communication, and industrial IoT applications. With its ultrahigh bandwidth, 6G can facilitate the use of AI and data analytics for predictive maintenance and optimization of production processes.

- **High-Fidelity Media and Entertainment**: 6G can enable the continuous streaming of high-fidelity media content, including 8 K and 16 K video, virtual concerts, and interactive gaming experiences. With its high data rates and low latency, 6G can provide an immersive and personalized media experience, allowing users to access and enjoy content continuously across various devices.
- **Environmental Monitoring and Sustainability**: 6G can be used for environmental monitoring and sustainability initiatives. It can support real-time monitoring of air quality, water management systems, precision agriculture, and enable data-driven decision-making for sustainable resource management.
- **Advanced Security and Privacy**: With the growing issues around cybersecurity and privacy, 6G can provide enhanced security mechanisms. It can employ advanced encryption techniques, blockchain-based authentication, and AI-powered anomaly detection to ensure secure and private communication in a hyper-connected world.

These are just a few potential applications and use cases of 6G. As the technology continues to evolve, it will likely unlock new possibilities and make the way for innovative solutions across industries, ultimately transforming the way we live, work, and interact with the world.

16.3.1 Enhanced Mobile Broadband (eMBB)

eMBB is one of the key pillars of 6G, focusing on delivering an exceptional mobile connectivity experience with significantly higher data rates, ultralow latency, and improved network capacity compared to previous generations. Here are some ways in which eMBB can be enhanced with 6G:

- **Gigabit Speeds**: 6G aims to provide unique data rates, potentially reaching multi-gigabit speeds. This ultra-fast connectivity will enable continuous streaming of high-definition and even 8 K/16 K video content, faster file downloads, and instantaneous cloud access. Users will experience ultra-responsive and lag-free interactions with applications and services.
- **Ultralow Latency**: 6G will significantly reduce latency, aiming for sub-millisecond response times. This low latency will enable real-time applications like cloud gaming, immersive VR experiences, remote robotics control, and AR applications, where instantaneous feedback and interactivity are important.
- **Large Device Connectivity**: 6G is expected to support a large number of connected devices, reaching a density of one million devices per square kilometer. This enhanced device connectivity will facilitate the IoT ecosystem, enabling continuous integration and communication among billions of connected

devices, ranging from smart home appliances and wearables to autonomous vehicles and industrial sensors.

- **Beamforming and MIMO Technologies**: 6G will build upon the advancements of previous generations in antenna technologies, like large MIMO and beamforming. These technologies enable the transmission and reception of data via multiple antennas, enhancing signal quality, increasing coverage, and improving network capacity.
- **Network Slicing and Dynamic Resource Allocation**: 6G will further refine network slicing capabilities, allowing the allocation of specific network resources tailored to different applications and user requirements. This dynamic resource allocation ensures optimal performance, prioritizing important services, and maximizing network efficiency.
- **AI-Enabled Network Optimization**: AI integration will play an important role in optimizing network performance in 6G. AI algorithms can analyze network data, predict traffic patterns, and dynamically allocate network resources to enhance efficiency, reduce congestion, and improve the overall user experience.
- **Holographic Communication**: With advancements in eMBB, 6G may enable holographic communication, where users can have immersive, three-dimensional video calls, creating a sense of presence and enhancing remote collaboration.
- **Cloud-Native Architecture**: 6G is expected to use cloud-native architecture, enabling more flexible and scalable network deployments. This architecture allows for the virtualization of network functions and the dynamic allocation of resources based on demand, leading to optimized network operations and improved cost-efficiency.

In summary, enhanced mobile broadband in 6G aims to deliver an unparalleled mobile connectivity experience, enabling ultra-fast speeds, ultralow latency, large device connectivity, and advanced applications and services. This will open up new possibilities for communication, entertainment, IoT, and other sectors, transforming the way we interact with mobile networks and the digital world. Figure 16.3 mentions few applications of 6G network in the near future.

16.3.2 Ultra-Reliable Low-Latency Communication (URLLC)

URLLC is an important feature in wireless communication systems, and it is expected to play a major role in the development of 6G networks. While 6G technology is still under development, it is anticipated to provide substantial improvements over existing networks, including higher data rates, lower latency, enhanced reliability, and support for large machine-type communications (mMTCs). Here are some key aspects of URLLC in the context of 6G:

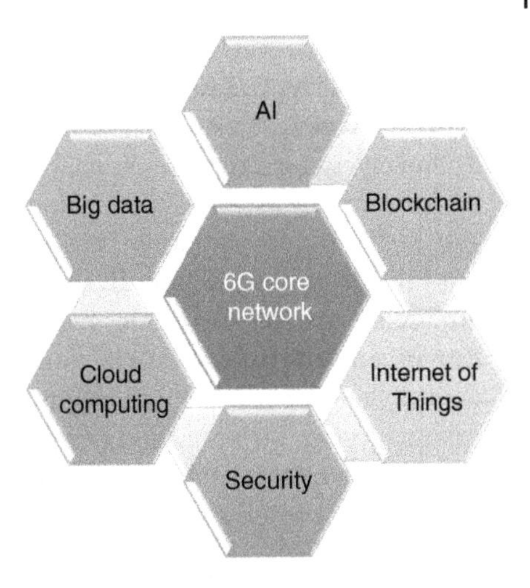

Figure 16.3 Application of 6G network.

- **Ultra-Reliability**: URLLC in 6G aims to provide highly reliable communication links with extremely low error rates. This is important for applications that require mission-important connectivity, like autonomous vehicles, remote surgery, industrial automation, and public safety services. 6G networks will use advanced error correction techniques, redundancy mechanisms, and robust channel coding to achieve ultra-reliable communication.
- **Low-Latency**: 6G networks will significantly reduce communication latency to support near real-time applications. URLLC in 6G aims to achieve end-to-end latency in the sub-millisecond range, enabling new applications that demand instantaneous response times. This is important for applications like haptic communication, AR, VR, gaming, and tactile Internet.
- **Network Slicing**: 6G will introduce network slicing as a fundamental architectural concept. It allows the partitioning of network resources into dedicated slices, each tailored to specific requirements. URLLC can benefit from network slicing by providing dedicated slices with ultra-reliable and low-latency characteristics. This ensures that important applications receive the necessary network resources and guarantees while coexisting with other types of traffic.
- **Edge Computing**: 6G will extensively use edge computing capabilities to reduce latency and improve reliability for URLLC. By distributing computing resources closer to the network edge, processing and decision-making can be performed in proximity to the data source, reducing the round-trip time. Edge computing enables efficient data offloading, real-time analytics, and localized processing for URLLC applications.
- **Advanced Antenna Technologies**: 6G is expected to incorporate advanced antenna technologies, like massive MIMO and beamforming, to enhance the reliability and capacity of URLLC. These technologies can improve signal

quality, mitigate interference, and enable precise directional communication, thereby reducing latency and improving overall network performance.
- **Hybrid Access Networks**: 6G networks may integrate multiple access technologies, including traditional cellular networks, satellite communications, terrestrial wireless systems, and even airborne platforms like drones and high-altitude platforms (HAPs). This heterogeneous access landscape can provide redundant and diverse connectivity options, contributing to the ultra-reliability and low-latency requirements of URLLC.
- **Enhanced Security**: As URLLC supports important applications and services, ensuring robust security mechanisms is of paramount importance. 6G networks will incorporate advanced security protocols, encryption algorithms, authentication mechanisms, and privacy-enhancing technologies to safeguard URLLC communications against potential threats.

Note that 6G is still in the early stages of research and development, and the specific features and technologies mentioned above may evolve over time. The actual implementation and standardization of 6G, including URLLC, will be subject to ongoing research, industry collaboration, and global standards bodies' decisions.

16.3.3 Large Machine-Type Communication (mMTC)

mMTC is another important aspect of 6G networks. It refers to the capability of connecting a large number of devices in a dense and scalable manner, catering to the needs of IoT and M2M communication. mMTC in 6G aims to provide efficient and reliable connectivity for a wide range of IoT devices and applications. Here are key issues for mMTC in the context of 6G:

- **Large Scalability**: 6G networks will be designed to support a significantly higher number of connected devices compared to previous generations. The mMTC paradigm in 6G will target connecting billions or even trillions of devices simultaneously, accommodating the exponential growth of IoT deployments and M2M communication.
- **Energy Efficiency**: To accommodate the large scale of IoT devices, 6G will prioritize energy-efficient communication protocols and network architectures. mMTC devices often have limited power resources, and optimizing energy consumption is important for prolonging device battery life and ensuring sustainable IoT deployments.
- **Low Device Cost**: 6G aims to reduce the cost of mMTC devices, making them more affordable and accessible for globally adoption. Lowering device costs can facilitate large-scale IoT deployments across various sectors, including smart cities, industrial automation, agriculture, and healthcare.

- **Enhanced Coverage and Connectivity**: 6G networks will provide improved coverage and connectivity options for mMTC devices. This includes extending coverage to remote and challenging environments, like underground areas, deep-sea locations, and rural regions. mMTC in 6G will use advanced techniques like device-to-device (D2D) communication, relay nodes, and mesh networking to enhance connectivity options.
- **Latency Optimization**: While mMTC generally does not require ultralow latency like URLLC, 6G will still aim to reduce latency for timely and responsive communication. Low-latency connectivity is important for certain mMTC applications, like real-time monitoring, control systems, and time-important IoT deployments.
- **Network Slicing and Resource Allocation**: Similar to URLLC, 6G's network slicing capabilities will be used to create dedicated slices for mMTC traffic. These slices will be optimized for the specific requirements of IoT devices, allowing efficient resource allocation, prioritization, and isolation for mMTC communications.
- **Integration of AI and Machine Learning**: 6G will incorporate advanced AI and ML techniques to optimize mMTC operations. AI-based algorithms can help with efficient device management, network optimization, predictive maintenance, and data analytics for IoT deployments. This integration enables intelligent decision-making, improved network performance, and enhanced user experience in mMTC scenarios.
- **Security and Privacy**: As mMTC involves the connection of a large number of devices, security and privacy become important issues. 6G will integrate robust security mechanisms, authentication protocols, and encryption algorithms to protect mMTC communications from unauthorized access, data breaches, and cyberattacks.

Note that 6G is still in the research and development phase, and the specific features and technologies for mMTC may evolve over time. Ongoing research, collaboration among industry users, and standardization efforts will shape the implementation and specifications of mMTC in 6G networks.

16.3.4 Internet of Things (IoT) and Industry 5.0

The IoT and Industry 5.0 are two key concepts that are expected to be greatly influenced by the advancements brought by 6G networks. Let us explore their relationship with 6G in more detail:

- **Internet of Things (IoT) and 6G**: The IoT is a network of interconnected devices, sensors, and objects that can collect, exchange, and act upon data. 6G is anticipated to provide the underlying infrastructure to support the large-scale and diverse requirements of IoT deployments. Some ways 6G can enhance the IoT ecosystem include:

- **Large Device Connectivity**: 6G networks will be designed to handle the large number of IoT devices by providing increased capacity and scalability. This will enable continuous connectivity and data exchange among billions or even trillions of IoT devices.
- **Enhanced Coverage and Range**: 6G aims to extend coverage to challenging environments, including remote areas, underground locations, and deep-sea regions. This expanded coverage will enable IoT deployments in a broader range of applications like agriculture, environmental monitoring, and infrastructure management.
- **Lower Latency and Faster Response Times**: 6G's reduced latency capabilities will enable near real-time communication between IoT devices and cloud-based applications. This is important for time-important applications like autonomous vehicles, industrial automation, and real-time monitoring.
- **Energy Efficiency**: 6G networks will prioritize energy-efficient communication protocols and mechanisms to optimize the power consumption of IoT devices. This will help extend device battery life and support sustainable IoT deployments.
- **Advanced Data Analytics**: 6G's integration of AI and ML capabilities will enable advanced data analytics at the edge, closer to IoT devices. This will facilitate real-time insights, predictive analytics, and intelligent decision-making within IoT applications.

- **Industry 5.0 and 6G**: Industry 5.0 represents the next phase of the Industrial Revolution, where humans and advanced technologies collaborate closely in manufacturing and production processes. 6G networks are expected to play an important role in enabling the vision of Industry 5.0 by providing advanced connectivity, automation, and intelligence. Here are some ways in which 6G can support Industry 5.0:
  - **Ultra-Reliable and Low-Latency Communication**: 6G's ultra-reliable and low-latency communication capabilities, like URLLC, will facilitate continuous and real-time communication between industrial machines, robots, and human operators. This will enable highly responsive and synchronized industrial processes, improving productivity and efficiency.
  - **Large Machine-Type Communication (mMTC)**: 6G's mMTC capabilities will allow large-scale connectivity of industrial sensors, actuators, and devices. This will enable extensive data collection, monitoring, and control, leading to intelligent and autonomous manufacturing systems.
  - **Edge Computing and AI Integration**: 6G's integration of edge computing and AI technologies will enable localized data processing, real-time analytics, and intelligent decision-making within the industrial environment. This

will reduce latency, enhance data security, and support real-time optimization and predictive maintenance in industrial operations.

- **Digital Twins and Virtual Reality**: 6G's high-speed and low-latency connectivity will enable the creation and synchronization of digital twins – virtual replicas of physical systems – in real time. Combined with VR technologies, Industry 5.0 can use 6G to enable immersive remote collaboration, training, and simulation for industrial applications.
- **Enhanced Security and Privacy**: Industry 5.0 demands robust security measures to protect important industrial processes, data, and intellectual property. 6G networks will incorporate advanced security protocols, encryption mechanisms, and privacy-enhancing technologies to ensure secure and trusted communication within the industrial ecosystem.

16.3.5 Augmented Reality (AR), Virtual Reality (VR), and Mixed Reality (MR)

AR, VR, and MR are immersive technologies that create interactive and realistic experiences by blending the digital and physical worlds. With the advancements brought by 6G networks, these technologies are expected to be significantly enhanced. Let us explore their relationship with 6G in more detail:

- **Enhanced Connectivity**: 6G networks will provide higher data rates, lower latency, and improved reliability, which are important for delivering high-quality AR, VR, and MR experiences. These immersive technologies require continuous and uninterrupted connectivity to ensure real-time interactions and content delivery. 6G's capabilities will enable a more immersive and responsive user experience.
- **Ultrahigh Data Rates**: AR, VR, and MR applications generate and consume large amounts of data, including high-resolution audio, video, and 3D content. 6G networks will provide ultrahigh data rates, enabling smooth streaming, low-latency content delivery, and high-fidelity experiences. This will eliminate buffering, reduce latency, and support the transfer of large amounts of data in real time.
- **Low Latency**: Latency is an important factor in AR, VR, and MR systems as even slight delays can disrupt the user experience and cause motion sickness. 6G's low-latency capabilities, potentially in the sub-millisecond range, will greatly reduce the delay between user actions and system response, ensuring a more natural and immersive experience. This is particularly important for real-time interactions, multiplayer experiences, and remote collaboration.
- **Edge Computing**: 6G will use edge computing technologies, which bring computation and data storage closer to the user and content source. This distributed

architecture reduces latency by processing AR, VR, and MR data locally, near the edge of the network. Edge computing in 6G will enable real-time rendering, content caching, and localized processing, improving the overall performance of immersive experiences.

- **Haptic Communication**: Haptic feedback, which provides a sense of touch and physical interaction, enhances the realism of AR, VR, and MR applications. 6G networks may introduce haptic communication capabilities, enabling the transmission of tactile feedback over the network. This can enhance user immersion and enable new types of interactive experiences, like touch-based interactions in virtual environments.
- **Multimodal Interfaces**: 6G networks will support multimodal interfaces, combining various sensory inputs like visual, auditory, and haptic feedback. This allows for more realistic and engaging AR, VR, and MR experiences. Users will be able to interact with the virtual world using gestures, voice commands, eye tracking, and other forms of input, creating a more intuitive and natural user interface.
- **Multiuser Experiences**: 6G's increased capacity and support for large connectivity will enable collaborative and multiuser AR, VR, and MR experiences. Multiple users will be able to interact and share virtual spaces simultaneously, opening up opportunities for remote collaboration, social interaction, and multiplayer gaming in immersive environments.
- **Edge-Based AI and Machine Learning**: 6G networks will integrate AI and ML capabilities at the network edge, enabling intelligent content delivery, real-time analytics, and personalized experiences. AI algorithms can optimize content rendering, adapt to user preferences, and enhance the realism of AR, VR, and MR applications.

Note that 6G is still in the research and development phase, and the specific features and technologies mentioned above may evolve over time. Ongoing research, industry collaboration, and standardization efforts will shape the implementation and specifications of AR, VR, and MR in 6G networks.

16.3.6 Smart Cities and Autonomous Systems

Smart cities and autonomous systems are two domains that are expected to greatly benefit from the advancements brought by 6G networks. Let us explore their relationship with 6G in more detail:

- **Smart Cities and 6G**: Smart cities use advanced technologies to improve the quality of life for citizens, enhance resource efficiency, and enable sustainable urban development. 6G networks will provide the underlying infrastructure to support the diverse requirements of smart city applications. Here are some ways in which 6G can enhance smart cities:

- **Large Device Connectivity**: 6G networks will support the connectivity of a large number of devices, sensors, and actuators deployed throughout the city. This allows for continuous data collection, monitoring, and control, enabling real-time insights for optimized resource management, traffic control, waste management, and environmental monitoring.
- **Ultra-Reliable and Low-Latency Communication**: 6G's ultra-reliable and low-latency communication capabilities, like URLLC, will facilitate real-time interactions between various components of a smart city ecosystem. This includes communication between autonomous vehicles, smart infrastructure, public safety systems, and citizen devices, ensuring efficient and coordinated operations.
- **Edge Computing and AI Integration**: 6G's integration of edge computing and AI technologies enables localized data processing, real-time analytics, and intelligent decision-making within the smart city environment. Edge computing in 6G can support real-time data analysis, predictive maintenance, and optimized resource allocation, improving the overall efficiency and responsiveness of smart city services.
- **Intelligent Transportation Systems**: 6G networks will support the connectivity requirements of autonomous vehicles, enabling reliable and low-latency communication between vehicles, infrastructure, and other road users. This promotes safer and more efficient transportation systems, reduces traffic congestion, and enables the implementation of advanced features like platooning and intelligent traffic management.
- **Environmental Monitoring and Sustainability**: 6G can enhance environmental monitoring and sustainability initiatives in smart cities. The network's large connectivity and low-power capabilities enable the deployment of a wide range of sensors for air quality monitoring, water management, waste management, and energy efficiency optimization. This facilitates data-driven decision-making and the implementation of sustainable practices.

- **Autonomous Systems and 6G**: Autonomous systems, including autonomous vehicles, drones, robots, and industrial automation, heavily depend on high-speed and reliable communication networks. 6G networks will provide the necessary connectivity and capabilities to support the rapid growth of autonomous systems. We can see that how 6G can benefit autonomous systems:
  - **Ultra-Reliable and Low-Latency Communication**: 6G's ultra-reliable and low-latency communication, like URLLC, is important for enabling real-time communication between autonomous systems and their control

centers. This ensures safe and efficient operation, remote monitoring, and immediate response to important events.

- **Edge Computing and AI Integration**: 6G's integration of edge computing and AI technologies brings computational capabilities closer to autonomous systems. This allows for real-time decision-making, reducing the reliance on centralized cloud infrastructure. Edge computing in 6G enables autonomous systems to process data locally, improving response times and reducing dependence on network connectivity.
- **Network Slicing**: 6G's network slicing capabilities enable the creation of dedicated network slices optimized for autonomous systems. These slices provide the required ultra-reliable and low-latency communication, ensuring that autonomous systems receive the necessary network resources for their operation without interference from other types of traffic.
- **Advanced Positioning and Localization**: 6G networks are expected to incorporate advanced positioning technologies, including highly accurate and reliable positioning systems like enhanced GPS or advanced satellite systems. These positioning capabilities are important for autonomous systems, enabling precise localization, navigation, and coordination in various environments.

16.3.7 Healthcare and Remote Surgery

Healthcare and remote surgery are areas that can greatly benefit from the advancements brought by 6G networks. Let us explore their relationship in more detail:

16.3.7.1 Telemedicine and Remote Healthcare

6G networks will enable advanced telemedicine and remote healthcare services, providing high-quality and real-time healthcare delivery regardless of geographical distance. Here's how 6G can enhance healthcare:

- **Ultra-Reliable and Low-Latency Communication**: 6G's ultra-reliable and low-latency communication capabilities, like URLLC, ensure that healthcare professionals can remotely interact with patients in real time. This enables timely diagnosis, remote consultations, and virtual care delivery, improving access to healthcare services.
- **High-Quality Video Streaming**: 6G's high data rates and low-latency communication will support high-quality video streaming, facilitating detailed visual examinations and accurate assessments during remote consultations. This enables healthcare providers to observe and analyze patient conditions in real time, leading to more effective diagnosis and treatment planning.
- **Remote Patient Monitoring**: 6G networks can support the continuous transmission of real-time patient data, including important signs, medical

images, and wearable device data. This allows healthcare providers to remotely monitor patients' health status, detect anomalies, and provide timely interventions.

- **Edge Computing and AI Integration**: 6G's integration of edge computing and AI technologies enables real-time data processing, analysis, and decision-making at the network edge. This can support real-time health monitoring, predictive analytics, and personalized treatment recommendations, improving patient outcomes and reducing the burden on centralized healthcare systems.
- **Enhanced Security and Privacy**: 6G networks will incorporate robust security and privacy measures to protect patient data during remote healthcare interactions. Advanced encryption, authentication protocols, and privacy-enhancing technologies will ensure the confidentiality and integrity of patient information.

16.3.7.2 Remote Surgery

Remote surgery, also known as telesurgery or teleoperated surgery, allows surgeons to perform operations on patients located at a distant location using robotic systems. 6G networks can play an important role in supporting the requirements of remote surgery:

- **Ultra-Reliable and Low-Latency Communication**: 6G's ultra-reliable and low-latency communication capabilities are essential for enabling real-time, high-precision control of surgical robots during remote procedures. This ensures that the surgeon's commands are executed without any noticeable delay, providing a continuous and responsive surgical experience.
- **High-Bandwidth and Low-Latency Video Streaming**: Remote surgery requires high-quality video streaming to provide the surgeon with a clear and detailed view of the surgical field. 6G's high data rates and low latency enable high-resolution video streaming, allowing surgeons to visualize the procedure in real time and make precise surgical decisions.
- **Haptic Feedback**: 6G networks may introduce haptic communication capabilities, enabling the transmission of tactile feedback from the surgical site to the surgeon's interface. This can provide the surgeon with a sense of touch, enabling more accurate manipulation of surgical instruments and enhancing the overall precision of the procedure.
- **Network Resilience**: Remote surgery demands a highly reliable and resilient network infrastructure. 6G networks can incorporate redundancy and fault-tolerant mechanisms to ensure continuous communication during important surgical procedures. This helps prevent interruptions and maintain the integrity of the surgical process.

- **Security and Privacy**: Remote surgery involves the transmission of sensitive medical and patient data. 6G networks will prioritize robust security measures, including encryption, authentication, and access control, to safeguard the privacy and confidentiality of patient information during remote surgical procedures.

Note that remote surgery is a complex and highly regulated field, and the adoption of such procedures requires extensive research, validation, and compliance with regulatory guidelines and safety standards.

16.3.8 Transportation and Intelligent Mobility

Transportation and intelligent mobility are areas that are expected to undergo a huge and rapid transformation with the advancements brought by 6G networks. Let us explore their relationship in more detail.

16.3.8.1 Connected and Autonomous Vehicles

6G networks will play an important role in supporting the communication and connectivity requirements of connected and autonomous vehicles. Here is how 6G can enhance transportation:

- **Ultra-Reliable and Low-Latency Communication**: 6G's ultra-reliable and low-latency communication capabilities, like URLLC, enable real-time communication between vehicles, infrastructure, and other road users. This ensures efficient coordination, improves traffic safety, and enables advanced features like cooperative collision avoidance and platooning.
- **Vehicle-to-Everything (V2X) Communication**: 6G networks will support V2X communication, allowing vehicles to exchange information with surrounding infrastructure, pedestrians, and other vehicles. This enables enhanced situational awareness, real-time traffic updates, and optimized routing, leading to improved traffic flow and reduced congestion.
- **Edge Computing and AI Integration**: 6G's integration of edge computing and AI technologies brings computational capabilities closer to the transportation environment. This enables real-time data processing, analytics, and decision-making, supporting functions like real-time route planning, traffic prediction, and intelligent traffic management.
- **Enhanced Positioning and Localization**: 6G networks may incorporate advanced positioning technologies, including highly accurate and reliable positioning systems like enhanced GPS or advanced satellite systems. This improves vehicle localization, navigation, and accuracy, enabling more precise and efficient transportation services.

- **Multimodal Connectivity**: 6G networks will provide continuous connectivity across different modes of transportation, enabling integrated and multimodal travel experiences. This facilitates smooth transitions between various transport modes, like public transportation, ride-sharing, and personal vehicles, providing more efficient and convenient transportation options for users.

16.3.8.2 Intelligent Mobility and Transportation Optimization

6G networks will support intelligent mobility solutions, enabling optimization and efficiency in transportation systems. Here is how 6G can enhance intelligent mobility:

- **Real-Time Data Collection and Analytics**: 6G's high-speed and low-latency communication enables real-time data collection from various sensors, infrastructure, and mobility platforms. This data can be analyzed and processed in real time to generate insights for optimizing transportation networks, managing traffic flow, and predicting demand patterns.
- **Demand-Responsive Mobility**: 6G networks can support demand-responsive mobility services, enabling dynamic and personalized transportation solutions. Real-time data and AI-driven algorithms can optimize routing, vehicle allocation, and scheduling based on demand patterns and user preferences, improving overall transportation efficiency.
- **Integrated Mobility Services**: 6G networks can facilitate the integration of different mobility services, like public transportation, ride-sharing, micro-mobility, and on-demand services. Continuous connectivity and interoperability enable users to access and switch between services, promoting efficient and sustainable travel options.
- **Smart Infrastructure and Traffic Management**: 6G's connectivity and real-time communication capabilities support smart infrastructure and traffic management systems. This includes intelligent traffic lights, adaptive signal control, dynamic routing, and congestion management, enabling optimized traffic flow, reduced travel time, and improved overall transportation efficiency.
- **Mobility as a Service (MaaS)**: 6G networks can enable the implementation of Mobility as a Service platforms, where users can access and continuously pay for various transportation modes through a single interface. This promotes sustainable transportation options, reduces private vehicle usage, and improves accessibility to mobility services.

Note that the full realization of intelligent mobility and the deployment of connected and autonomous vehicles depend not only on technological advancements but also on regulatory frameworks, industry collaborations, and public acceptance. The deployment of 6G networks and related technologies will require careful planning.

16.4 Challenges and Issues in 6G Development

The development of 6G technology comes with several challenges and issues that need to be addressed [41–43]. These challenges include:

- **Technical Complexity**: Developing 6G technology involves overcoming technical challenges like achieving ultrahigh data rates, ultralow latency, and large device connectivity. Innovations in network architecture, RF engineering, signal processing, and antenna design are required to meet these ambitious goals.
- **Spectrum Availability**: 6G will require a huge amount of spectrum to deliver its promised capabilities. However, the available spectrum is limited, and regulatory frameworks need to be established to allocate and manage the spectrum efficiently. Harmonization of spectrum globally will be important for continuous global connectivity.
- **Infrastructure Requirements**: Deploying 6G networks will require huge investments in infrastructure, including the installation of new BS, small cells, and network densification. Building the necessary infrastructure to support 6G technology faces challenges in terms of cost, scalability, and environmental impact.
- **Energy Efficiency**: As the demand for connectivity and data increases, ensuring energy efficiency in 6G networks becomes important. Balancing high data rates and low latency with energy consumption requires innovations in energy-efficient network design, power management, and energy harvesting techniques.
- **Security and Privacy**: With the increasing connectivity and data exchange in 6G networks, there is a need to address security and privacy issues [39, 44–50]. Developing robust security mechanisms, encryption protocols, and privacy-enhancing technologies will be essential to protect user data and prevent cyber threats.
- **Standardization and Interoperability**: The development of 6G requires global standardization to ensure interoperability and continuous connectivity across different networks and devices. Collaborative efforts among industry players, standardization bodies, and regulatory agencies are necessary to establish common standards and protocols.
- **Regulatory and Policy Frameworks**: Governments and regulatory bodies need to establish frameworks that promote innovation, fair competition, consumer protection, and address the challenges associated with 6G technology.

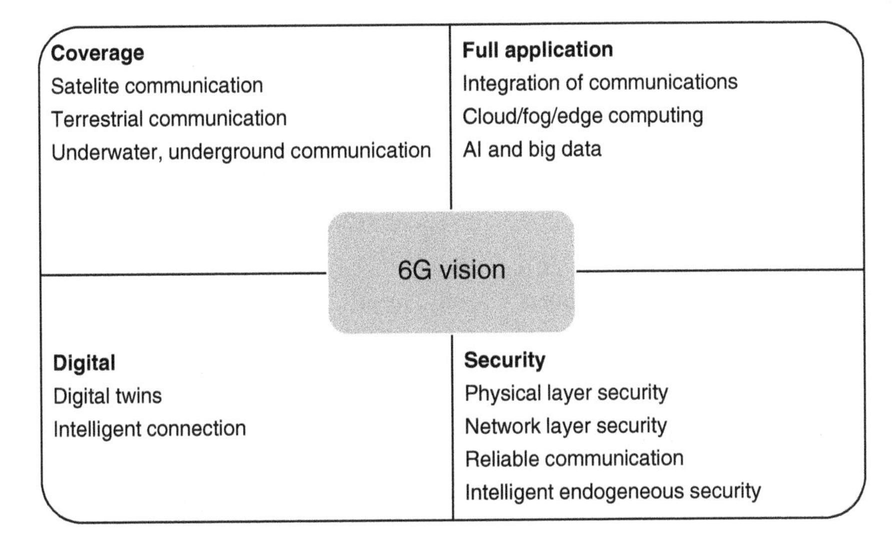

Figure 16.4 6G vision.

Policies regarding spectrum allocation, infrastructure deployment, data governance, and privacy regulations will shape the development and deployment of 6G.

Hence, addressing these challenges and issues requires collaboration among users, including governments, industry players, academia, standardization bodies, and civil society. It is essential to ensure that 6G technology is developed and deployed in a manner that maximizes its benefits while mitigating potential risks and ensuring the well-being of individuals and society. Figure 16.4 provides detailed information for 6G vision.

16.4.1 Spectrum Allocation and Regulation

Spectrum allocation and regulation play an important role in the development and deployment of 6G networks. As 6G aims to support a wide range of applications and services with diverse requirements, efficient and well-regulated spectrum allocation is essential. Here are some key issues related to spectrum allocation and regulation in 6G development:

- **Spectrum Availability**: The availability of suitable spectrum is important for the successful deployment of 6G networks. As demand for wireless communication continues to grow, regulators and policymakers need to identify and allocate frequency bands that can accommodate the increased capacity and data rates required by 6G. This may involve repurposing existing bands, allocating new frequency ranges, or exploring innovative spectrum-sharing approaches.

- **Spectrum Harmonization**: Harmonization of spectrum allocation across different countries and regions is important for global interoperability and roaming capabilities in 6G networks. Harmonized spectrum enables continuous connectivity and supports consistent service experiences for users across international borders. International coordination and collaboration among regulatory bodies and standardization organizations are necessary to ensure harmonized spectrum usage.
- **Spectrum Efficiency**: Spectrum efficiency is a key issue to maximize the utilization of available frequency bands. In 6G, advanced technologies, like dynamic spectrum access, cognitive radio, and beamforming, can be employed to improve spectral efficiency. Regulatory frameworks should encourage the adoption of these technologies and promote spectrum-sharing strategies to optimize spectrum utilization.
- **Spectrum Sharing**: With the increasing demand for spectrum, regulators are exploring new approaches for spectrum sharing to ensure efficient spectrum utilization. Dynamic spectrum access, spectrum-sharing databases, and spectrum-sharing frameworks enable multiple users to access the same frequency bands under specified conditions. These techniques can facilitate the coexistence of different services and help alleviate spectrum scarcity challenges.
- **Spectrum Auctions and Licensing**: Governments often conduct spectrum auctions to allocate spectrum licenses to operators and service providers. Spectrum auctions ensure fair access to limited resources and provide incentives for efficient spectrum usage. Regulatory frameworks need to establish transparent, competitive, and inclusive processes for spectrum auctions, taking into account factors like market competition, service diversity, and coverage obligations.
- **Spectrum for Vertical Industries**: 6G is expected to serve a wide range of vertical industries, including healthcare, transportation, manufacturing, and agriculture. To cater to their specific requirements, regulators need to consider allocating dedicated spectrum bands or providing spectrum access mechanisms tailored to support the unique needs of these industries. This could involve spectrum licensing frameworks that prioritize vertical industry use cases or the establishment of dedicated frequency bands for specific vertical applications.
- **International Spectrum Coordination**: Spectrum allocation is a global issue, and international coordination is important to avoid interference and ensure efficient spectrum usage. Regulatory bodies, standardization organizations, and industry users collaborate to harmonize spectrum usage across borders, facilitating continuous global connectivity and interoperability of 6G networks.
- **Regulatory Frameworks for Emerging Technologies**: 6G networks will introduce new technologies and use cases, like THz communication, satellite integration, and non-terrestrial networks. Regulators need to adapt and develop

regulatory frameworks that address the unique characteristics, challenges, and opportunities presented by these emerging technologies, ensuring appropriate spectrum allocations, standards, and licensing policies.

In summary, regulators and policymakers play a central role in shaping the regulatory landscape for 6G development. They need to balance the need for efficient spectrum utilization, promote competition, support innovation, and address societal and environmental issues while fostering a conducive environment for investment and deployment of 6G networks. Continuous dialogue and collaboration between regulators, industry users, and standardization bodies are essential for achieving effective spectrum allocation and regulation in 6G.

16.4.2 Network Architecture and Infrastructure

The network architecture and infrastructure of 6G development are expected to undergo rapid advancements to meet the diverse requirements of future applications and services. While specific details are still being researched and defined, here are some key aspects that are being considered for the network architecture and infrastructure of 6G:

- **Hyperconnected and Intelligent Networks**: 6G aims to create hyperconnected networks that continuously integrate various communication technologies, including terrestrial networks, satellite networks, aerial platforms (like drones), and possibly even non-terrestrial networks like stratospheric platforms or low Earth orbit satellite constellations. This integration will enable ubiquitous and uninterrupted connectivity, providing high-quality services in diverse scenarios.
- **Terahertz Communication**: 6G is expected to use THz frequencies to achieve extremely high data rates and ultralow latency. THz communication can enable multi-terabit-per-second data transmission, unlocking new possibilities for applications like holographic communication, immersive AR, and ultrahigh-definition video streaming. However, THz communication requires advancements in hardware technologies and overcoming propagation challenges associated with this frequency range.
- **Integrated Access and Backhaul**: 6G networks will likely integrate access and backhaul functions, enabling more efficient and flexible network deployments. By combining access and backhaul, 6G networks can optimize resource utilization, reduce infrastructure costs, and enable rapid network deployment and scalability.
- **Heterogeneous Networks and Network Slicing**: 6G is expected to support heterogeneous networks that integrate various access technologies, like cellular networks, Wi-Fi, and short-range communication systems. Network slicing,

a technique introduced in 5G, will continue to evolve in 6G, allowing the creation of dedicated virtual networks optimized for specific use cases, industries, or applications. Network slicing enables the customization of network resources, performance characteristics, and security parameters to meet diverse requirements.
- **Edge Computing and Distributed Intelligence**: 6G networks will heavily use edge computing capabilities to bring computational resources closer to the network edge. This enables real-time processing, low-latency services, and distributed intelligence for applications like real-time analytics, AI-driven decision-making, and localized data processing. Edge computing in 6G will enable new services, reduce network congestion, and enhance user experiences.
- **AI-Enabled Network Management and Orchestration**: 6G networks will incorporate AI and ML techniques for intelligent network management and orchestration. AI algorithms can optimize resource allocation, predict network demands, detect anomalies, and enable self-healing capabilities. These AI-driven functionalities will enhance network efficiency, adaptability, and resilience.
- **Security and Privacy**: 6G networks will prioritize robust security and privacy mechanisms. As the scale and complexity of connected devices increase, security challenges will intensify. 6G networks will employ advanced encryption algorithms, authentication mechanisms, privacy-preserving techniques, and secure communication protocols to ensure the confidentiality, integrity, and privacy of data and communications.
- **Sustainable and Energy-Efficient Solutions**: 6G networks will consider sustainable and energy-efficient solutions to reduce the environmental impact of network infrastructure. This includes the adoption of energy-efficient hardware, intelligent power management, and green network practices to minimize energy consumption and carbon footprint.
- **Standardization and Interoperability**: Standardization bodies like the International Telecommunication Union (ITU) and industry consortia play an important role in defining the specifications and protocols for 6G networks. Interoperability between different vendors and technologies is essential to ensure continuous connectivity and enable a diverse ecosystem of devices and applications.

Note that 6G is still in the early stages of research and development, and the exact network architecture and infrastructure will evolve over time. Researchers, industry users, and standardization bodies are actively working on defining the requirements and characteristics of 6G.

16.4.3 Security and Privacy

Security and privacy are important issues in the development of 6G networks. As 6G aims to support a wide range of applications and services, including sensitive and important domains, robust security and privacy measures must be integrated into the network architecture and protocols [39, 44–51]. Here are key aspects related to security and privacy in 6G development:

- **Enhanced Encryption and Authentication**: 6G networks will employ advanced encryption algorithms and authentication mechanisms to secure communications between devices, users, and network elements. Strong encryption protocols ensure the confidentiality and integrity of data, protecting it from unauthorized access and tampering.
- **Secure Network Infrastructure**: 6G networks will focus on securing the underlying network infrastructure, including BS, access points, and core network elements. Implementing security measures like secure bootstrapping, firmware integrity checks, and secure management interfaces will help prevent unauthorized access and protect against attacks targeting the network infrastructure.
- **Privacy-Preserving Technologies**: Privacy is a major issue in 6G networks, given the vast amount of personal data generated by connected devices and applications. Privacy-preserving technologies, like differential privacy, secure multiparty computation, and federated learning, can be employed to protect users' personal information while enabling data-driven services and applications.
- **Trustworthy Identity Management**: Robust identity management is important in 6G networks to ensure the authenticity and integrity of users and devices. The use of digital certificates, secure key management, and multifactor authentication mechanisms can establish trust and prevent unauthorized access.
- **Security for Vertical Industries**: 6G networks will cater to diverse vertical industries, like healthcare, transportation, and industrial automation. Each industry has specific security requirements and regulations. 6G must incorporate industry-specific security measures, like secure data exchange in healthcare, secure communication in autonomous vehicles, and protection against cyber-physical attacks in industrial environments.
- **Threat Intelligence and Real-Time Security**: 6G networks will use AI and ML techniques to analyze network traffic, detect anomalies, and identify potential security threats in real time. This proactive approach enhances the network's ability to respond to emerging security risks promptly.
- **Resilience and Disaster Recovery**: 6G networks will incorporate resilience and disaster recovery mechanisms to ensure continuity of services in the event of natural disasters, network failures, or cyberattacks. Redundancy, backup

systems, and rapid restoration capabilities will be essential to minimize service disruptions and recover from incidents quickly.

- **User Empowerment and Control**: 6G networks should empower users with control over their data and privacy preferences. Privacy settings, consent mechanisms, and transparent data usage policies will allow users to make informed decisions about sharing their data and enable them to exercise control over their privacy.
- **Collaboration and Standards**: Collaboration among users, including network operators, vendors, researchers, and regulators, is essential to establish common security and privacy standards for 6G. Industry consortia, standardization bodies, and regulatory agencies play an important role in defining security guidelines, best practices, and compliance requirements to ensure a consistent and secure 6G ecosystem.

As 6G networks evolve, it is important to recognize that security and privacy are ongoing issues. Continuous research, security audits, and updates to security protocols will be necessary to address emerging threats and vulnerabilities in the rapidly evolving technology landscape.

16.4.4 Interoperability and Standardization

Interoperability and standardization are important aspects of 6G development to ensure continuous connectivity, compatibility, and harmonization across different networks, devices, and services. Here is a closer look at the importance of interoperability and standardization in the context of 6G:

- **Interoperability Enables Continuous Connectivity**: Interoperability ensures that diverse devices, networks, and applications can work together smoothly, regardless of their manufacturers or service providers. In the context of 6G, interoperability allows for continuous connectivity between various communication technologies, like cellular networks, Wi-Fi, satellite networks, and other emerging wireless technologies. It enables users to access services and applications continuously across different devices and networks.
- **Harmonization of Standards Promotes Global Compatibility**: Standardization is the process of establishing common technical specifications, protocols, and guidelines to ensure compatibility and uniformity across the ecosystem. In the context of 6G, harmonized standards enable global compatibility, roaming, and consistent service experiences across different regions and countries. International standardization bodies, like the ITU and other industry consortia, play an important role in developing and promoting these standards.
- **Efficient Resource Utilization and Optimization**: Standardized interfaces, protocols, and network architectures enhance resource utilization and optimization in 6G networks. By defining common specifications and protocols, different

vendors and technologies can continuously integrate and interoperate, resulting in more efficient network deployments and resource allocation. This leads to improved overall network performance, reduced complexity, and enhanced scalability.

- **Promoting Innovation and Market Competition**: Standardization provides a level playing field for technology vendors, promotes healthy market competition, and encourages innovation. When standards are well-defined, vendors can focus on developing innovative products, services, and applications that adhere to those standards. This fosters a competitive environment that benefits end users by providing a wider range of choices and driving technological advancements.
- **Ecosystem Development and Collaboration**: Interoperability and standardization promote collaboration among industry users, including network operators, device manufacturers, service providers, and researchers. These users collaborate to define common specifications, share best practices, and address technical challenges. Such collaboration facilitates the development of a vibrant ecosystem where different players can collaborate, innovate, and create new solutions.
- **Regulatory Compliance and Policy Alignment**: Standardization helps ensure regulatory compliance and policy alignment in the deployment of 6G networks. By adhering to standardized specifications and protocols, network operators and service providers can meet regulatory requirements, security guidelines, and privacy regulations. Standardization also supports policy alignment across different regions, facilitating the adoption of consistent practices and regulations related to 6G deployment.
- **Future-Proofing and Long-Term Investment Protection**: Standardization efforts in 6G take into account future requirements and anticipated technological advancements. By establishing flexible and adaptable standards, the industry can better accommodate future innovations and avoid premature obsolescence of infrastructure investments. Standardization enables long-term investment protection for both network operators and end users.
- **Continuity and Evolution of Technology**: As 6G networks evolve, standardization ensures continuity and continuous transition between different generations of technology. Standardization bodies work on defining specifications and protocols that enable backward compatibility and smooth migration paths from earlier generations like 5G to 6G. This allows for the gradual deployment and adoption of 6G technologies without disrupting existing services and infrastructure.

The standardization process involves close collaboration between industry users, standardization bodies, regulatory agencies, and research communities. It is an ongoing effort that adapts to emerging technologies, addresses new use cases, and incorporates feedback from the industry. The goal is to develop a comprehensive set of standards that support the successful deployment and operation of 6G networks, enabling a globally connected and interoperable ecosystem.

16.4.5 Ethical and Social Implications

The development of 6G networks brings with it various ethical and social implications that need to be considered. While 6G technology holds the potential to enable transformative applications and services, it is essential to address these implications to ensure responsible and beneficial deployment. Here are some key ethical and social issues related to 6G development:

- **Digital Divide**: 6G networks may exacerbate existing digital divides, leading to unequal access to advanced connectivity and services. It is important to address affordability, availability, and accessibility barriers to ensure that underserved communities and regions are not left behind. Efforts should be made to bridge the digital divide, promote digital inclusion, and provide equal opportunities for all individuals and communities.
- **Privacy and Data Protection**: The extensive connectivity and data-intensive nature of 6G networks raise major privacy issues. Collecting, storing, and processing vast amounts of user data may risk privacy breaches and unauthorized use of personal information. Strong data protection measures, transparency, informed consent, and user control mechanisms should be in place to safeguard individuals' privacy rights and prevent misuse of personal data.
- **Security and Cybersecurity**: 6G networks present new security challenges due to their complex and interconnected nature. The potential risks include data breaches, cyberattacks, and the compromise of important infrastructure. Robust security measures, threat intelligence, and resilient architectures should be implemented to protect networks, devices, and users from malicious activities. Collaboration between industry, government, and research communities is important to address evolving cybersecurity threats effectively.
- **Algorithmic Bias and Fairness**: AI-driven technologies in 6G may introduce biases and unfairness in decision-making processes. ML algorithms can reflect the biases present in training data, leading to discriminatory outcomes in areas like employment, finance, and healthcare. Ensuring fairness, transparency, and accountability in algorithmic systems is essential to mitigate biases and promote equitable outcomes.
- **Social Impact and Job Displacement**: The deployment of 6G networks, along with automation and AI technologies, may have implications for the workforce. While new job opportunities may emerge, there is a possibility of job

displacement and changes in employment patterns. Ensuring a just transition and providing reskilling and upskilling opportunities for affected workers can help mitigate negative social and economic impacts.

- **Ethical AI and Autonomous Systems**: 6G networks may facilitate the deployment of autonomous systems, like autonomous vehicles and drones, that depend on AI algorithms for decision-making. Ensuring ethical behavior, transparency, and accountability of these systems is important to prevent harm to individuals, society, and the environment. Ethical issues, like safety, privacy, and accountability, should be embedded in the design and deployment of AI-enabled autonomous systems.
- **Environmental Sustainability**: The increasing energy consumption and carbon footprint associated with advanced wireless networks raise environmental issues. The development of 6G should prioritize energy-efficient technologies, green infrastructure, and sustainable practices to minimize its impact on the environment. This includes optimizing energy consumption, promoting RES, and reducing electronic waste.
- **Social Implications of Emerging Applications**: 6G networks enable a wide range of applications, like telemedicine, VR, and smart cities. These applications may have implications for healthcare delivery, human interaction, and urban environments. It is important to consider the social, cultural, and ethical implications of these applications to ensure they align with societal values, respect human rights, and promote social well-being.

Addressing these ethical and social implications requires a multidisciplinary approach involving collaboration among policymakers, industry users, academia, civil society organizations, and the public. Engaging in transparent and inclusive discussions, establishing regulatory frameworks, and implementing responsible practices are essential for the ethical and socially responsible deployment of 6G technologies.

16.5 Impact of 6G on Various Industries and Sectors

The deployment of 6G-enabled technologies is expected to have a transformative impact across various industries and sectors. While the specific applications and benefits will depend on technological advancements and specific use cases [44, 45], here are some potential impacts of 6G on different industries:

- **Healthcare**: 6G can revolutionize healthcare delivery by enabling real-time remote diagnostics, telemedicine, and remote surgery. High-speed, low-latency connections will facilitate continuous communication between doctors and patients, regardless of geographical distances, leading to improved access to healthcare services.

- **Transportation and Autonomous Systems**: The ultralow latency and high reliability of 6G networks can significantly enhance the safety and efficiency of transportation systems. It will support the global deployment of autonomous vehicles, enabling real-time communication between vehicles, infrastructure, and pedestrians. This can reduce accidents, optimize traffic flow, and enable new mobility services.
- **Manufacturing and Industrial Automation**: 6G's large device connectivity and ultra-reliable communication will drive advancements in industrial automation. It will facilitate the deployment of smart factories, where machines, robots, and sensors can communicate and collaborate continuously, leading to increased productivity, predictive maintenance, and optimized supply chains.
- **Energy and Utilities**: 6G can play an important role in the transformation of energy and utilities sectors. It can enable smart grid systems that provide real-time monitoring and control of energy distribution, leading to more efficient energy management, reduced carbon footprint, and improved integration of RES.
- **Entertainment and Media**: 6G's high data rates and low latency will revolutionize the entertainment and media industry. It will enable immersive experiences, including VR, AR, and high-quality streaming services. Additionally, 6G networks will support real-time interactive gaming, live events, and personalized content delivery.
- **Agriculture**: In the agricultural sector, 6G can enable precision farming by providing real-time data on soil moisture, crop health, and weather conditions. This data can be utilized to optimize irrigation, fertilizer usage, and pest control, leading to increased crop yields, reduced resource wastage, and improved sustainability.
- **Smart Cities**: 6G will be a catalyst for the development of smart cities, where various sectors like transportation, energy, healthcare, and public safety are continuously connected. It will enable efficient urban planning, real-time monitoring of infrastructure, intelligent traffic management, and improved citizen services.
- **Education**: 6G-enabled technologies can transform education by enabling immersive and personalized learning experiences. Virtual classrooms, interactive simulations, and remote collaboration will become more continuous, providing new opportunities for distance education and lifelong learning.

These are just a few examples of how 6G can impact various industries and sectors. The ultrahigh data rates, ultralow latency, and large device connectivity of 6G networks will unlock innovative applications and drive digital transformation across multiple domains, leading to increased efficiency, improved services, and enhanced user experiences.

16.5.1 Communication and Connectivity

6G is expected to have a profound impact on communication and connectivity, revolutionizing the way we connect, communicate, and interact with the world. Here are some key impacts that 6G is anticipated to have:

- **Ultra-Fast Speeds**: 6G is projected to provide unique data speeds, potentially reaching terabits per second. This ultra-fast speed will enable near-instantaneous downloads, continuous streaming of high-resolution content, and real-time communication with minimal latency. It will significantly enhance user experiences and support data-intensive applications.
- **Ultra-Reliable Low-Latency Communication (URLLC)**: 6G aims to provide ultra-reliable and low-latency communication, enabling mission-important applications and services that require instant responsiveness and high reliability. This will be important for applications like autonomous vehicles, remote surgery, and industrial automation, where any delay or disruption could have severe consequences.
- **Large Connectivity**: 6G will support large mMTC, enabling a vast number of connected devices and sensors to communicate simultaneously. This will make the way for the IoT to expand further, connecting billions or even trillions of devices, and facilitating continuous communication between them. It will unlock new possibilities for smart cities, smart homes, and various industrial applications.
- **Enhanced Coverage and Availability**: 6G is expected to extend coverage and availability to remote and underserved areas, addressing connectivity gaps and bridging the digital divide. By using innovative technologies like satellite networks and aerial platforms (like drones), 6G can provide connectivity in previously inaccessible regions, enabling more inclusive and equitable access to communication services.
- **Immersive Experiences**: 6G will enable immersive experiences through advanced AR, VR, and MR applications. The high data rates, ultralow latency, and continuous connectivity of 6G networks will provide users with immersive and interactive virtual experiences, transforming industries like gaming, entertainment, education, and healthcare.
- **Intelligent Connectivity**: 6G networks will integrate AI and ML capabilities to enable intelligent connectivity. These networks will have the ability to dynamically adapt to user requirements, environmental conditions, and network congestion. AI algorithms will optimize network resources, predict user behavior, and personalize services, enhancing the overall connectivity experience.

- **Terahertz Communication**: 6G is expected to use THz frequencies for communication, providing ultrahigh data rates and capacity. THz communication can enable applications like holographic communication, high-definition video streaming, and ultrahigh-speed data transmission. However, THz communication also faces technical challenges related to propagation and hardware development.
- **Ubiquitous Connectivity**: 6G aims to provide ubiquitous connectivity, ensuring that users can stay connected regardless of their location or mobility. Continuous handovers between different networks and technologies, like cellular networks, Wi-Fi, and satellite systems, will enable uninterrupted connectivity, supporting users' needs even in highly dynamic environments.
- **Green and Sustainable Connectivity**: With increasing focus on sustainability, 6G development highlights energy efficiency and green networking. Efforts are being made to optimize energy consumption, reduce the carbon footprint, and minimize electronic waste associated with network infrastructure. Green and sustainable practices will be incorporated into the design and deployment of 6G networks.

In summary, the impact of 6G on communication and connectivity is expected to be transformative, enabling new applications, enhancing user experiences, and driving innovation across industries. It will create a hyperconnected world where communication is faster, more reliable, and more immersive, facilitating advancements in areas like healthcare, transportation, education, entertainment, and beyond.

16.5.2 Manufacturing and Industry

6G is anticipated to have a major impact on the manufacturing and industry sectors, bringing forth transformative changes and unlocking new opportunities for efficiency, automation, and connectivity. Here are some key impacts of 6G on manufacturing and industry:

- **Industrial Automation and Robotics**: 6G will enable advanced industrial automation and robotics by providing ultra-reliable and low-latency communication. It will facilitate real-time control, remote monitoring, and coordination of industrial processes, leading to increased productivity, precision, and efficiency. Robotic systems will be able to communicate continuously and collaborate with each other, enabling complex and synchronized manufacturing operations.
- **Internet of Things (IoT) in Manufacturing**: 6G's large mMTC capabilities will enhance the implementation of IoT in manufacturing. A multitude of connected sensors, devices, and machinery will enable real-time monitoring, data

collection, and analysis across the production chain. This will enable predictive maintenance, optimized resource allocation, and improved quality control, leading to reduced downtime, cost savings, and enhanced product quality.

- **Digital Twins and Simulation**: 6G's ultra-fast speeds and low latency will support the deployment of digital twins and simulation technologies in manufacturing. Digital twins are virtual replicas of physical assets, products, or processes, allowing for real-time monitoring, analysis, and optimization. Simulation technologies powered by 6G will enable virtual testing and optimization of production systems, leading to reduced development costs, accelerated time-to-market, and improved product performance.
- **Artificial Intelligence and Machine Learning**: 6G networks will use AI and ML capabilities to enable intelligent manufacturing systems. Real-time data analytics, predictive algorithms, and autonomous decision-making will optimize production processes, enable adaptive manufacturing, and improve overall operational efficiency [39, 44–51]. AI-powered systems will learn from data generated by connected devices and continuously optimize manufacturing operations.
- **Supply Chain Optimization**: 6G will enhance supply chain management by providing real-time visibility and traceability of goods and materials throughout the entire supply chain. Connected devices, sensors, and blockchain technology will enable secure and transparent tracking of products, ensuring authenticity, reducing counterfeit risks, and streamlining logistics operations. This will result in improved inventory management, reduced costs, and enhanced customer satisfaction.
- **Remote Operations and Maintenance**: 6G's ultra-reliable low-latency communication will enable remote operation and maintenance of industrial equipment and facilities. Experts will be able to remotely monitor, diagnose, and troubleshoot machinery, reducing the need for on-site visits and minimizing downtime. This remote capability will improve operational efficiency, reduce costs, and enhance worker safety by minimizing exposure to hazardous environments.
- **Augmented Reality (AR) in Manufacturing**: 6G's high-speed connectivity and low latency will facilitate the global adoption of AR in manufacturing. AR overlays virtual information onto the physical world, enabling workers to access real-time instructions, guidance, and visualizations during production processes. This will improve worker productivity, reduce errors, and support training and knowledge transfer in manufacturing operations.
- **Energy Efficiency and Sustainability**: 6G development highlights energy efficiency and sustainability. The integration of green and sustainable practices in manufacturing processes, supported by 6G technologies, will enable optimized resource utilization, reduced waste, and lower energy consumption. This

will contribute to environmental sustainability and help industries meet their sustainability goals.

In summary, 6G's impact on manufacturing and industry will drive digital transformation, automation, and connectivity. It will enable smarter, more efficient, and sustainable manufacturing processes, leading to increased productivity, improved product quality, and enhanced competitiveness in the global market.

16.5.3 Healthcare and Life Sciences

6G technology is expected to have a transformative impact on healthcare and life sciences, revolutionizing the way healthcare is delivered and advancing medical research. Here are some key impacts of 6G on healthcare and life sciences:

- **Telemedicine and Remote Healthcare**: 6G's ultra-reliable low-latency communication will enable real-time, high-quality telemedicine services. Healthcare professionals will be able to remotely diagnose and treat patients, monitor important signs, and provide personalized care. Remote healthcare consultations, remote surgeries, and remote patient monitoring will become more feasible and accessible, especially in underserved areas.
- **Remote Surgery and Robotic-Assisted Procedures**: With 6G's ultralow latency and high reliability, remote surgery and robotic-assisted procedures will be significantly enhanced. Surgeons will be able to perform complex operations with the assistance of robotic systems, even from remote locations. This will expand access to specialized surgical expertise, minimize the need for patient travel, and improve surgical outcomes.
- **Internet of Medical Things (IoMT)**: 6G's large mMTC capabilities will advance the IoMT. Wearable devices, biosensors, and implantable devices will continuously communicate with healthcare providers, enabling continuous monitoring of patients' health conditions. This real-time data collection and analysis will support preventive healthcare, early detection of diseases, and personalized treatment plans.
- **AI-Assisted Diagnostics and Precision Medicine**: 6G's high-speed connectivity and data processing capabilities will facilitate the use of AI and ML algorithms in healthcare. AI systems will analyze vast amounts of patient data, medical images, and genetic information to assist in diagnosis, predict diseases, and recommend personalized treatment plans. Precision medicine approaches will be enhanced, enabling tailored treatments based on an individual's genetic profile and health data.
- **Real-Time Health Monitoring and Emergency Response**: With 6G's ultra-fast speeds and low latency, real-time health monitoring systems will

become more effective. Patients with chronic conditions, elderly individuals, and those requiring constant monitoring will benefit from wearable devices and remote monitoring solutions. In emergency situations, timely communication and coordination between healthcare providers, emergency responders, and patients will be facilitated, leading to faster response times and improved outcomes.

- **Virtual and Augmented Reality in Medical Training**: 6G's high-speed connectivity and low latency will support the integration of VR and AR in medical training and education. Healthcare professionals and students will have access to immersive, realistic simulations for surgical training, anatomy education, and complex medical procedures. This will enhance skills acquisition, reduce training costs, and improve patient safety.
- **Collaborative Research and Data Sharing**: 6G's secure and reliable communication will enable continuous collaboration and data sharing among researchers, healthcare institutions, and life sciences organizations. Large-scale genomic data, clinical trial data, and medical research findings can be shared and analyzed more efficiently, accelerating medical breakthroughs and advancing drug discovery processes.
- **Personal Health Management and Wellness**: 6G will empower individuals to take charge of their health through personal health management apps, wearable devices, and real-time health monitoring. Users will have access to personalized health information, wellness recommendations, and behavior change interventions. This will promote preventive healthcare, self-care, and overall well-being.

In summary, 6G's impact on healthcare and life sciences will enable the delivery of more accessible, personalized, and efficient healthcare services. It will enhance patient outcomes, facilitate remote care, accelerate medical research, and support the shift toward proactive and personalized approaches to healthcare.

16.5.4 Transportation and Logistics

6G technology is expected to have a profound impact on transportation and logistics, revolutionizing the way goods and people are transported and improving efficiency, safety, and sustainability. Here are some key impacts of 6G on transportation and logistics:

- **Intelligent Mobility and Connected Vehicles**: 6G will enable intelligent mobility by providing ultra-fast and reliable connectivity to vehicles. Connected vehicles will communicate with each other, with traffic infrastructure, and with the cloud in real time. This will enable advanced driver assistance systems, cooperative collision avoidance, and traffic optimization, leading to safer and more efficient transportation.

- **Autonomous Vehicles and V2X Communication**: 6G's ultra-reliable low-latency communication will be instrumental in supporting autonomous vehicles. The high-speed and low-latency connectivity will enable real-time communication between autonomous vehicles (V2V) and between vehicles and infrastructure (V2X). This will enhance the coordination and decision-making capabilities of autonomous vehicles, improving safety and efficiency on the roads.
- **Traffic Management and Optimization**: 6G's connectivity and real-time data processing capabilities will revolutionize traffic management and optimization systems. Intelligent algorithms and AI-powered traffic management systems will analyze real-time data from connected vehicles, sensors, and infrastructure to optimize traffic flow, reduce congestion, and improve overall transportation efficiency.
- **Supply Chain and Logistics Optimization**: 6G's large mMTC capabilities will enable real-time tracking, monitoring, and optimization of goods throughout the supply chain. Connected sensors, RFID tags, and IoT devices will provide accurate and up-to-date information on inventory, shipment status, and delivery routes. This will enable more efficient supply chain management, reduced costs, and improved logistics planning.
- **Enhanced Fleet Management**: 6G's connectivity and data analytics capabilities will enhance fleet management operations. Real-time monitoring of vehicles' health, fuel consumption, and driver behavior will enable proactive maintenance, fuel optimization, and improved driver safety. Fleet managers will have access to real-time data and analytics to optimize routes, schedules, and resource allocation, leading to cost savings and increased operational efficiency.
- **Smart Infrastructure and Traffic Safety**: 6G will enable the development of smart infrastructure, including smart traffic lights, road sensors, and adaptive signage. These intelligent infrastructure elements will communicate with vehicles, providing real-time traffic information, hazard alerts, and road condition updates. This will enhance traffic safety, improve navigation, and provide a continuous travel experience.
- **Last-Mile Delivery and Drones**: 6G's low latency and high data rates will support the deployment of drones and UAVs for last-mile delivery. Drones will be able to communicate with each other, with the delivery infrastructure, and with centralized control systems in real time. This will enable efficient and rapid delivery of packages, especially in urban areas, reducing congestion and improving delivery speed.

- **Sustainability and Energy Efficiency**: 6G development highlights energy efficiency and sustainability. By optimizing routes, reducing traffic congestion, and enabling more efficient logistics operations, 6G will contribute to reducing carbon emissions and environmental impact. Additionally, smart charging infrastructure for electric vehicles can be facilitated, promoting the adoption of sustainable transportation solutions.
- **Passenger Experience and Entertainment**: 6G's ultra-fast speeds and low latency will enhance the passenger experience during travel. Passengers will have access to high-definition entertainment, AR and VR experiences, and continuous connectivity for remote work and communication. This will transform the travel experience and create new business opportunities in the transportation industry.

In summary, 6G's impact on transportation and logistics will drive increased efficiency, safety, and sustainability in the movement of goods and people. It will enable intelligent mobility, autonomous vehicles, real-time traffic management, and optimized logistics operations, transforming the way transportation systems.

16.5.5 Entertainment and Media

As of my last knowledge update in September 2021, 6G technology was still in the early stages of development, and there were limited details available about its specific capabilities and features. However, based on the general trajectory of technological advancements and the impact of previous generations of wireless networks, we can speculate on some potential impacts of 6G on the entertainment and media industry. Note that these predictions may not be entirely accurate, as the actual implementation and features of 6G are subject to change.

- **Faster and More Reliable Streaming**: 6G is expected to provide significantly faster data transfer speeds compared to its predecessors. This means that streaming services for video, music, and other media content could experience a substantial boost in performance. Users may be able to enjoy ultrahigh-definition and immersive media experiences with minimal buffering or latency issues.
- **Enhanced Virtual and Augmented Reality**: 6G's advancements in bandwidth, latency, and capacity may lead to major improvements in VR and AR technologies. This could result in more realistic and immersive experiences for users, enabling new forms of entertainment and media consumption. For example, live sports events could be streamed in VR, allowing viewers to feel as though they are physically present at the venue.
- **Improved Gaming Experiences**: The gaming industry is expected to benefit greatly from 6G. With faster and more stable connections, cloud gaming could become more mainstream, eliminating the need for powerful local hardware.

Multiplayer experiences could also be enhanced, with lower latency enabling smoother and more responsive gameplay. Additionally, the potential for AR and VR integration could open up new possibilities for immersive gaming experiences.

- **Enhanced Content Creation and Distribution**: 6G's capabilities may have a profound impact on content creators and distributors. Higher bandwidth and improved connectivity could facilitate real-time content creation and collaboration, allowing media professionals to work together continuously, regardless of their physical locations. This could lead to more efficient production processes and the creation of innovative, interactive content.
- **Advanced Advertising and Personalized Content**: With the increased data transfer speeds and improved connectivity of 6G, targeted advertising and personalized content delivery could become even more precise and efficient. Advertisers and content providers could use real-time user data and analytics to deliver highly tailored experiences, leading to more engaging and relevant content for consumers.
- **Internet of Things (IoT) Integration**: 6G is expected to play an important role in connecting and managing the vast number of IoT devices [39, 45–50]. This integration could have implications for the entertainment and media industry by enabling continuous interactions between devices and services. For instance, smart homes could use 6G's capabilities to provide personalized media experiences across multiple devices and platforms.

Note that these predictions are speculative and based on assumptions about the potential capabilities of 6G technology. The actual impact on the entertainment and media industry will depend on how 6G is implemented, adopted, and utilized by various users.

16.5.6 Education and Research

While 6G technology is still in its early stages of development and there is limited information available about its specific features and capabilities, we can anticipate some potential impacts on education and research based on the trends observed with previous generations of wireless networks. Note that these predictions are speculative and may not align with the actual implementation and features of 6G.

- **Enhanced Connectivity and Accessibility**: 6G is expected to provide significantly faster data transfer speeds, lower latency, and improved network capacity compared to previous generations. This could lead to improved connectivity and accessibility for educational institutions and researchers, particularly in remote or underserved areas. Students and researchers would be able to access online resources, participate in virtual classrooms, and collaborate on research projects more continuously.

- **Immersive Learning Experiences**: The high data transfer speeds and low latency of 6G may enable the development of immersive learning experiences. VR and AR applications could be more widely integrated into educational curricula, allowing students to engage with complex subjects in a more interactive and immersive manner. This could enhance understanding, retention, and overall learning outcomes.
- **Remote Learning and Collaboration**: 6G's improved connectivity and reliability could significantly enhance remote learning and collaboration. With faster and more stable connections, students and researchers could participate in virtual classrooms, attend lectures, and engage in group projects without the limitations of traditional distance learning. Real-time collaboration and communication among students and researchers across different locations could become more continuous and efficient.
- **Advanced Research Capabilities**: 6G's high-speed and low-latency connections could empower researchers with more advanced capabilities. They would be able to access and analyze vast amounts of data in real time, enabling more sophisticated research in various fields. This could accelerate the pace of scientific discoveries, facilitate data-intensive research, and enable remote experiments and simulations.
- **Internet of Things (IoT) Integration**: 6G is expected to play an important role in supporting the integration and management of IoT devices. This integration could have implications for education and research by enabling smart classrooms, campus automation, and data-driven insights. IoT-enabled sensors and devices could be used to create interactive and adaptive learning environments, optimize resource utilization, and facilitate data collection for research purposes.
- **Lifelong Learning and Personalized Education**: The combination of 6G's connectivity, low latency, and IoT integration could support personalized and adaptive learning experiences. AI-powered platforms could use real-time data to tailor educational content and methodologies to individual students' needs, preferences, and learning styles. Lifelong learning could be facilitated through personalized learning paths and continuous access to educational resources.

Note that these predictions are based on assumptions and general trends observed with previous generations of wireless networks. The actual impact of 6G on education and research will depend on the specific implementation, adoption, and utilization of the technology by educational institutions, researchers, and policymakers.

16.6 Analysis and Future Scope for 6G-Enabled Technologies

16.6.1 Analysis

6G-enabled technologies are still in the early stages of development, and concrete specifications and standards have yet to be defined. However, based on the envisioned capabilities and goals, several analysis points can be identified:

- **Enhanced Connectivity**: 6G aims to provide ultrahigh data rates, ultralow latency, and large device connectivity. This will revolutionize communication, enabling faster and more reliable connections for a wide range of applications.
- **Transformative Applications**: The capabilities of 6G will drive advancements in various sectors, including IoT, smart cities, healthcare, transportation, entertainment, and education. It will enable innovations like autonomous systems, immersive experiences, and intelligent automation.
- **Convergence of Technologies**: 6G will likely integrate technologies like AI, ML, quantum computing, and edge computing. This convergence will enable real-time decision-making, enhanced security, and personalized services, opening up new possibilities across industries.
- **Socioeconomic Impact**: The deployment of 6G will have a huge socioeconomic impact. It will create new job opportunities, spur innovation and economic growth, and improve the quality of life through enhanced services and connectivity.

16.6.2 Future Scope

The future scope of 6G-enabled technologies is vast and holds tremendous potential. Here are some key aspects that may shape the future of 6G:

- **Technology Advancements**: Continued research and development efforts will lead to advancements in 6G technology, including improvements in data rates, latency, energy efficiency, and network architecture. These advancements will expand the capabilities and applications of 6G networks.
- **Global Standardization**: International standardization bodies and industry alliances will work toward defining the specifications and standards for 6G technology. Global collaboration will be important to ensure interoperability, continuous roaming, and global adoption of 6G networks.
- **Network Infrastructure**: The deployment of 6G will require huge investment in network infrastructure, including the deployment of small cells, mmWave technology, and network densification. The development of new wireless technologies and deployment strategies will shape the scalability and coverage of 6G networks.

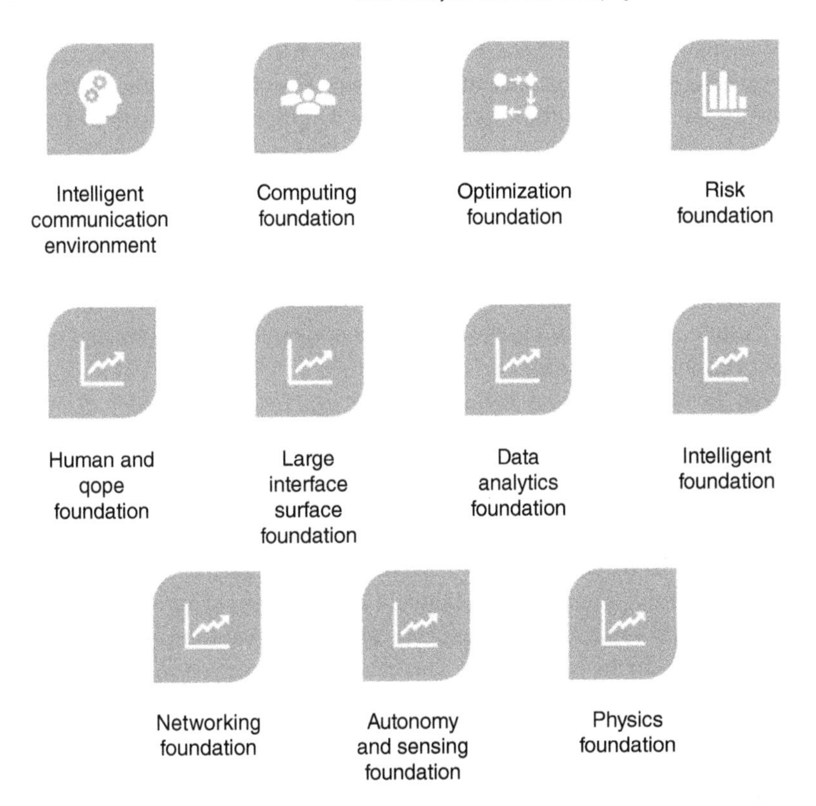

Figure 16.5 Analysis and future scope for 6G-enabled technologies.

- **Policy and Regulatory Frameworks**: Governments and regulatory bodies will need to establish policies and regulations that foster innovation, protect user privacy, ensure fair competition, and address ethical issues associated with 6G-enabled technologies.

Figure 16.5 provides analysis and future scope for 6G-enabled technologies in detail.

In summary, the future scope of 6G-enabled technologies is promising, with the potential to revolutionize communication, enable transformative applications, and drive socioeconomic progress. Continued research, global collaboration, policy development, and responsible deployment will be key to unlocking the full potential of 6G in the coming years and decades.

16.6.3 Market Potential and Economic Impact

6G-enabled technologies have the potential to bring about huge market opportunities and economic impacts. While it is important to note that 6G is still in the early stages of development, we can discuss some general aspects that could contribute to its market potential and economic impact. It is worth noting that the actual outcomes will depend on the specific implementation, adoption, and utilization of 6G technologies.

- **New Business Models and Industries**: 6G is expected to enable a wide range of new business models and industries. The high-speed, low-latency connectivity provided by 6G can make the way for innovative services and applications across sectors like entertainment, healthcare, transportation, manufacturing, and agriculture. This can lead to the emergence of new market players and the creation of entirely new industries.
- **Increased Productivity and Efficiency**: The enhanced capabilities of 6G, including faster data transfer speeds and improved connectivity, can contribute to increased productivity and efficiency in various sectors. Industries can use 6G-enabled technologies to streamline processes, optimize resource allocation, and automate tasks, leading to cost savings and improved overall efficiency.
- **Advancements in Automation and Artificial Intelligence**: 6G's capabilities can significantly impact the adoption and advancement of automation and AI. The low latency and high data rates of 6G can support real-time data processing and enable AI-driven applications. This can lead to increased automation in industries, ranging from autonomous vehicles and robotics to smart cities and intelligent manufacturing, with potential benefits in terms of cost reduction, enhanced safety, and improved decision-making processes.
- **Internet of Things (IoT) Expansion**: 6G is expected to play an important role in supporting the large-scale deployment of IoT devices. With its high capacity and connectivity, 6G can facilitate the continuous integration and management of a vast number of IoT devices, enabling a wide range of applications in smart homes, smart cities, industrial IoT, and more. This expansion of IoT can lead to new business opportunities, improved operational efficiency, and data-driven insights.
- **Job Creation and Economic Growth**: The adoption and deployment of 6G-enabled technologies can generate huge job opportunities across various sectors. The development and maintenance of 6G infrastructure, the creation of new services and applications, and the demand for skilled professionals in areas like data analytics, AI, and cybersecurity can contribute to job creation. Additionally, the economic growth stimulated by 6G technologies can have positive ripple effects on related industries and the overall economy.

Note that it is important to recognize that the market potential and economic impact of 6G-enabled technologies will depend on various factors, including regulatory frameworks, infrastructure investments, industry collaborations, and market demand. The successful realization of these impacts will require coordinated efforts from governments, industry players, and other users to use the full potential of 6G technologies.

16.6.4 Timelines for Deployment and Adoption

A general idea of the expected timelines for the deployment and adoption of 6G-enabled technologies based on the typical progression of wireless communication technologies can be given as:

- **Research and Development (R&D) Phase**: The initial phase involves extensive research, experimentation, and development of key technologies and concepts for 6G networks. This phase includes exploring advanced communication techniques, spectrum allocation, energy efficiency, and potential use cases.
- **Standardization**: After the R&D phase, international standardization organizations, like the ITU, will start developing standards for 6G networks. This process involves defining technical specifications, protocols, and requirements to ensure compatibility and interoperability across different vendors and networks.
- **Trial Deployments**: Once the standards are established, trial deployments of 6G networks and technologies may begin. These trials help validate the theoretical concepts and test the practical implementation of 6G in real-world scenarios. Initially, these trials may be limited to specific locations or use cases.
- **Commercial Deployment**: After successful trial deployments and refinement of the technology, commercial deployments of 6G networks are expected to start. The timeline for commercial deployment will depend on various factors, including the readiness of infrastructure, availability of compatible devices, regulatory approvals, and market demand.
- **Adoption and Expansion**: As 6G networks become commercially available, adoption will gradually increase. Initially, the adoption rate may be slower due to the limited coverage and availability of compatible devices. Over time, as the infrastructure expands and device manufacturers introduce 6G-enabled devices, the adoption rate is expected to accelerate.

Note that the timelines for deployment and adoption can vary significantly based on several factors, including technological advancements, regulatory environment, investment levels, and market demand. As 6G technology is still in its early stages, the precise timeline for its deployment and globally adoption is uncertain. It is anticipated that the deployment of 6G networks and the adoption of 6G-enabled technologies will occur over the next decade or longer.

16.6.5 Collaboration and Research Initiatives

Several collaboration and research initiatives were underway to advance the development of 6G-enabled technologies. These initiatives involve academia, industry players, and standardization bodies. While I cannot provide real-time updates, here are some notable examples of collaborative efforts at that time:

- **University-Led Research**: Universities and research institutions globally have been actively engaged in 6G research initiatives. These include initiatives like the University of Oulu's 6G Flagship Program in Finland, which aims to drive the development of 6G technologies and applications through multidisciplinary research.
- **Industry Consortiums**: Various industry consortiums and alliances have been formed to facilitate collaboration and research in 6G technology. For example, the Next G Alliance, launched by the Telecommunications Industry Association (TIA) in the United States, brings together industry users to drive the development of 6G in North America.
- **Standardization Bodies**: International standardization organizations play an important role in defining the technical standards for 6G. The ITU and other bodies like 3GPP (Third Generation Partnership Project) are involved in shaping the global standards for 6G technology.
- **Public–Private Partnerships**: Governments and private entities are collaborating to foster 6G research and development. For instance, in the European Union, the Horizon Europe program includes funding opportunities for 6G research projects, fostering public–private partnerships and cross-border collaborations.
- **Industry Collaboration**: Companies in the telecommunications and technology sectors are collaborating to drive 6G research and development. Partnerships and collaborations between companies like Nokia, Ericsson, Huawei, Samsung, and others are expected to contribute to the advancement of 6G-enabled technologies.

Note that the landscape of collaboration and research initiatives for 6G technology is rapidly evolving. New partnerships and initiatives may have emerged since my knowledge cutoff, and it is recommended to refer to the latest information from academic institutions, industry organizations, and standardization bodies for the most up-to-date developments in the field of 6G-enabled technologies.

16.6.6 Ethical, Legal, and Policy Issues

The deployment of 6G-enabled technologies will bring about various ethical, legal, and policy issues that need to be addressed. While we can provide a general overview, it is important to note that the specific issue and regulations will vary across countries and jurisdictions. Here are some key areas of issue:

- **Privacy and Data Protection**: With the increased connectivity and data exchange in 6G networks, privacy becomes an important issue. There will be a need to establish robust data protection regulations, consent mechanisms, and safeguards to ensure that personal information is collected, stored, and used in a secure and transparent manner.

- **Security and Cybersecurity**: As connectivity expands with 6G, the vulnerability to cyber threats also increases. There will be a need for enhanced security measures to protect important infrastructure, prevent unauthorized access, and ensure the integrity and confidentiality of data transmitted over 6G networks.
- **Digital Divide**: The deployment of advanced technologies like 6G may exacerbate the existing digital divide, where certain populations or regions have limited access to connectivity. It is important to address this divide by implementing policies that ensure equitable access to 6G networks, especially in underserved areas.
- **Ethical Use of Technology**: As 6G-enabled technologies enable new applications like autonomous vehicles, smart cities, and AR, ethical issues arise. These include issues related to AI ethics, algorithmic transparency, fairness, accountability, and the impact on employment.
- **Spectrum Allocation and Regulation**: 6G networks will require a huge amount of radio spectrum to operate effectively. Governments and regulatory bodies will need to develop policies and regulations to ensure efficient spectrum allocation, prevent interference, and promote fair competition among service providers.
- **Environmental Impact**: The increasing energy consumption of advanced wireless networks, including 6G, raises issues about the environmental impact. Policies and regulations may need to address energy efficiency requirements, responsible e-waste management, and the use of RES to mitigate the environmental footprint of 6G networks.
- **Intellectual Property Rights**: The development and deployment of 6G technologies may involve intellectual property issues. Companies and users need to navigate licensing agreements, patents, and standards development to ensure fair and equitable access to technology while protecting intellectual property rights.

Note that addressing these ethical, legal, and policy issues requires close collaboration among governments, industry users, researchers, and regulatory bodies. It is essential to strike a balance between innovation, societal benefits, and protecting individual rights and interests as 6G-enabled technologies continue to evolve.

16.6.7 Predictions for 6G's Role in Future Technological Advancements

While the specific role and impact of 6G technology in future technological advancements are uncertain, there are several areas where 6G is expected to play a major role based on the potential capabilities and goals envisioned for the technology. Here are some predictions for 6G's role in future technological advancements:

- **Hyperconnectivity and Ubiquitous Communications**: 6G is expected to provide even higher data rates, lower latency, and greater capacity compared to previous generations. This will enable hyperconnectivity, allowing for continuous communication between devices and enabling new applications that require real-time, high-bandwidth connections.
- **Internet of Things (IoT) and Smart Cities**: With its ability to support large device connectivity, 6G is likely to drive advancements in IoT applications. It will enable the deployment of a vast number of connected devices and sensors in various domains, including smart cities, agriculture, healthcare, transportation, and industrial automation, leading to more efficient and sustainable systems.
- **Immersive Experiences and Extended Reality (XR)**: 6G is expected to enhance immersive technologies like VR, AR, and MR. The high data rates, low latency, and high reliability of 6G networks will enable continuous and immersive XR experiences, revolutionizing areas like entertainment, education, remote collaboration, and telemedicine.
- **Artificial Intelligence (AI) and Machine Learning (ML)**: 6G networks can facilitate the deployment of AI and ML algorithms at the network edge, enabling real-time data processing and decision-making. This will lead to advancements in autonomous systems, intelligent automation, and personalized services, where devices can use AI capabilities while minimizing latency and reliance on cloud computing.
- **Holographic Communication and Telepresence**: 6G's ultrahigh data rates and low latency may enable the development of holographic communication and telepresence applications. Users may be able to project and interact with 3D holographic representations of people or objects in real time, enhancing remote collaboration and communication.
- **Sustainable and Energy-Efficient Networks**: Energy efficiency and sustainability are important issues for future networks. 6G aims to optimize energy consumption by using technologies like energy harvesting, intelligent power management, and dynamic resource allocation. This can contribute to reducing the environmental impact of communication networks.
- **Quantum Communications and Computing**: As quantum technologies advance, 6G may incorporate quantum communication protocols for enhanced security and privacy. Furthermore, 6G networks may facilitate distributed quantum computing, enabling the processing of complex algorithms and unlocking new possibilities in cryptography, optimization, and simulations.

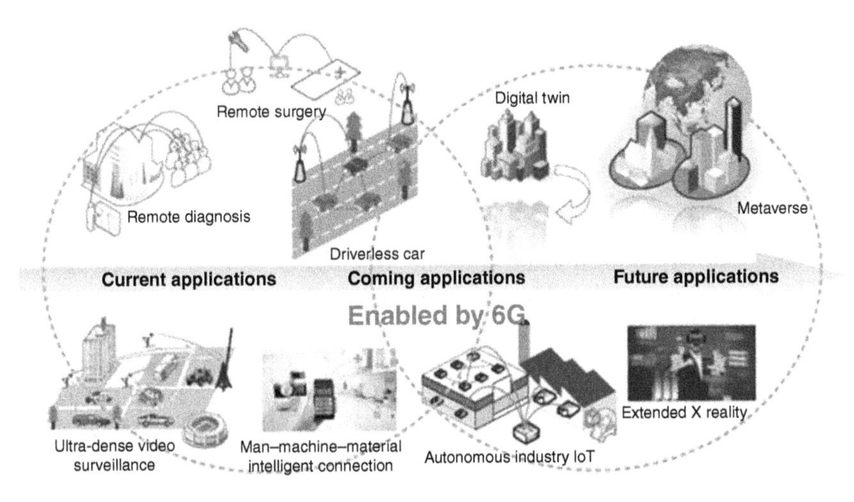

Figure 16.6 Coming and future applications enabled by 6G.

Figure 16.6 depicts few coming and future applications enabled by 6G. Note that these predictions are speculative and based on the potential capabilities of 6G technology. The actual realization and impact of these advancements will depend on various factors, including technological advancements, regulatory frameworks, market demands, and societal acceptance. The full potential of 6G technology will likely emerge as it progresses from research and development to deployment and globally adoption.

16.7 Conclusion

6G-enabled technologies have huge potential to shape the next generation of communication and drive technological advancements. With its unique features of ultrahigh data rates, ultralow latency, and large device connectivity, 6G aims to provide hyperconnectivity and enable groundbreaking applications across various domains. The fundamentals of 6G technology are built upon extensive research and collaboration efforts, involving academia, industry players, and standardization bodies. Through these initiatives, the development of 6G is progressing, although it is still in the early stages. Note that key issues in the deployment and adoption of 6G include privacy, security, ethical use of technology, digital divide, and environmental impact. Addressing these issues will be essential to ensure responsible and equitable deployment of 6G networks. Looking ahead in the future, the applications of 6G technology are expected to be diverse and transformative.

From IoT and smart cities to immersive experiences and XR, 6G will enable new levels of connectivity and innovation. The integration of AI, ML, and quantum technologies with 6G networks will further amplify its capabilities and enable intelligent automation, holographic communication, and quantum communication and computing. However, huge challenges lie ahead. The standardization of 6G, regulatory frameworks, spectrum allocation, and infrastructure development are complex tasks that require international collaboration and coordination. Moreover, the ethical and societal implications of 6G technologies, as well as addressing the digital divide, need to be carefully addressed to ensure equitable access and responsible use. In summary, 6G-enabled technologies represent a promising future where hyperconnectivity, immersive experiences, and advanced applications are the norm. As research and development continue, and collaboration across users strengthens, the full potential of 6G will gradually explain, bringing forth a new era of connectivity and technological advancements with both opportunities and challenges to address.

References

1 Imoize, A.L., Adedeji, O., Tandiya, N., and Shetty, S. (2021). 6G enabled smart infrastructure for sustainable society: opportunities, challenges, and research roadmap. *Sensors* 21 (5): 1709.

2 Zhang, F., Zhang, Y., Weidang, L. et al. (2022). 6G-enabled smart agriculture: a review and prospect. *Electronics* 11 (18): 2845.

3 Xiao, Z. and Zeng, Y. (2022). An overview on integrated localization and communication towards 6G. *Science China Information Sciences* 65: 1–46.

4 Ahmadi, H., Nag, A., Khar, Z. et al. (2021). Networked twins and twins of networks: an overview on the relationship between digital twins and 6G. *IEEE Communications Standards Magazine* 5 (4): 154–160.

5 Wild, T., Braun, V., and Viswanathan, H. (2021). Joint design of communication and sensing for beyond 5G and 6G systems. *IEEE Access* 9: 30845–30857.

6 Gustavsson, U., Frenger, P., Fager, C. et al. (2021). Implementation challenges and opportunities in beyond-5G and 6G communication. *IEEE Journal of Microwaves* 1 (1): 86–100.

7 Nawaz, F., Ibrahim, J., Muhammad, A.A. et al. (2020). A review of vision and challenges of 6G technology. *International Journal of Advanced Computer Science and Applications* 11, no. 2.

8 Viswanathan, H. and Mogensen, P.E. (2020). Communications in the 6G era. *IEEE Access* 8: 57063–57074.

9 Yap, K.Y., Chin, H.H., and Klemeš, J.J. (2022). Future outlook on 6G technology for renewable energy sources (RES). *Renewable and Sustainable Energy Reviews* 167: 112722.

10 Hakeem, Shimaa A.A., Hanan H. Hussein, and HyungWon Kim. "Vision and research directions of 6G technologies and applications." *Journal of King Saud University-Computer and Information Sciences* 34, no. 6 (2022): 2419–2442.

11 Nayak, S. and Patgiri, R. (2021). *6G Communication Technology: A Vision on Intelligent Healthcare*, 1–18. Health Informatics: A Computational Perspective in Healthcare.

12 Giordani, M., Polese, M., Mezzavilla, M. et al. (2020). Toward 6G networks: use cases and technologies. *IEEE Communications Magazine* 58 (3): 55–61.

13 Lu, Y. and Zheng, X. (2020). 6G: a survey on technologies, scenarios, challenges, and the related issues. *Journal of Industrial Information Integration* 19: 100158.

14 Chen, S., Liang, Y.-C., Sun, S. et al. (2020). Vision, requirements, and technology trend of 6G: how to tackle the challenges of system coverage, capacity, user data-rate and movement speed. *IEEE Wireless Communications* 27 (2): 218–228.

15 Wikström, G., Peisa, J., Rugeland, P. et al. (2020). Challenges and technologies for 6G. In: *2020 2nd 6G Wireless Summit (6G SUMMIT)*, 1–5. IEEE.

16 Zong, B., Fan, C., Wang, X. et al. (2019). 6G technologies: key drivers, core requirements, system architectures, and enabling technologies. *IEEE Vehicular Technology Magazine* 14 (3): 18–27.

17 Abdel Hakeem, Shimaa A., Hanan H. Hussein, and HyungWon Kim. "Security requirements and challenges of 6G technologies and applications." *Sensors* 22, no. 5 (2022): 1969.

18 Ji, B., Han, Y., Liu, S. et al. (2021). Several key technologies for 6G: challenges and opportunities. *IEEE Communications Standards Magazine* 5 (2): 44–51.

19 Chowdhury, M.Z., Shahjalal, M., Ahmed, S., and Jang, Y.M. (2020). 6G wireless communication systems: applications, requirements, technologies, challenges, and research directions. *IEEE Open Journal of the Communications Society* 1: 957–975.

20 Nayak, S., and Patgiri, R. (2020). 6G communication: Envisioning the key issues and challenges. *arXiv preprint arXiv:2004.04024.*

21 6G Flagship, Key Drivers and Research Challenges for 6G Ubiquitous Wireless Intelligence, White Paper, Sep. 2019. [Online]. Available: https://www.mobilewirelesstesting.com/wp-content/uploads/2019/10/5G-evolution-on-the-path-to-6G-wpen3608%E2%80%933326-52v0100.pdf.

22 Key R&D projects 2019: Wideband communications and new networks. https://service.most.gov.cn/u/cms/static/201812/12164952skqa.pdf (accessed 10 July 2022).

23 Artificial Intelligence Aided D-band Network for 5G Long Term Evolution. https://cordis.europa.eu/project/id/871464 (accessed 24 July 2022).

24 Research and Development on Satellite-Terrestrial Integration Technology for Beyond 5G. https://www2.nict.go.jp/spacelab/en/pjstit.html (accessed 24 July 2022).

25 Guidelines for the declaration of 2021 projects for the Multimodal Networks and Communications, key project (in Chinese). http://gdstc.gd.gov.cn/attachment/0/422/422266/3296387.pdf (accessed 27 July 2022).

26 6G SENTINEL — The next generation of mobile communications. https://www.fraunhofer.de/en/research/lighthouse-projects-fraunhofer-initiatives/fraunhofer-lighthouse-projects/6g-sentinel.html (accessed 11 July 2022).

27 Project RISE-6G. https://rise-6g.eu (accessed 10 July 2022) [12] REINDEER — REsilient INteractive applications through hyper Diversity in Energy Efficient RadioWeaves technology. https://reindeer-project.eu (accessed 11 July 2022).

28 Project Hexa-X. https://hexa-x.eu (accessed 10 July 2022).

29 RINGS. https://beta.nsf.gov/funding/opportunities/resilient-intelligent-nextg-systems-rings IEEE Communications Surveys and Tutorials, Vol. Xx, No. Xx, February 2023 (accessed 24 July 2022).

30 KT and Hanwha Systems are Jointly Developing 6G Quantum Cryptography Technology. https://www.kedglobal.com/cn/6g/newsView/ked202207120033 (accessed 24 July 2022).

31 Open6GHub – 6G for Society and Sustainability. https://www.open6ghub.de/en# (accessed 21 October 2022).

32 6GStart: Starting the Sustainable 6G SNS Initiative for Europe. https://5g-ppp.eu/6gstart (accessed 26 October 2022).

33 6GTandem: A Dual-frequency Distributed MIMO Approach for Future 6G Applications. https://security-link.se/6gtandem (accessed 21 October 2022).

34 Hexa-X-II, the Second Phase of the European 6G Flagship Initiative. https://hexa-x.eu/hexa-x-ii-the-second-phase-of-the-european-6g-flagship-initiative (accessed 21 October 2022).

35 TERA6G. https://www.hhi.fraunhofer.de/en/departments/pc/projects/tera6g.htm (accessed 21 October 2022).

36 Gupta, R., Reebadiya, D., and Tanwar, S. (2021). 6G-enabled edge intelligence for ultra-reliable low latency applications: vision and mission. *Computer Standards & Interfaces* 77: 103521.

37 Adhikari, M., Hazra, A., Menon, V.G. et al. (2021). A roadmap of next-generation wireless technology for 6G-enabled vehicular networks. *IEEE Internet of Things Magazine* 4 (4): 79–85.

38 Mohsan, S.A., Hassnain, A.M., Malik, W. et al. (2020). 6G: envisioning the key technologies, applications and challenges. *International Journal of Advanced Computer Science and Applications* 11 (9).

39 Ajanthaa, L., Seranmadevi, R., Sree, P.H., and Tyagi, A.K. (2024). Engineering applications of artificial intelligence. In: *Enhancing Medical Imaging with Emerging Technologies*. https://doi.org/10.4018/979-8-3693-5261-8.ch010.

40 Mohsan, S.A., Hassnain, N.Q., Othman, H. et al. (2021). A vision of 6G: technology trends, potential applications, challenges and future roadmap. *International Journal of Computer Applications in Technology* 67 (2–3): 275–288.

41 Banafaa, M., Shayea, I., Din, J. et al. (2023). 6G mobile communication technology: requirements, targets, applications, challenges, advantages, and opportunities. *Alexandria Engineering Journal* 64: 245–274. https://doi.org/10.1016/j.aej.2022.08.017.

42 Whig, P., Velu, A., and Naddikatu, R.R. (2022). The economic impact of AI-enabled blockchain in 6G-based industry. In: Dutta Borah, M., Singh, P., Deka, G.C. (eds.), *AI and Blockchain Technology in 6G Wireless Network, Blockchain Technologies*. Springer, Singapore. https://doi.org/10.1007/978-981-19-2868-0_10.

43 Ziegler, V. and Yrjola, S. (2020). 6G indicators of value and performance. In: *2020 2nd 6G Wireless Summit (6G SUMMIT)*, 1–5. IEEE.

44 Nair, M.M. and Tyagi, A.K. (2023). Blockchain technology for next-generation society: current trends and future opportunities for smart era. In: *Blockchain Technology for Secure Social Media Computing*. https://doi.org/10.1049/PBSE019E_ch11.

45 Tyagi, A.K., Kumari, S., Chidambaram, N., and Sharma, A. (2024). Engineering applications of blockchain in this smart era. In: *Enhancing Medical Imaging with Emerging Technologies*. https://doi.org/10.4018/979-8-3693-5261-8.ch011.

46 Tyagi, A.K., Swetta Kukreja, R., and Sivakumar, P. (2024). Role of blockchain technology in smart era: a review on possible smart applications. *Journal of Information & Knowledge Management* https://doi.org/10.1142/S0219649224500321.

47 Tyagi, A.K. and Tiwari, S. (2024). The future of artificial intelligence in Blockchain applications, in book title: machine learning algorithms using Scikit and TensorFlow environments. *IGI Global* https://doi.org/10.4018/978-1-6684-8531-6.ch018.

48 Kumari, S., Thompson, A., and Tiwari, S. (2024). 6G-enabled internet of things-artificial intelligence-based digital twins: cybersecurity and resilience, in the book titled: emerging technologies and security in cloud computing. *IGI Global* https://doi.org/10.4018/979-8-3693-2081-5.ch016.

49 Nair, M.M. and Tyagi, A.K. (2023). 6G: technology, advancement, barriers, and the future. In: *6G-Enabled IoT and AI for Smart Healthcare*. CRC Press.

50 Nair, M.M. and Tyagi, A.K. (2021). Privacy: history, statistics, policy, laws, preservation and threat analysis. *Journal of Information Assurance & Security.* 16 (1): 24–34.

51 Wang, C.-X., You, X., Gao, X. et al. (2023). On the road to 6G: visions, requirements, key technologies and testbeds. *IEEE Communications Surveys & Tutorials* 25 (2): 905–974.

17

Conclusion

This chapter is summarized with several discussions and interesting remarks for future readers or researchers. This can be discussed as follows.

17.1 Move to a New Era – A Glimpse into 6G

The world of mobile communication is on its major revolution. Although we are still getting acquainted with the capabilities of fifth generation (5G), researchers are already peering into the future, envisioning the next giant leap: sixth generation (6G). Expected to arrive sometime in the 2030s, 6G promises to be a game changer, pushing the boundaries of speed, reliability, and connectivity to previously unimaginable heights [1].

Imagine downloading a movie in an instant, holding a virtual meeting with lifelike holographic projections, or controlling machines in real time with your mind. These futuristic scenarios might seem like science fiction today, but 6G has the potential to make them a reality. With terabit-per-second (Tbps) data rates and near-zero latency (the time it takes for data to travel), 6G will usher in an era of hyper-connectivity, empowering entirely new applications and transforming the way we live, work, and interact with the world around us.

However, the road to 6G is not without its challenges. Operating at much higher frequencies (terahertz spectrum) compared to current generations, 6G signals have a shorter range and are more easily disrupted. Developing energy-efficient components and network infrastructure to handle the massive data flow is also important [2]. Security issues also need to be addressed, as a more connected world necessitates robust safeguards for data privacy and network security.

Despite these difficulties, research and development in 6G is brimming with exciting possibilities. Novel materials and innovative antenna designs are being

6G-Enabled Technologies for Next Generation: Fundamentals, Applications, Analysis and Challenges, First Edition. Amit Kumar Tyagi, Shrikant Tiwari, Shivani Gupta, and Anand Kumar Mishra.

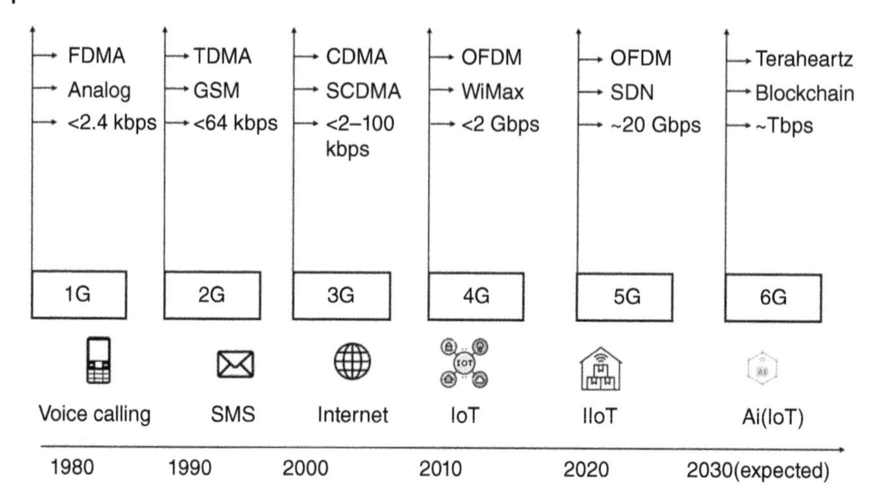

Figure 17.1 Overview of technology.

investigated to overcome the limitations of the terahertz spectrum. Artificial intelligence (AI) is used to play an important role, in optimizing network management, resource allocation, and security protocols. Additionally, intelligent surfaces capable of manipulating radio waves to improve signal propagation hold promise for extending network coverage and enhancing efficiency.

The potential societal impact of 6G is truly transformative (refer Figure 17.1). It can empower the Internet of Things (IoT) to its full potential, enabling seamless communication between billions of devices. Industries such as manufacturing, healthcare, and transportation can use 6G to create intelligent factories, implement remote surgery with haptic feedback, and develop autonomous vehicles with ultrareliable communication. Additionally, 6G has the potential to bridge the digital divide by providing high-speed internet access to remote and underserved areas.

17.2 Unveiling the Power of 6G – A Leap Toward Hyper-connectivity

The future of communication is used for a quantum leap with the arrival of 6G technology. This next-generation network promises to revolutionize the way we interact with the digital world, offering connectivity so fast and reliable that it seems almost instantaneous.

The true power of 6G lies in its ability to deliver data at microsecond speeds, a staggering speed of 1000 times faster than its predecessor, 5G. Imagine downloading an entire movie in a blink of an eye or attending a virtual meeting where participants appear as life-sized holograms [3]. With 6G, buffering, lags, and frustrating disconnections become relics of the past.

One of the most essential advancements of 6G resides in its ability to enhance machine-to-machine communication. This translates to seamless interoperability within the burgeoning IoT ecosystem. Billions of devices, from smart appliances to self-driving cars, will communicate with each other efficiently, paving the way for a truly intelligent and connected world.

The speed difference between 5G and 6G is truly remarkable. Although 5G delivers 20 gigabytes of data in a millisecond (1000 microseconds), 6G accomplishes the same feat in a mere microsecond, transferring a staggering one terabyte (1000 gigabytes) of data. This exponential leap in speed unlocks a plethora of groundbreaking applications. Imagine networked vehicles navigating city streets in perfect harmony or smart factories operating with unparalleled efficiency, all made possible by the near-instantaneous communication offered by 6G.

Another exciting feature of 6G is the integration of AI and machine learning directly into the network itself. This empowers the network to make autonomous decisions at every layer, optimizing performance and adapting to changing demands. This makes the way for a new wave of applications that leverage AI capabilities [4]. Public safety systems could react quicker to emergencies, health monitoring could become more responsive, and facial recognition technology could operate with enhanced accuracy.

Experts anticipate the deployment of 6G networks sometime in the late 2020s, potentially as early as 2028. Today's governments, telecommunication giants, infrastructure providers, and leading academic institutions are all actively involved in the development of 6G technology.

17.3 The Road to 6G – Issues and Challenges

The growth of 6G promises future growth with many possibilities. Imagine instantaneous data transfer, seamless communication across the globe, and the ability to connect billions of devices in a symphony of information exchange. However, this revolutionary technology is not without its difficulties. Note that the path to 6G necessitates addressing important challenges to ensure its successful implementation.

One of the most important issues surrounding 6G is the sheer scale of infrastructure required. Unlike its predecessors, 6G will depend on a dense network of small cells – low-powered base stations with limited coverage areas. Deploying this intricate web, especially in remote locations, presents an essential logistical and financial challenge. The complete number of small cells needed will also lead to a considerable increase in energy consumption by wireless networks.

Another potential roadblock lies in the realm of network interference. 6G operates at a much higher frequency than previous generations, making it more susceptible to interference from other networks and devices. This can lead to disruptions in call quality, data transfer speeds, and overall network performance. Imagine the frustration of an essential video conference being interrupted due to signal problems.

As discussed above, security also emerges as a major issue. The rapid-fast speeds and minimal latency of 6G make it an attractive target for malicious actors. Additionally, the increased integration of AI and machine learning into the network creates new vulnerabilities that require robust defense mechanisms. As we lock our doors at night, robust security protocols are more important to safeguard the integrity of 6G.

On the other hand, privacy is an aspect that demands careful consideration. The vast amount of data flowing through 6G networks raises issues about mass surveillance and potential abuse of personal information [5]. Imagine a world where every step we take is tracked and monitored – 6G necessitates the development of strong privacy frameworks to ensure our digital lives remain secure.

The cost of 6G technology also presents a lot of challenges. The development of new devices and infrastructure will likely be expensive, potentially excluding a significant portion of the population, particularly in developing countries. This could exacerbate the digital divide, creating a world where access to the benefits of 6G becomes a privilege rather than a right.

Finally, the digital divide itself emerges large. The deployment of 6G networks could widen the gap between those who have access to cutting-edge technology and those who remain disconnected. Bridging this divide necessitates proactive measures to ensure equitable access to the benefits of 6G.

17.4 Navigating the Road to 6G: Unveiling the Challenges

The impending arrival of 6G, the next generation of wireless communication, promises to revolutionize how we interact with the world. Yet, the path toward this technological leap is not without its difficulties. We need to discuss the significant challenges associated with 6G, using real-world examples to illustrate each point.

One of the most pressing challenges lies in achieving the massive leap in data transmission speeds and capacity expected from 6G. We are looking at data rates reaching 1 Tbps, a staggering 100 times faster than 5G. To achieve this, significant advancements are needed in wireless communication technology. This includes utilizing higher frequency bands, deploying advanced antenna systems, and developing novel modulation techniques.

Millimeter wave (mmWave) technology, operating between 24 and 100 GHz, is a prime candidate for 6G due to its immense bandwidth and potential for high data rates. However, mmWave signals have limitations. Their shorter range and increased susceptibility to interference, compared to lower frequencies, pose significant challenges for signal propagation and network coverage. Imagine trying to shout across a vast canyon – the message might not get through!

Another major difficulty for 6G is ensuring ultralow latency and high reliability. Real-time applications such as autonomous vehicles, remote surgery, and industrial automation rely on 6G, requiring latency (data travel time) below 1 millisecond and near-perfect reliability (99.999%) [6]. Meeting these stringent demands necessitates significant advancements in network architecture. Note that techniques such as edge computing, network slicing, and software-defined networking come into play.

Edge computing involves processing data closer to its source, like having a local translator instead of depending on one far away. This approach reduces latency and improves reliability. However, deploying these distributed computing resources can be complex and expensive, similar to setting up mini data centers across a vast network.

Security and privacy are essential issues in the age of 6G. As the amount of data transmitted over wireless networks explodes, the risk of cyberattacks and data breaches rises. Robust security mechanisms are important to safeguard against these threats. Think of building a high-security vault for this valuable data. Encryption, blockchain technology, and AI-based intrusion detection systems are potential solutions.

Blockchain offers a secure and decentralized way of managing network resources, similar to a shared ledger accessible to all authorized parties. Meanwhile, AI-based intrusion detection systems can act as vigilant guards, constantly monitoring and responding to cyberattacks in real time.

Finally, navigating the regulatory and standardization landscape presents a challenge. Global collaboration and coordination among governments, industry leaders, and academic institutions are vital for 6G's development [7]. Establishing new regulatory frameworks and industry standards is important for ensuring interoperability (seamless communication between different systems) and fair competition. Imagine having different power plugs in every country – it would not be very convenient! Organizations like the International Telecommunication Union (ITU) are actively working on developing 6G standards, creating a common blueprint for designing and deploying 6G networks.

17.5 6G: Ushering in a New Era of Business Transformation

The digital landscape is constantly evolving, and businesses must adapt to stay competitive. Enter 6G technology, which is used to be a game changer, not just offering faster internet but a complete major shift in how we perceive digital innovation. It is a leap beyond mere speed improvements; it is an invitation to discuss how ultrafast connectivity can transform every facet of business innovation.

17.5.1 6G: A Gateway to Business Transformation

Imagine instant communication, seamless collaboration, and breakthroughs in AI and IoT – all facilitated by 6G. This next-generation technology promises to redefine business operations, paving the way for future success. A McKinsey report underscores this vision, highlighting 6G's potential to spark innovation, attract investment, and revitalize the telecommunications sector [8]. Reports even suggest a 6G market reaching trillions, signifying its pivotal role in driving both technological and economic growth.

17.5.2 Demystifying 6G

6G utilizes previously untapped radio frequencies and harnesses cognitive technologies such as AI and IoT. This translates to high-speed, low-latency communication, exceeding current 5G networks by a significant margin. This future-proof network caters to diverse applications, from enhanced mobile broadband to mission-important communication and vast IoT deployments. Supporting the Fourth Industrial Revolution (or Industry 4.0) [9–15], acting as a bridge between humans, machines, and the environment.

With real-time processing capabilities, 6G unlocks possibilities in smart cities, autonomous driving, and immersive virtual reality, paving the way for a truly interconnected world. As the successor to 5G, its development is well underway, with standards expected to be finalized by 2028 and widespread deployment by 2030 (as per a Bloomberg report).

17.5.3 The Business Case for 6G

This new wave of technology promises seamless, immersive experiences across various domains. Imagine devices capable of tasks previously hindered by technological limitations. From remote healthcare monitoring to automated manufacturing, 6G's potential to revolutionize business operations is vast.

17.5.4 Preparing for the 6G Revolution

To reap the benefits of 6G, businesses must be proactive. This involves strategizing for integration, discussing innovative business models, and building partnerships to ensure a smooth transition into this 6G-driven digital ecosystem [16].

17.5.5 6G and Industry 4.0

Industry 4.0, characterized by automation and digitalization of manufacturing, relies heavily on robust communication infrastructure. 6G emerges as a game changer, providing a high-speed, low-latency network that facilitates real-time data exchange, machine-to-machine communication, and remote monitoring. Here is how 6G can revolutionize Industry 4.0 [9–15]:

- **Predictive Maintenance:** Real-time equipment monitoring allows for preventative measures and reduced downtime.
- **Enhanced Quality Control:** Augmented and virtual reality facilitate more accurate and efficient inspections.
- **Streamlined Supply Chain Management:** Real-time tracking and monitoring of goods through connected devices within the supply chain network.

Although 6G technology is still in its early stages, its potential to reshape our world is undeniable. The hyper-connectivity it promises will redefine how we live, work, and interact with technology. From revolutionizing industries to fostering entirely new applications, 6G presents a future brimming with possibilities. However, significant challenges lie ahead. Developing the necessary infrastructure and ensuring robust security measures will be important in new industry revolutions [17].

The journey toward 6G is a collaborative effort. Governments, academia, and the private sector must work together to navigate the technical difficulties and ethical issues. International collaboration will be important in establishing global standards and ensuring equitable access to this transformative technology.

Hence, the growth of 6G marks an important moment in human history. As we move toward this exciting journey, it is vital to remember that technology is a tool. The true power of 6G lies in our ability to harness it responsibly and ethically, shaping a future that benefits all of humanity.

References

1 Loscri, V., Chiaraviglio, L., and Vegni, A.M. (2024). The Road towards 6G: Opportunities, Challenges, and Applications. *A Comprehensive View of the Enabling Technologies*, Springer Book, doi: 10.1007/978-3-031-42567-7.

2 Alhammadi, A., Shayea, I., El-Saleh, A.A. et al. (2024). Artificial intelligence in 6G wireless networks: opportunities, applications, and challenges. *International Journal of Intelligent Systems* 1: 2024. https://doi.org/10.1155/2024/8845070.
3 Chen, H., Keskin, M.F., Sakhnini, A. et al. (2024). 6G localization and sensing in the near field: features, opportunities, and challenges. *IEEE Wireless Communications* 31 (4): 260–267. doi: 10.1109/MWC.011.2300359.
4 Chataut, R., Nankya, M., and Akl, R. (2024). 6G networks and the AI revolution-exploring technologies, applications, and emerging challenges. *Sensors (Basel)* 24 (6): 1888.
5 Kazmi, S.H.A., Qamar, F., Hassan, R. et al. (2024). Security of federated learning in 6G era: a review on conceptual techniques and software platforms used for research and analysis. *Computer Networks* 110358.
6 Liwen, Z., Qamar, F., Liaqat, M. et al. (2024). Towards efficient 6G IoT networks: A perspective on resource optimization strategies, challenges, and future directions. *IEEE Access* 1–1.
7 Wang, C.-X., You, X., Gao, X. et al. (2023). On the road to 6G: visions, requirements, key technologies, and testbeds. *IEEE Communications Surveys and Tutorials* 25 (2): 905–974.
8 Qadir, Z., Le, K.N., Saeed, N., and Munawar, H.S. (2023). Towards 6G internet of things: recent advances, use cases, and open challenges. *ICT Express* 9 (3): 296–312.
9 Hemamalini, V., Tyagi, A.K., Vasuki, P., and Kumari, S. (2024). Industrial automation in drug discovery: the emerging of smart manufacturing in Industry 5.0. In: *Converging Pharmacy Science and Engineering in Computational Drug Discovery*. https://doi.org/10.4018/979-8-3693-2897-2.ch008.
10 Tyagi, A.K., Priya, R.L., Mishra, A.K., and Balamurugan, G. (2023). Industry 5.0: potentials, issues, opportunities, and challenges for society 5.0. In: *Privacy Preservation of Genomic and Medical Data*. https://doi.org/10.1002/9781394213726.ch17.
11 Tyagi, A.K., Arumugam, S.K., Raghavendra Prasad, P., and Sharma, A. (2024). The position of digital society, Healthcare 5.0, and Consumer 5.0 in the Era of Industry 5.0. In: *Advancing Software Engineering Through AI, Federated Learning, and Large Language Models*. IGI Global. doi: 10.4018/979-8-3693-3502-4.ch017.
12 Tyagi, A.K. (2024). Transformative effects of ChatGPT on the modern era of education and society: from society's and industry's perspectives. In: *Machine Learning Algorithms Using Scikit and TensorFlow Environments*. IGI Global. https://doi.org/10.4018/978-1-6684-8531-6.ch019.
13 Shrikant Tiwari, R., Reddy, R., and Tyagi, A.K. (2024). Position of blockchain: Internet of Things-Based Education 4.0 in Industry 5.0 – A discussion of issues

and challenges. In: *Architecture and Technological Advancements of Education 4.0*. IGI Global. doi: 10.4018/978-1-6684-9285-7.ch013.

14 Gomathi, L., Mishra, A.K., and Tyagi, A.K. (2023). Industry 5.0 for Health-care 5.0: opportunities, challenges and future research possibilities. In: *2023 7th International Conference on Trends in Electronics and Informatics (ICOEI), Tirunelveli, India*, 204–213. https://doi.org/10.1109/ICOEI56765.2023.10125660.

15 Singh, R., Tyagi, A.K., and Arumugam, S.K. (2024). Imagining the sustainable future with Industry 6.0: a smarter pathway for modern society and manufacturing industries. In: *Machine Learning Algorithms Using Scikit and TensorFlow Environments*. IGI Global. doi: 10.4018/978-1-6684-8531-6.ch016.

16 Salahdine, F., Han, T., and Zhang, N. (2023). 5G, 6G, and beyond: recent advances and future challenges. *Annals of Telecommunications* 78 (9): 525–549.

17 Chen, W., Lin, X., Lee, J. et al. (2023). 5G-Advanced toward 6G: past, present, and future. *IEEE Journal on Selected Areas in Communications* 41 (6): 1592–1619. doi: 10.1109/JSAC.2023.3274037.

Index

6G-Enabled Technologies for Next Generation: Fundamentals, Applications, Analysis and Challenges, First Edition. Amit Kumar Tyagi, Shrikant Tiwari, Shivani Gupta, and Anand Kumar Mishra.

f

g

k

l

r

s